Crustaceans
Endocrinology, Biology and Aquaculture

Editor

Valerio Zupo

Department of Ecosustainable of Marine Biotechnology
Stazione Zoologica Anton Dohrn, Napoli, Italy

CRC Press
Taylor & Francis Group
Boca Raton London New York

CRC Press is an imprint of the
Taylor & Francis Group, an **informa** business

A SCIENCE PUBLISHERS BOOK

Cover credit: Photos by V. Zupo, M. Lorenti and A. Siciliano

First edition published 2023
by CRC Press
6000 Broken Sound Parkway NW, Suite 300, Boca Raton, FL 33487-2742

and by CRC Press
4 Park Square, Milton Park, Abingdon, Oxon, OX14 4RN

Library of Congress Cataloging-in-Publication Data (applied for)

ISBN: 978-0-367-42070-3 (hbk)
ISBN: 978-1-032-39072-7 (pbk)
ISBN: 978-0-367-85342-6 (ebk)

DOI: 10.1201/9780367853426

Typeset in Times New Roman
by Radiant Productions

Preface

Crustaceans are a huge group of arthropods and they play key ecological roles in any aquatic environment, while only a few species are adapted to sub-aerial and humid environments. Their evolutionary success is not only due to a wide set of morphological and biological adaptations, but also to some key features, such as their peculiar endocrinology. Due to a remarkable physiological plasticity, their incredible diversity, and an impressive simplicity of their morpho-functional adaptations, crustaceans have contributed fundamentally to our knowledge of cell, molecular, and developmental biology. For example, a few key glands produce dramatic effects in any crustacean, even if ascribed to distant taxa. In this book we will stress how their sex determination is due to a single gland, the Androgenic Gland, and this feature favours the evolution of several sexual strategies, ranging from gonochorism, to proterandry to partial or total hermaphroditism. Other fundamental processes for crustaceans, such as moulting cycles and circadian rhythms, are driven by hormonal effectors and respond to clear ecological constraints. In addition, crustaceans are characterized by a series of super-versatile chemical and optical sensors deserving attention, because they play important biological roles, linked to chemical ecology issues, and their functioning is going to be transformed due to the global changes and particularly to ocean acidification. Considering the importance of crustaceans for the aquatic ecology (e.g., copepods in the marine plankton, amphipods and isopods in the benthos), the biotechnological importance of model species for scientific research and the central role of decapod crustaceans for aquaculture practices, it is worth understanding their physiology and ecology, both for theoretical and practical issues. In this book, chief scientists pool their expertise to explain fundamental biological and physiological mechanisms of the main taxa of crustaceans. Prior to start the discussion of these key topics, however, let's briefly introduce what a crustacean is, from a morphological and evolutionary point of view. Here we will avoid introducing a taxonomic description of this group of invertebrates: classifications are in a continuous progress and new web-based tools are extremely more efficient than a printed book to provide clear and updated information on this topic. Consequently, we invite the reader to refer to affordable websites to obtain all the necessary information on the taxonomy of crustaceans and the tools needed for their correct identification: this will permit to avoid wastes of space and contribute useful information on the physiology, ecology and biotechnology of crustaceans. On the other hand, the huge biodiversity of crustaceans is reflected in the diversity of morpho-functional adaptations observed in each taxonomic group. According to [1], 75,000 species of crustaceans are present in the world.

However, other authors [2] estimated a lower number of species (between 45,000 and 50,000 according to [3]). Various scientists have questioned the monophyletic origin of the group, because of the huge variability of life forms [1], and also because crustaceans "*lack distinct sets of apomorphies, as observed in other arthropods*" [2]. Nowadays Crustacea are considered a paraphyletic group and part of the Pancrustacea/ Tetraconata, including both "Crustacea" and Hexapoda [4]. However, when we examine the diagnostic characters usually associated with crustaceans, we constantly find a range of evolutionary adaptations that may be summarized as follows (according to a very exhaustive categorization proposed by [2], that we will strictly follow, to describe the next points) to illustrate the main features of crustaceans:

1. **Head made of five somites**, each bearing a set of appendages consisting of two pairs of antennae, a pair of mandibles, and two pairs of maxillae. This should be a distinctive feature of crustaceans. However, not all crustaceans have a head containing five somites (e.g., Remipedia have six). Furthermore, insects and myriapods bear as well a head containing five somites, along with a pair of mandibles and two pairs of maxillae. The only feature distinctive for crustaceans is the presence of a second set of sensory antennae, since the head of myriapods and insects bears only a single pair of antennae. However, second antennae, that are primarily sensory, are not a characteristic feature of all groups of crustaceans and, in addition, a limb in this position is not exclusive to crustaceans, but shared with merostomes, arachnids and pycnogonids. Consequently, this first "*diagnostic character*" of crustaceans is not sufficiently informative!

2. **Body consisting of three regions**, head, thorax, and abdomen. The tagmosis (body divided into a head, thorax and abdomen), characterizing most arthropods, is certainly distinctive, but also this feature is shared with insects. Thus, regarding this "feature" we find some similarities [5] between crustaceomorphs and insects or, more formally [6], within the monophylum of Pancrustacea (or Tetraconata). In addition, this diagnostic character finds exceptions even within crustaceans; for example, remipedes have a long trunk with no differentiation between anterior and posterior sectors and mystacocarids have five thoracomeres while branchiurans have four, and malacostracans have seven or eight. Even the definition of "abdomen" has a different evolutionary meaning among crustaceans, since the total number of segments composing this region, as well as the molecular expression of two homeobox family genes [7], defining the abdomen and the pleon, may vary among crustaceans. According to the innervation and the role of these structures, Schram and Koenemann [8] concluded that the term *pleon*, applied to the posterior region of the malacostracan body, is not an alternative for the term *abdomen*, because the pleon of malacostracans and the abdomen of other crustaceans show different developmental pathways. Consequently, also this characterization of crustaceans is not a unique feature, nor an informative statement.

3. **Trunk appendages primitively multi-ramous** are commonly considered a characteristic feature of Arthropoda. However, divergent development strategies are observed in various crustaceans, as in the case of the multi-ramous limb of branchiopods, having quite a different development as compared to crustaceans

Figure 1. Biodiversity of amphipod crustaceans in a historical printing by Antonio Della Valle, *Gammarini del Golfo di Napoli* (Fauna & Flora, 20). Berlin, Friedländer, 1893. Print test of plate 4. Archivio storico Stazione Zoologica Anton Dohrn, Ua.IV.1. For gentle permission by the Curator of the Stazione Zoologica historical archives Christiane Groeben agreed by the President Prof. R. Danovaro.

bearing biramous or uniramous limbs [9]. In conclusion, even this feature seems to be not sufficiently informative to distinguish the body structure of crustaceans from other arthropods.

4. **Development consisting of a series of discrete larval and/or juvenile stages**, initiated by a *nauplius* larva. This character apparently distinguishes crustaceans, since most of them hatch as a larva which not only has fewer segments than adults, but also exhibits a distinctive shape, absence of key structures (e.g., pedunculated eyes, characterizing only adults) and different morphologies of appendages (e.g., adapted to swimming, while post-larvae are adapted to benthic environments). Further molts not only add segments, but also metamorphose the shape of the body, of various appendages, eyes and swimming appendages. However, even this feature does not represent an apomorphy for crustaceans. In fact, almost all arthropods have larvae (from trilobites on) and nauplii are known in fossil records of several arthropods [10]. Various groups of larvae characterize and distinguish the taxa of crustaceans, such as a distinctive *nauplius* (a larva characterized by only three sets of limbs: the first and second antennae and the mandibles) with fronto-lateral horns and a post-naupliar *cypris* larva in the Cirripedia, or a characteristic *nauplius* with a rnaupliar process on the antennae for Branchiopoda. A *zoea* larva distinguishes malacostracan decapods. Thus, the *nauplius* represents a phylotypic stage characterizing crustaceans. However, this larval phase may be passed within embryogenesis. Consequently, some crustaceans do not begin independent life as a *nauplius* but rather as a *metanauplius* (a stage with more than the three sets of naupliar limbs and three naupliar segments). In addition, *nauplii* show quite different structures and functions according to crustacean taxa, with or without *orthonaupliar* stages which may be non-feeding, until the *metanaupliar* stages. Based on the observations of Scholtz [11], we may conclude [2] that the *nauplius* larva cannot be a phylotypic stage for all crustaceomorphs. Although the *nauplius* larva is regarded as a derived form, the main problem is that some taxa may have lost it, while other taxa probably never developed it [2]. Notwithstanding this, the developmental patterns and the *nauplius* larva offer an unambiguous set of apomorphic descriptors to diagnose a monophyletic crustacean, because we can assume that the *nauplius* has independently evolved several times and, consequently... it has been lost several times! Thus, at least from the point of view of reproduction and larval development [8] we can find diagnostic features suggesting crustaceans as a monophyletic group, and this finding shows how important the reproductive biology of crustaceans is, at least from the point of view of evolutionary biologists. With this in mind, we will continue our morpho-functional overview of crustaceans by describing their lives under various points of view, taking into account mainly their physiology (first part of this book), ecology (second part), and biotechnological (third part) aspects.

In **the first part of this book**, we afford in Chapter 1 the basic structure of crustacean endocrinology, to stimulate new ideas to search for animal models. In Chapter 2 their sexual biology is presented in cooperation with Alan Hodgson, through a comprehensive description, useful for reproductive, evolutionary and developmental biologists, as well as for invertebrate biologists, comparative endocrinologists and aquaculture specialists, by adopting schematic representations to synthesize the huge amount of available information. In Chapter 3, Ulrich Hoeger and Sven Schenk offer a delicious analysis of the structure, function and diversity, evolution and origin of crustacean yolk proteins, as an exclusive and valuable contribution of this book. In Chapter 4, Hardege, Fletcher and Breithaupt show how crustaceans communicate with others and with their environment in a splendid analysis of the methods they use to take advantage of their chemical sensors to interact and evolve. Finally, in Chapter 5, Maria Costantini and co-workers afford the intriguing field of crustaceans as "models" for the scientific research, taking into account the possible contribution of each key species to the omics approaches and to the investigation of stress responses, as a useful and concrete tool contributing to the activities of each experimental scientist.

In **the second part**, the ecological and taxonomical approach starts with Chapter 6, where the biodiversity of crustacean decapods is analyzed to answer basic questions about the factors triggering world's biodiversity trends, also according to ocean acidification (OA). In Chapter 7, Maurizio Lorenti offers an ecological perspective of isopod crustaceans as consumers of seagrass tissues. In Chapter 8, Felicita Scapini presents the ecology and the intriguing ethology of littoral amphipods, including behavioral adaptations and daily rhythms to avoid dehydration, along with special techniques for space orientation. In Chapter 9, Mazza and Tricarico show how ethology and ecology interact to determine the destiny of a species, as a contribution to nature and species conservation, as well as to sustainable and ethical farming. In Chapter 10, Hardege and Fletcher analyze the ecology of crustaceans according to climate changes, demonstrating that ocean acidification and other pressures influence their physiology, and showing emerging evidence that behavior, fitness and ultimately reproduction and survival of coastal crustaceans are negatively altered under climate change conditions. Finally, in Chapter 11, Jeff Ram and co-workers afford the issue of the biodiversity of freshwater crustaceans, analyzing a range of environments, from ubiquitous groups found in the frigid waters of the Antarctic to the completely light-deprived subterranean caves of the tropics. From these, we move to **the third part of the book**, dedicated to the biotechnological applications made possible by the study of crustaceans, with special attention to aquaculture and biomonitoring. In the Chapter 12, Francesca Carella analyzes the role of crustaceans as parasites of cultured organisms (mainly fishes) and, *vice-versa*, she analyzes the causative agents of crustacean's diseases when they are, in their turn, the subjects of aquaculture practices. This chapter is viewed as a practical tool for anybody involved in culture productions and contains tips about pathogen classification, life cycle, host-pathogen interaction and aetio-pathogenesis. A very special practical approach for scientists is contained in Chapter 13, where Marco Guida and co-workers show the biomonitoring techniques made possible by the use of crustaceans as bioindicators. The chapter integrates information across various levels of organization, from genetic sequencing to ecological and environmental traits for *Daphnia* spp. In Chapter 14, we analyze an emerging field of research, represented by the use of natural products to control the invasions of fish parasites, including the sea lice (copepods) responsible for mass mortalities of salmon cultured in various areas of the world. The control of the expansion of crustaceans, in this case, is vital to save a very lucrative market. In Chapter 15, Glaviano and Mutalipassi present a new automatic device for the continuous culture of crustaceans, framed in the field of techniques for the smart control of invertebrate culture, adequate both for scientific and aquaculture purposes and useful also for the culture of crustaceans for the ornamental market. In the Chapter 16, Paolucci, Coccia, Polese and Di Cosmo cooperate to explain the current issues on freshwater crayfish aquaculture, with special focus on welfare that has reached in the last decade a key importance in any activity of culture both for science and aquaculture. The chapter has been recently updated according to the latest rules for animal welfare, now including crustaceans. In Chapter 17, Maria Costantini and co-workers show the importance of molecular approaches to the practices of crustacean aquaculture and this last contribution completes our vision of crustaceans, as organisms to culture, protect, fight, adopt, investigate, respect and, finally… love!

Ischia, January the 30th, 2022   Valerio Zupo
President of the International Society
of Invertebrate Reproduction and Development

Literature cited

1. P.A. Meglitsch, and F.R. Schram. (1991). *Invertebrate zoology.* Oxford Univ. Press. New York, 623 pp.
2. F.R. Schram. (2013). pp. 1–33. *In*: Watling, L., and Thiel, M. (eds.). *Functional Morphology and Diversity.* Oxford Univ. Press. New York.
3. P. Bouchet. (2006). pp. 33–62. *In*: Duarte, C.M. (ed.). *The Exploration of Marine Biodiversity: Scientific and Technical Challenges.* Fundacion BBVA, Bilbao, Spain.
4. C. Wolff, and M. Gerberding. (2015). Chapter 4. *In*: Andreas Wanninger (ed.). *Evolutionary Developmental Biology of Invertebrates.* Springer-Verlag. Wien.
5. W. Wheeler, G. Giribet, and G.D. Edgecombe. (2004). pp. 281–295. *In*: Cracraft, J., and Donoghue, M. (eds.). *Assembling the Tree of Life. Arthropod Systematics: The Comparative Study of Genomic, Anatomical, and Paleontological Information.* Oxford Univ. Press, New York.
6. G. Giribet, S.Richter, G.D. Edgecombe, and W.Wheeler. (2005). *Crust. Issues* **16**: 307–352.
7. A. Abzhanov, and T.C. Kaufman. (2004). *Crust. Issues* **15**: 43–74.
8. F.R. Schram, and S. Koenemann. (2004). *Crust. Issues* **15**: 75–92.
9. T.A. Williams. (1998). *Dev. Genes Evol.* **207**: 427–434.
10. K.J. Müller, and D. Walossek. (1986). *Trans. R. Soc. Edinburgh, Earth Env. Sci.* **77**: 157–179.
11. G. Scholtz. (2000). *J. Zool Syst. Evol. Res.* **38**: 175–187.

Acknowledgements

I have a profound debt of gratitude towards all authors of individual chapters of this book. They are prominent scientists in various fields of the crustacean biology, physiology, evolution, ecology and biotechnology and their prompt answer to our request to write specialistic chapters was enthusiastic. This is a demonstration of how enthusiasm and scientific excellence synergistically proceed! I am indebted with Thomas Viel for his kind help in the final revision of texts. We thank Christiane Groeben and the President of Stazione Zoologica Anton Dohrn, Prof. R. Danovaro, for contributing splendid historical images helping our imagination and enthusiasm about crustacean biodiversity. Finally, I must acknowledge the indispensable contribution of Dr. M. Costantini, for her continuous help in the production of the final version of this volume. As a matter of fact, she contributed as an editor of this publication.

Contents

Preface   iii

Acknowledgements   ix

PART 1: Physiology Issues

1. **Crustacean Endocrinology: Fascinating Topic for Biologists or a   3
   Peculiar Opportunity for Biotechnologies? A Historical View with
   Functional Perspectives**
   Valerio Zupo and *Penny M. Hopkins*

2. **Sexual Biology and Reproduction**   18
   Valerio Zupo and *Alan N. Hodgson*

3. **Crustacean Yolk Proteins: Structure, Function and Diversity**   40
   Ulrich Hoeger and *Sven Schenk*

4. **Infochemicals Recognized by Crustaceans**   72
   Joerg D. Hardege, *Nicky Fletcher* and *Thomas Breithaupt*

5. **Crustaceans as Good Marine Model Organisms to Study Stress**   82
   Responses by –Omics Approaches
   Maria Costantini, *Roberta Esposito* and *Nadia Ruocco*

PART 2: Ecology and Taxonomy

6. **Crustacean Decapods are Models to Describe the General Trends of**   109
   Biodiversity According to Ocean Acidification
   Valerio Zupo and *Emanuele Somma*

7. **Isopod Crustaceans as Seagrass Consumers: A Mediterranean**   121
   Perspective
   Maurizio Lorenti

8. **Ecology and Ethology of Littoral Amphipods**   134
   Felicita Scapini

9. **Ethology of Crustaceans Influencing their Ecology**   146
   Giuseppe Mazza and *Elena Tricarico*

10. **Crustacean Ecology in a Changing Climate**   157
Jörg D. Hardege and *Nicky Fletcher*

11. **The Biodiversity of Freshwater Crustaceans Revealed by**   166
Taxonomy and Mitochondrial DNA Barcodes
Adrian A. Vasquez, Brittany L. Bonnici, Donna R. Kashian,
Jorge Trejo-Martinez, Carol J. Miller and *Jeffrey L. Ram*

PART 3: Aquaculture and Biotech

12. **Crustaceans as Pathogens and Most Common Pathogens of**   185
Crustaceans
Francesca Carella

13. **Biotechnologies Linked to Crustaceans**   204
Antonietta Siciliano, Giovanni Libralato and *Marco Guida*

14. **Copepods vs. Salmons: Environmental Treats for Crustaceans**   213
or Possible Eco-Sustainable Solutions?
Valerio Zupo, Valerio Mazzella, Patrick Fink, Mahasweta Saha,
Ylenia Carotenuto and *Mirko Mutalipassi*

15. **Automatic Culture of Crustaceans as Models for Science**   225
Francesca Glaviano and *Mirko Mutalipassi*

16. **Current Issues on Freshwater Crayfish Aquaculture with a Focus**   241
on Crustacean Welfare
Marina Paolucci, Elena Coccia, Gianluca Polese and *Anna Di Cosmo*

17. **Advanced Molecular Biology Techniques Applied to Crustacean**   274
Aquaculture
Maria Costantini, Roberta Esposito, Serena Federico and *Valerio Zupo*

Index   293

PART 1
Physiology Issues

CHAPTER 1

Crustacean Endocrinology
Fascinating Topic for Biologists or a Peculiar Opportunity for Biotechnologies? A Historical View with Functional Perspectives

Valerio Zupo[1],* and *Penny M. Hopkins*[2]

1.1 Introduction

About a decade has passed since Penny Hopkins [*1*] wrote a splendid review on crustacean endocrinology, to celebrate the 50th anniversary of publications for General and Comparative endocrinology, starting the discussion in 1961. This review, however, followed the one that the same author, in cooperation with David Borst, published in 2001 for American Zoologist [*2*]. The latter review, in its turn, was proposed to celebrate the career of Milton Fingerman, who published in 1997 a general analysis of crustacean endocrinology [*3*]. It appears that the history of crustacean endocrinology is enriched by periodic reconsiderations of this topic, since it contains valuable elements to comprehend general biology concepts.

It is generally considered that crustacean endocrinology started in the middle of 1920', when a series of studies were published [*4*], confirming that colour changes in crustaceans are under hormonal control [*5*], although some earlier observations [*6*] had already revealed evidence of endocrine regulation in moulting, coloration, and secondary sex characteristics [*1*]. The first anatomical structure with possible endocrine role discovered in crustaceans was the pericardial organ [*7*]. As a matter of fact, the first discovery of a hormone in crustaceans, with a clear indication of its nature and physiological role, is dated 1921, when Courrier defined for the first time the factors influencing the appearance of sexual characters in crustaceans. Further, Carlisle and Knowles [*8*] discovered a hormone having a pigmentary effect

---
[1] Stazione Zoologica Anton Dohrn, Department of Ecosustainable Marine Biotechnology, Villa Comunale. 80121 Naples, Italy.
[2] Department of Biology, University of Oklahoma, Norman, OK 73019.
* Corresponding author: vzupo@szn.it

in the body of crustaceans and they considered 1928 to be the year when the study of crustacean hormones began. This finding was part of efforts by the two authors to provide a complete model for Crustacean endocrinology that is… still missing, nowadays! However, it was preceded by various studies performed all over the world, indicating the role of sexual hormones [9], the anatomy of reproductive systems [10], the position and the role of glands involved in the moulting process [11]. After these key steps, i.e., after 1959, researches were marked by the discovery that most crustacean hormones are neuro-hormones, produced by specialized neuroendocrine cells. In fact, besides the first indication of endocrine activities in crustaceans related to pigmented cells, most of the studies that followed focused on reproduction.

The presence of steroid hormones in crustaceans was discovered in 1964 by Hampshire and Horn [12]. It was initially called "*Crustecdysone*", but later it was shown to be identical to the ecdysone of insects. Later studies suggested that the source of the crustacean ecdysone were the paired y-organs located in the anterior thorax of most crustaceans. Later Sagi et al. [13] showed that the endogenous gland (AG) of some male crustaceans produced a protein hormone necessary and sufficient for establishing maleness.

Several studies aimed at understanding the role of parasites on the biology, physiology and ecology of various species (e.g., *Sacculina*, [6], or Bophirids, [14,15]). Even more striking is the fact that, during a historical period mainly characterized by morphological studies, the attention of scientists was attracted by functional investigations, in the case of crustaceans. Crustaceans were subjects of pioneering studies in the field of physiology and in general of functional investigations, including regeneration of limbs. More recently, the endocrinology of crustaceans has moved from rudimentary techniques of extirpation and transplant to the modern tools aimed at elucidating gene structure and molecular functions of hormones, often coupled with evolutionary views of their effects in a range of organisms, including vertebrates.

Why does the endocrinology of crustaceans call for our attention? There are several reasons for studying this intriguing subject but among them, the most important may be summarized as:

(1) **A common architecture** characterizes such diverse taxa as early malacostracans and decapods, indicating that the general structure of the endocrine system has been established quite early in the evolution. For example, the AG provides a similar hormone and similar functions in amphipods, decapods and most crustaceans, as well as the sinus glands, y-organs and other fundamental endocrine structures probably established in less evolved arthropods.

(2) **Common basic rules** characterize the whole taxon, as in the determination of the sex and the moulting cycles, within a complex architecture of endocrine effectors, comprising dozens of different hormones still not completely identified. As a matter of fact, crustaceans share a very complex endocrine system and their endocrinology provides basic biology information, but it also has the potential for enhancing our ability to culture important species, as in the case of decapods [16].

(3) **Single glands may independently produce critical individual effects,** although a complex architecture is envisaged, as above stated. This is the case of sex differentiation, for example, that in crustaceans is controlled by a single gland, the AG: its presence produces a male individual, while its surgical removal produces immediately the shift to a female individual, with equilibrated physiology and functions [*13*]. All this simplicity, within a very complex physiological architecture, increases the possibility for testing and biotechnologies.

(4) **Several hormones share important functions** in different organisms, as in the case of diabetogenic hormones [*17*], juvenile and steroid hormones [*18*], hyperglycemic activities [*19*] and insulin-like hormones [*20*]. These similarities permit to tailor specific biotechnologies in various fields of medicine and aquaculture.

(5) **Neurosecretion** is widespread in crustaceans, as well as in vertebrates and in other invertebrates. The concept that nerves may produce and release hormones into the blood or hemolymph is quite exciting. The crustacean X-Organ/ Sinus Gland system (XO/SG) resembles the functioning of hypothalamic/ neurohypophysis system of vertebrates and represents a model for electric signals transformed into hormones, released from axonal endings into the circulation, to reach target tissues [*21*]. Also, hormones from the XO/SG are controlled by such neurotransmitters as dopamine and serotonin [*22*], and this opens clear opportunities for investigations in this field, using model crustaceans. In this chapter, some essential news about the structure and the physiology of crustacean's glands will be given, to facilitate the research of those willing to apply their functions and roles to understand basic biology mechanisms, or to develop novel biotechnologies.

1.2 The "families" of neuro-hormones

Besides the anatomical distribution and the physiology of glands, neuro-hormones of crustaceans have historically been organized into four main "families" in order to facilitate the comprehension of their complex relationships, synthesis and effects on target organs. A summary of their relationships and effects is given below based on Hopkins [*1*] in order to re-evaluate the discipline from a biochemical point of view.

The CHH family takes its name from the Crustacean Hyperglycemic Hormone and contains other immunoreactive peptides detected in the sinus gland and in the pericardial organs (PO) of decapods, as well as in extra-eyestalk nervous tissues in a wide variety of species including the gut, various CNS ganglia and other tissues. Various metabolic functions, including moulting, osmoregulatory functions, and ovarian control were demonstrated. CHH extracted from the eyestalks inhibit the levels of ecdysteroids and may raise the emolymph sugar, inhibiting the production of ecdysteroids by Y-organ, but isoforms extracted from the PO are inactive, indicating that different isoforms produced in different neuroendocrine glands may have a range of functions. The identification of hormones within this family started from an individual important hormone [*17*], the diabetogenic factor (DF) previously described. The DF is produced in the central nervous system and is released in response to various stresses clues. Further it was noted that the DF is part of a family of peptides with a variety of functions and sizes. The CHH peptides are structurally related, pleiotropic, and unique to crustaceans, although there are some CHH orthologues found in insects and involved in ion transport. The pleiotropic effects of CHH hormones include elevation of blood glucose, osmoregulation, control of female reproduction, control of the moult cycle, and probably various additional, still unknown functions. The CHH family is further divided into two types, I and II, according to the functions on target tissues and on the post-translational processes involved in their release.

The MIH family takes its name from the very important and first discovered Moult-Inhibiting Hormone. As indicated above, the ablation of eyestalks immediately triggers physiologic adaptations to start a moult process, and this demonstrated the presence of a "moult-inhibiting factor" in the eyestalk ganglia. Further, the production of moulting hormone (ecdysteroid) from the Y-organs (YO) located in the anterior body cavity was demonstrated. The onset of proecdysis and ecdysis represents a trigger for the cessation of MIH release in the hemolymph. MIH also inhibits other neurosecretory compounds involved in the stimulation of the production of ecdysteroids. While the synthesis of ecdysteroids is quite complex in insects and based on various peptides, in crustaceans it appears simpler and different.

The MIH peptide consists of five alpha-helices. MIH genes are expressed in the eyestalk ganglia as well as in optic nerves, ventral nerve cords, and thoracic ganglia. There are multiple forms, multiple sites of synthesis and several possible functions of MIH on blood glucose and on the moult cycle. Two MIH isoforms were identified in the eyestalks and neuroendocrine glands of shrimps, classified as MIH-A and MIH-B. The first acts as a moult-inhibitor. The second influences the gonad physiology. MIH-B may significantly prolong the moult cycle of shrimps with ablated eyestalks, but it was not as effective as MIH-A. However, various decapod groups share and are characterized by different isoforms of MIH. In eyestalk-less crustaceans a rapid decline in circulating ecdysteroids is observed, following ecdysis. This decline has been attributed to steroid negative feedback on the Y-organs, but the exact role of MIH has been debated within the complex mechanisms of moult control in crustaceans, due to the synergistic activity of inhibitory and stimulatory neuropeptides.

The VIH family takes its name from the Vitellogenin-Inhibiting Hormone that controls ovarian maturation by inhibition of the production of vitellogenin (Vg) synthesis. Vg is synthesized in ovarian and hepatopancreatic tissues and it is the precursor of the primary yolk protein—the vitellin, as reported and detailed in specific chapters of this book. VIH inhibits Vg synthesis and in fact, gene knockdown, using double stranded Vg RNA treatment in *Penaeus monodon*, resulted in premature ovarian maturation. However, the molecular structure of VIH is similar to that of MIH.

The PDH family takes its name from pigmentary-effector hormones and contains highly conserved peptides that range from eight to eighteen amino acids with effects on a variety of pigment-containing cells. Neuropeptides in the PDH family were the first to be found in the sinus gland of eyestalks. As above mentioned, the first neuropeptide sequenced was the red pigment concentrating hormone (RPCH). In fact, stellate cells in the epidermis, containing pigments, can change colour in response to the influence of the background. Four different chromatophores, white, red, black, and yellow are controlled in decapods to match background colours. The pigmentary effector family of hormones consists of pigment concentrating hormones, pigment dispersing hormones, and related peptides. Pigment concentrating hormones are octapeptides and they concentrate pigment in chromatophores and cause dark adaptation in eyes of various crustaceans, but it has no known effect on other metabolic functions. However, there is some overlap in the effects and species specificity of the PDH family in crustaceans, and RPCH is related to a similar peptide found in insects, the adipokinetic hormone (AKH). Finally, pigment-dispersing hormones have some neuromodulatory effects and may control circadian rhythms of crustaceans, having a conserved homology and anatomical location to circadian control cells of insects.

1.3 The steroid hormones

Sterols and steroid compounds are obtaining an increasing attention by scientists and biotechnologists, due to their key roles in several biological processes [23]. From the chemical point of view, steroids are a family of natural terpenoid lipids characterized by four fused alicyclic rings named "gonane". The biological functions of each individual steroid are determined by the oxidation state of their nucleus rings and the presence of distinctive functional groups [24]. Since steroids play important biological roles at various hierarchical levels (e.g., signalling mechanisms, cell membrane stabilization, regulation of cell proliferation and tissue differentiation) they may be key compounds for various biotechnological applications. In crustaceans, steroids represent a well-known and fundamental class of hormones [25]. All crustaceans produce endogenous sex steroid hormones similar to those of vertebrates [26]. The cytochrome P450 system rules synthesis of steroid hormones and their metabolism (including the production of hydroxylated or dehydrogenated metabolites) in crustaceans, but other enzymes are also involved in ecdysteroidogenesis. However, the biosynthetic pathways are different from those known in vertebrates [27].

Some steroids as estradiol and testosterone are present in the body homogenates of various crustaceans [28]. This is also the case of testosterone and other monohydroxylated and nonpolar testosterone metabolites [29]. The production of steroid hormones is largely influenced by the determination of the sex, both in gonochoristic and in hermaphroditic species. For example, testosterone levels in the hemolymph of crustaceans are largely higher in males and some categories, as androstenedione, are not detectable in females [29]. However, hemolymph levels of testosterone, 17α-hydroxytestosterone and 17β-estradiol did not significantly vary in females of crustaceans during the ovarian cycle [30] and this finding calls for more detailed information on the actual roles of steroid hormones in the reproduction of crustaceans. Consequently, although several studies documented the presence of steroids in the haemolymph of crustaceans and their vertebrate-like sex hormones, the mechanisms of their action in crustaceans (and Protostomes in general) are only poorly understood. Similarly, some steroid hormones collectively called "ecdysteroids" may play a gonadotropic role in crustaceans and hemolymph trends of ecdysteroids are correlated with those of sequestered ecdysteroids in ovaries [*31*]. However, it is still unknown if they may influence ovarian development.

The arthropod moulting hormone was first isolated from insects and it was called ecdysone (Butenandt and Karlson 1954), then identified as well in crustaceans, with the first isolation of 20-hydroxyecdysone (20E) from spiny lobsters [*32*]. The moulting hormones produced by Y-organ in all ecdysozoans (i.e., insects and crustaceans) are steroids as well. From the biosynthetic point of view, cholesterol (a steroid hormone precursor) is delivered to Y-organ, where steroidogenesis takes place on mitochondria and microsomes, similarly to the biosynthetic pathways observed in all vertebrates [*33*].

Various categories of hormones may interact in the physiology of crustaceans. For example, it has been proposed that the genesis of ecdysteroids in the Y-organ is suppressed by the Moult Inihbiting Hormone (MIH) during most time during the moult cycle, to be relieved from the suppression during the pre-molt stage, when the levels of the MIH are low [*34*]. Altogether, all published data strongly suggest that MIH acts directly on the Y-organs to suppress ecdysteroidogenesis. Ecdysteroids by Y-organ are considered to influence both the duration of the moulting cycle and the onset of reproduction [*35*].

It is worth to note that a specific class of sesquiterpenoids actively influence the metabolism of lipidic compounds in the plasma of crustaceans. Methylfarnesoate (MF) is a sesquiterpenoid structurally related to the juvenile hormone III of insect [*36*]. This is a secretory product of the mandibular organ, which participates in controlling growth and reproduction of crustaceans [*37*]. The existence of an inhibitory factor that negatively regulates MF synthesis was proposed [*38*] based on the observation that eyestalk ablation caused an increase in the MF levels in the hemolymph [*39*], while injections of eyestalk extracts reversed the effect of eyestalk ablation [40]. MF also indirectly affects moulting, probably via signaling of 20-hydroxyecdysone (20-HE).

On the overall, moulting is regulated by ecdysteroid hormones in close cooperation with the development and reproduction cycles in crustaceans (cross-links with methyl-farnesoate signalling), but their susceptibility to toxicants still needs to be studied. Other biochemical targets for xenobiotics were recently discovered in crustaceans and these should be explored by further ecotoxicological studies to obtain new molecular biology information about ecdysteroid receptors. Some sex steroid hormones known from vertebrates, as testosterone and progesterone, have been reported in crustaceans, but the knowledge about their targets (crustacean steroid receptors) and signalling is incomplete. The determination of the sex in developing juveniles (affecting the sex ratio in populations, as it will be described in further chapters) is a parameter sensitive to various xenobiotics (including endocrine disruptors) and external influences [*41*] but its modulation by general environmental stress and non-specific toxicity is still largely unknown. In parallel, steroid receptors (SR) were detected initially in Deuterostomes [*42*] (Figure 1).

However, the existence of ancestral steroid receptor (AnSR) was postulated also in their predecessors (including Protostomes) and set at about 600–1200 million years ago [*43*]. An estrogen-related receptor (ERR) was isolated from *Drosophila melanogaster*, and its presence was paralleled in females of *Gammarus fossarum*

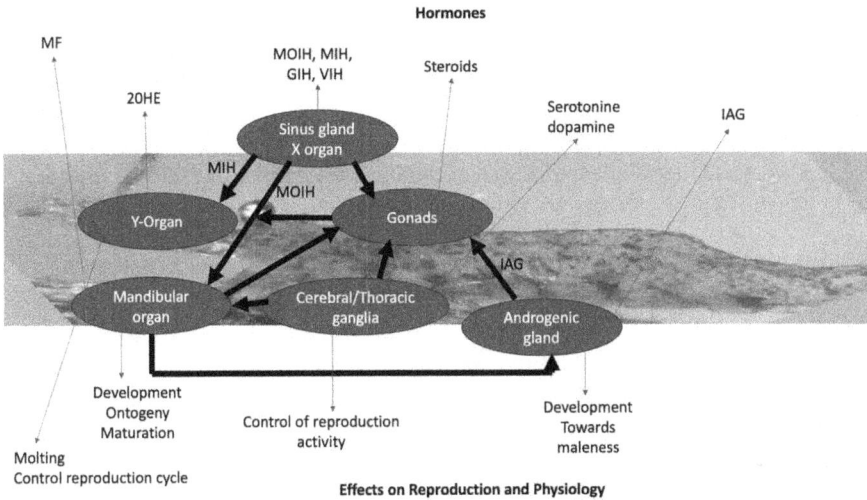

Figure 1. The structure of the endocrine system of crustaceans. The main glands are indicated in ovals and the hormones they produce are reported in the upper part of the scheme. Some of the main relationships among glands, and some of the hormones involved in such relationships, are indicated in the figure. The principal effects on reproduction and physiology are indicated in the lower part of the scheme.

using a cross-reactivity assay with antibodies raised against highly conserved DBD domain of vertebrate ERα ortholog.

1.4 The main hormonal axes: eyestalks, sinus gland, and neurosecretory mimicry

As above stressed, first investigations on crustaceans indicated that the bulk of the endocrine system was located in the central nervous system (CNS) [*44*], including thoracic ganglia and glands located close to the gut and the hearth, like the pericardial organs (PO), located on the lateral walls of the pericardial cavity. The anterior cardiac plexus (ACP) is a neuroendocrine structure too, located in the stomatogastric nervous system, as a branch of the CNS controlling the gut, and it comprises the stomatogastric ganglion (STG), which releases important neuro-hormones too. The commissural organs consist of nerve trunks that run between connectives, lateral to the oesophagus, and the stomach, as part of the stomatogastric nervous system [*45*]. However, probably due to their distal position, but also to the complexity of the neuroendocrine arches therein located, eyestalks were among the most studied structures in all arthropods since Zeleny [*46*]. Eyestalks are *de facto* considered as an extension of crustacean's brain and they produce two families of neuropeptides, i.e., the hyperglycemic hormones (CHH) and the pigmentary effectors (PDH), which will be further described. Studies by Koller [*47*] demonstrated that extracts of the eyestalks produce immediate blanching in the body of *Crangon crangon* and, given the ease of detaching and testing extracts of eyestalks, this was one of the first body region investigated for hormonal production. In 1933, Hanstrom highlighted the importance of a "blood gland" in the eyestalk, whose role was further clarified by the same author in 1937 (and the name shifted to "sinus gland") when Hanstrom [*48*] noted that the gland is located next to a large hemolymph sinus, in the eyestalk, especially in decapods exhibiting well-developed pedunculated eyes. In sessile-eyed crustaceans, as amphipods and isopods, the sinus gland migrates in the head, close to the optic centres [49], but it is always present and quite important for various physiologic activities [50], including mimicry. Evidently, the fact that changes of colour (often produced in response to the colours of the surrounding environment) are related to the neuroendocrine activity of a gland located within the eyes and the optic centres definitely makes sense. Thus, Brown [51] demonstrated in his early studies that some large "neurons" previously detected by invertebrate anatomists close to the central nervous organs of crustaceans, were actually glandular organs, secreting substances with clear hormonal functions, because they controlled the physiology of chromatophores. The importance of neurosecretory products for the biology of crustaceans was further established by Bliss (1956) and Kleinholtz (1966).

The chromatophores of crustaceans are a fascinating topic of research [52], since most crustaceans are characterized by evident mimicry [41] and, as above mentioned, pigment cells primed the studies of crustacean endocrinology yet in 1921. Aggregation of chromatophores dispersed in the integument of crustaceans is responsible for the change of colour of crustaceans, although their mimicry and the speed of change may vary according to the species, as firstly indicated by studies on colours changes performed by Koller [53] in *Crangon Crangon*. In addition, crustaceans change the position of pigmented cells in the retina of ommatidia, in their compound eyes, and the first experiments indicated that the transfusion of hemolymph from a pale specimen to a dark one produced an impressive change of colour. Further studies [5] demonstrated that chromatophores are not directly innervated, thus their colour had to be ruled by hemolymph-born compounds and this scientist also discovered that the eyestalk is an important and central endocrine organ influencing (among other activities) also the pigmentation of crustaceans. Crustaceans have normally various types of chromatophores (red, black, yellow, white, etc.) and the final colour of their body depends on the state of expansion/contraction of each type of chromatophore. In species able to adapt the colour of the integument according to the background, as exemplified by the shrimp *Hippolyte inermis* the dispersion or concentration of pigments of different colours produces a pattern that may precisely resemble the image of the substratum (Figure 2).

Thus, various pigments must be controlled independently [54] and it is still debated if various hormones may produce these effects or if a nervous effector may play locally a role. As well, retinal pigments (both distal, and reflecting) are under hormonal control [55], based on a light-adapting neuro-hormone (LAH) and a dark-adapting hormone (DAH) [56], while the proximal pigments of the retina are neutrally controlled [57]. Definitely, the adaptation to light and dark and the physiology of eye pigments represents a fascinating matter since it comprises various subsequent hormonal and non-hormonal arches, which represent a unique case in the field of invertebrate biology, at the best of our knowledge.

After the first discovery of the activity of eyestalks in the control of mimicry functions, further studies [58] demonstrated that they must contain other glands, besides the sinus gland. In fact, the ablation of a sinus gland produces changes in the colour of crustaceans, but also precocious moulting, as observed by ablating

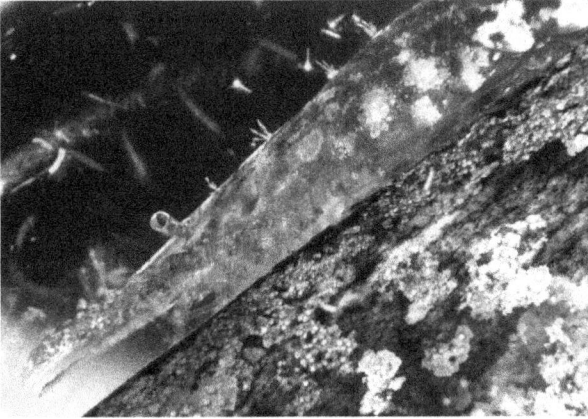

Figure 2. An adult female of *Hippolyte inermis* on a leaf of *Posidonia oceanica* repeats the patterns of colors and shapes of the leaf surface with its epiphytes.

whole eyestalks in several decapods [*59*]. As a consequence of these studies, a complex cluster of neurosecretory axonal terminations was identified in the *medulla terminalis* of eyestalks and it was named "X-organ", because its functions were still unknown [*60*]. It was then demonstrated (quite interestingly) that these axons are simply pipelines for neuro-hormones produced in distant nerve cell bodies and function merely as sites of storage-release [*61*]. The complex of the *medulla terminalis*-X organ resembles the functions of hypothalamic-neurophypophysis complex in vertebrates: this is another awesome feature of crustaceans!

It was very early observed that the ablation of both eyestalks brings on precocious moulting activity in *Uca pugilator* and less precipitously, but still evident, when one single eyestalk is ablated. The sinus gland contains a moult-inhibiting hormone (MIH, that is a peptidic neuro-hormone) and consequently, the partial (1 eyestalk) or total (2 eyestalks) removal of this hormone induces immediate moulting [*62*]. In contrast the Y-organ, located in the anterior thorax of most crustaceans, secretes ecdysteroids [*11*], i.e., the moulting hormones characterizing all arthropods. MIH inhibits the release and/or synthesis of ecdysteroids. In addition, the complex medulla terminalis-X-organ produces some compounds (as 3-hydroxyl-L-kiurenine (3-OH-K) and xanthurenic acid (XA)) that inhibit the synthesis of ecdysteroids by the Y-organ. An interesting interaction with these hormones is the one involving the so-called juvenile hormone, methylfarnesoate (MetF) [*63*], because it was demonstrated that MetF stimulates ecdysteroid secretion by the Y-organ of decapods.

The ablation of eyestalks also produces maturation of the ovaries in decapod crustaceans, because the sinus gland contains a gonad-inhibiting neuro-hormone (GIH), both in males and in female crustaceans. For this reason, the ablation of eyestalks (with a series of accurate operations performed by experts to avoid physiologic damage to the animals) is applied in shrimp farms all over the world, to prime reproduction. To this end, normally a single eyestalk is ablated in females, to avoid dramatic changes of the prawn physiology and keep high survival rates, and it is sufficient to prime egg maturation and reproduction. In addition, a gonad stimulating hormone (GSH) has been identified, produced in the brain and the thoracic ganglia of males and females of prawns and other crustaceans [*64*]. Both GIH and GSH are peptidic neuro-hormones and they both act, in males, on the physiology of the AG, whose hormone (an insulin-like compound) controls the physiology of testes and maintains the secondary sexual characters. Here there is another important material for biotechnologies, since the manipulation of AG in aquaculture decapods may lead to the production of monosex stocks of shrimps and prawns, with very interesting results on the productions [*65*]. This technique will be described more in detail in further chapters. In addition, the AG may be affected (and destroyed) by environmental influences and this produces interesting ecological interactions and further biotechnologies in the field of human medicine [*66*]. This topic will be discussed in further chapters but it is evident that the physiology of AG in crustaceans is a very interesting topic [*67*]. In fact, in male crustaceans, unlike other invertebrates and vertebrates, the endocrine and gametogenic functions are separated in different organs, i.e., the AG and the testis [*68*].

In addition, the sinus gland produces a substance (a diabetogenic factor also known as hyperglycemic hormone, CHH) that triggers an increase in the concentration of glucose of the hemolymph [*69*]. It is worth noting that MIH and CHH exhibit similar activities and possibly interact with the same receptors on Y-organs, thus, the role of these multifunctional peptides deserves further investigation. However, we still miss complete information about the structure and the functioning of receptors for both steroid and peptide hormones of crustaceans [*2*].

1.5 Other neuroendocrine glands

Over 100 neuropeptides, belonging to several families, were identified in decapods, euphausids and copepods. Such an abundance and diversity of peptides in the nervous system indicates the complexity of crustacean hormonal pathways and the possibilities for hormonal crosstalks. Several peptides are neurotransmitters or neuromodulators controlling the cascade release of other neuro-hormones. In some instances, particularly for those produced in neurosecretory tissues and neurohemal organs, they may directly act as neuro-hormones on target tissues. In 1935, evidence of neurosecretion in the central nervous ganglia of *Palaemonetes* sp. was obtained [*51*]. However, further studies [*70*] indicated quite a complex picture, since the post-esophageal commissure is another important source of hormones, producing effects on the mimicry of *Crangon septemspinosa*. Two post-commissural nerves, in fact, bear neurosecretory axons [*44*] further demonstrated in various decapods [*71*] involved in the physiology of crustacean chromatophores. Further, some nerves located in the pericardial cavity of crustaceans (pericardial organs) produce neuro-hormones influencing the physiology of the heart [*7*] and based on these observations, the first concept of "neurohemal organs" was proposed [*8*] and applied to various animal groups, including vertebrates.

Molecular evidence revealed the presence of cardioactive peptides, corozonin, allostatin, and neuroparsins. The functions of these peptides are not totally known. However, it was demonstrated that the heart of *Maya squinado* actively responds to extracts of pericardial organ, containing aminergic and peptidergic compounds, with increases in amplitude and decrease of frequency. Interestingly, two cardioprotective peptides (Proctolin and Nonapeptide) have been isolated from the pericardial organs of crustaceans [73] and this finding triggers novel biotechnologies made available by the investigation of crustacean physiology. Besides these important features, the neuroendocrine centres, including eyestalks, brain and thoracic ganglia, produce hormones that regulate salt and water balance, through ion transportation, both in freshwater and in marine species [73]. The research on neuroendocrine systems of crustaceans is still open, but it is evident how the endocrinology of these invertebrates can be extended to our human physiology, and may be, as well, useful for biotechnological approaches. For example, several studies have shown the effects of neurotransmitters (e.g., biogenic amines as 5-HT and CHH) on the hormonal releases from neuroendocrine centres [74]. Remarkably, 5-HT also induces the release of GSH both in males and females. Dopamine (DA) inhibits gonad maturation (blocking the release of GSH and contemporaneously triggering the release of GIH). Also 5-HT triggers the release [22] of red pigment dispersing hormone (RPDH). Consequently, individual hormones are involved in various functions and this explains why the ion equilibria, the body colour and the maturation of gonads are strictly interconnected in crustaceans. For example, in several species, males become reddish and change their behaviour during the reproductive period, and they may change their exposure to light, since these multiple activities are modulated by the same hormones [75]. The rhythms of moults and the final metamorphosis of decapod larvae are often dependent on light intensity and this finding is key to the production of post-larvae by valuable aquaculture species.

As above stated, several neuro-hormones are peptidic in structure and some of them have been finally identified and sequenced. In addition, peptides are important to assure the activity of steroid hormones (the so-called ecdysteroid binding proteins) [76]. The first crustacean neuro-hormone completely sequenced by Fernlund and Josefsson [77] was RPCH. It is an octapeptide showing some differences in various crustaceans. Other research identified the structure of PDH, and of the "crustacean cardioactive peptide" (CCAP, a nonapeptide). Likewise, GSH is a peptide and the determination of its structure has biotechnological value, since it is involved both in hyperglicemic processes and as a stimulant for the reproductive system. Finally, the peptidic structures of the gonad-inhibiting hormone (GIH), the molt-inhibiting hormone (MIH) and crustacean hyperglycemic hormone (CHH) were identified, showing that they all share similar length (71–78 aminoacids) and that isomeric forms of single aminoacids lead to dramatic changes of activity. The AG insulin-like hormone (IAG) has been recently identified [20] and this represents an additional important step for biotechnologies because it inhibits vitellogenenesis by inhibiting leucine incorporation into ovaries and produces immediate masculinization in treated females of crustaceans [78].

1.6 Non-neural endocrine glands

In the anterior sectors of the body of crustaceans, are found the steroid-producing Y-organs named as a counterpoint of the neuro-hormonal X-organs. The Y-organs are a pair of glands positioned in the maxillary segment of the head of crustaceans, mimicing the function of the prothoracic gland of insects, and involved in the crustacean moulting processes [*11*]. As a matter of fact, various diuretic hormones, ecdysis-triggering hormones, and the eclosion hormone, among several others, are neuro-hormones with well-established endocrine roles in insects, but still undefined in crustaceans. The two key glands involved in important processes, as the mimicry and the moulting cycle [*79*], were named X-organ and Y-organ, respectively, because their roles were initially hypothesized but not demonstrated. Research identified their functions, but the names are still used to indicate these glands, having quite different histological characterization (neuroendocrine the first, non-neural the second). Overall, ecdysteroids may also have a role in the control of molting and the synthesis of yolk. In amphipods and isopods [*80*] Methylfarnesoate (MF) is produced by mandibular organs [*18*]. MF is a terpenoid hormone [*63*] that is structurally related to the juvenile hormone (JH) of insects. It is produced in the mandibular organ (MO) which is located on the posterior surface of the lower apodeme and in the vicinity of the posterior mandibular adductor muscle.

The red pigment concentrating hormone (RPCH), besides its role in the control of pigmentation, stimulates the synthesis of methylfarnesoate [*81*]. In contrast, the pigment-dispersing hormone (PDH) inhibits the synthesis of methylpharnesoate [*82*]. This may at least partially explain the above-mentioned relationship between light irradiance and photoperiod, with reproduction of crustaceans and the larval growth, since hormones linked to the light perception of crustaceans (PDH and RPCH) also have a direct influence on the synthesis of other hormones involved in growth and reproduction. This topic deserves further in-dept studies to be completely understood and explained.

Proceeding caudally in the body of crustaceans we detect another endocrine gland, present only in males and thus named Androgenic Gland (AG) after Cronin (1947). It is a ductless small gland, commonly attached to the sub-terminal ejaculatory region of the *vas deferens* or the sperm-ducts of male crustaceans (with a few exceptions). In 1954 Charniaux-Cotton identified its actual function and proposed a name. This gland is responsible for the appearance of male characters, such as the presence of an *appendix masculina* and the development of testes and *vas deferens*. Masculinization of the external characters of female decapods was observed after the implantation of an androgenic gland [*83*]. However, a range of abnormal development of gonads was observed in andrectomized males, depending on the species and the age at which the ablation was performed. In adult males of several species the gland doesn't seem to be necessary for sex differentiation [*65*]. There is an innate "femaleness" of crustaceans, and the actual role of the AG is to prevent feminine development in the early stages of post-larvae, by an inhibition of differentiation of the juvenile gonad into an ovary. In fact, the administration of diatom extracts (known to produce a destruction of AG in the shrimp *Hippolyte inermis*) triggers a complete shift of sex in young shrimps if conducted in the first 10 days of post-larval development [*84*]. In contrast, a late administration of diatom extracts after the second week of post-larval development produces incomplete females, scarcely developed males, or it has no effects [*41*]. However, in further stages of development, in several crustaceans, the ablation of the AG (or its natural decay, which occurs in protandric or hermaphroditic species) leads to the development of an ovotestis, which further transforms into an ovary (in sex shifting species). These differences were related to the tissue stability characterizing the AG of different species of decapods, which leads to gonochoristic species when it is high, or to complete hermaphroditism when it is low, with all the intermediate stages characterizing different groups of crustaceans. These interesting relationships will be detailed in the chapter of this book dedicated to the reproductive biology of crustaceans.

From a histological point of view, the AG takes the shape of thin chords of cells [85] or of compact lobes [86] as observed in *Hippolyte inermis* [87], or also a combination of the two structures, as found in *Macrobrachium rosenbergii* [88]. Hormones produced by the AG also induce regression and total destruction (by apoptosis processes) of an ovary, if implanted in adult female crustaceans, and an immediate transformation of ovarian tissues into testes.

In genetic female crustaceans this gland is absent (from hatching, in gonochoristic species, or after partial or total regression, in hermaphroditic species) but ovaries are a source of another hormone (the so called "ovarian hormone") triggering the development of secondary sexual characters in females, as the ovigerous setae on the appendix interna of pleopods [9]. Interestingly, the testes of adult crustaceans seem to be unable to produce any hormones, since the simple presence of an AG and its hormone (IAG) assure all the necessary functions of maleness.

1.7 External influences on hormonal control of crustaceans

Besides the effects of light and temperature on reproduction and growth, previously discussed and easily explained by analysing the multiple effects of neuroendocrine effectors, various environmental pressures may modify the complex functioning of the endocrine systems of crustaceans. In this category we must ascribe various pollutants, whose individual and direct influence on endocrine axes has been demonstrated. For example, heavy metals influence the limb regeneration and the moulting process of decapod crustaceans [89]. Methylmercury and Cadmium inhibit these processes (in particular, Cadmium inhibits BPDH synthesis), while Naphtalene induces ovary athresia by producing an accumulation of neurosecreted hormones, including GSH, in the brain and BPDH in the eyestalks. Contemporaneously, both Cadmium and Naphtalene induce hyperglicemia in the emolymph of decapod crustaceans [90]. All these relationships bear a tremendous biotechnological power in various fields. First of all, given the relationships between body colour and hormonal control, the possibility of using crustaceans as bioindicators for pollutants (since it is easy to detect their colour changes) may be envisaged. Of course, the same relationships may be applied for conservation issues and to rapidly check the status of health in natural populations of various crustaceans. Finally, in aquaculture practices, some compounds could be used to improve the colour (that is directly influencing the market price) of valuable species of decapods.

References

1. P.M. Hopkins. (2012). *Gen. Comp. Endocrinol.* **175**: 357–66.
2. P.M. Hopkins, and D.W. Borst. (2001). *Amer. Zool.* **41**: 361–363.
3. M. Fingerman. (1997). *Physiological Zoology* **70**: 257–269.
4. G. Koller. (1927). *Z. Vergleichende Physiol.* **5**: 192–246.

5. E. Perkins. (1928). *J. Exp. Zool.* **50**: 71–105.
6. R. Courrier. (1921). *CR Hebd Acad. Sci.* **173**: 668–71.
7. F. Jolyet, and H. Viallanes. (1893). *Ann. Sci. Nat. Zool. Biol. Anim.* **14**: 387–404.
8. D.B. Carlisle, and F.G.W. Knowles. (1959). *Endocrine control in Crustaceans.* Cambridge University Press, Cambridge.
9. H. Charniaux-Cotton. (1952). *CR Hebdomadaires Acad. Sci.* **234**: 2570–2572.
10. L.E. Cronin. (1947). *J. Morphol.* **81**: 209–239.
11. M. Gabe. (1953). *CR Hebdomadaires Acad. Sci.* **237**: 1111–1113.
12. F. Hampshire. (1966). *Horn. Chem. Commun.* **1966**: 37–38.
13. A. Sagi, E. Snir, and I. Kalahila. (1997). *Inv. Repr. Dev.* **31**: 1–3.
14. G. Reverberi. (1942). *Pubbl Stazione Zoologica Napoli* **19**: 225–316.
15. G. Reverberi, and M. Pitotti. (1942). *Pubbl. Stazione Zoologica Napoli* **19**: 111–184.
16. G.M. Hoover, S.H. Sokolow, J. Kemp et al. (2019). *Nat. Sustain.* **2**: 611–620.
17. A. Abramowitz, F. Hisaw, and D. Papandrea. (1944). *Biol. Bull.* **86**: 1–5.
18. D.W. Borst, H. Laufer, M. Landau et al. (1987). *Insect Biochem.* **17**: 1123–1127.
19. E.S. Chang, G.D. Prestwich, and M.J. Bruce. (1990). *Biochem. Biophys. Res. Commun.* **171**: 818–826.
20. T. Ventura, and A. Sagi. (2012). *Biotechnol. Adv.* **30**: 1543–50.
21. R. Andrew, and A. Saleuddin. (1978). *Can. J. Zool.* **56**: 423–430.
22. M. Fingerman. (1995). *Amer. Zool.* **35**: 68–78.
23. L. Fernández-Cabezón, B. Galán, and J.L. García. (2018). *Front. Microbiol.* **9**: 958–962.
24. D. Lednicer. (2011). *Steroid Chemistry at a Glance.* Chichester, Wiley.
25. E.S. Chang, S.A. Chang, and E.P. Mulder. (2001). *Am. Zool.* **41**(5): 1090–1097.
26. W.S. Baldwin, D.L. Milam, and G.A. LeBlanc. (1995). Environ. *Toxicol. Chem.* **14**: 945–952.
27. M.O. James, and S.M. Boyle. (1998). *Comp. Biochem. Physiol.* **121**: 157–172.
28. D.J. Huang, and S.Y. Wang. (2004). *Chen. Chemosphere* **57**: 1621–1627.
29. T. Verslycke, K. De Wasch, H.F. De Brabander, and C.R. Janssen. (2002). *Gen. Comp. Endocrinol.* **126**: 190–199.
30. J. Martins, K. Riberio, T. Rangel-Figueiredo, and J. Coimbra. (2007). *J. Crustac. Biol.* **27**: 220–228.
31. T. Subramoniam. (2000). *Comp. Biochem. Physiol.* **125**: 135–156.
32. D.H.S. Horn, E.J. Middleton, J.A. Wunderlich, and F. Hampshire. (1966). *Chem. Com.* **1966**: 339–340.
33. E. Spaziani, M.P. Mattson, W.N.L. Wang, and H.E. McDougall. (1999). *Am. Zool.* **39**: 496–512.
34. T. Nakatsuji, C.Y. Lee, and R.D. Watson. (2009). *Comp. Biochem. Physiol. A Mol. Integr. Physiol.* **152**: 139–48.
35. E. Mazurová, K. Hilscherová, R. Triebskorn, H.R. Koehler, B. Maršálek, and L. Bláha. (2008). *Biologia* **63**: 139–150.
36. E. Homola, and E.S. Chang. (1997). *Comp. Biochem. Physiol. Biochem. Mol. Biol.* **117**: 347–56.
37. H. Laufer, N. Demir, X. Pan, J.D. Stuart, and J.S. Ahl. (2005). *J. Insect Physiol.* **51**: 379–84.
38. H. Laufer, D. Borst, F.C. Baker, C. Carrasco, M. Sinkus et al. (1987). *Science* **235**: 202–205.
39. D.W. Borst, J. Ogan, B. Tsukimura, T. Claerhout, and K.C. Holford. (2001). *Amer. Zool.* **41**: 430–41.
40. B. Tsukimura, and D.W. Borst. (1992). *Gen. Comp. Endocrinol.* **86**: 287–303.
41. V. Zupo. (2000). *Mar. Ecol. Prog. Series* **201**: 251–259.
42. J.W. Thornton, E. Need, and D. Crews. (2003). *Sci. Total Environ.* **301**: 1714–1717.
43. J.W. Thornton. (2004). *Nature Rev. Genet.* **5**: 366–375.
44. F. Knowles. (1953). *Nature* **171**: 131–132.
45. Y.W. Hsu, D. Messinger, J. Chung, S. Webster, H. de la Iglesia, and A. Christie. (2006). *J. Exp. Biol.* **209**: 3241–3256.
46. C. Zeleny. (1905). *J. Exp. Zool.* **2**: 1–102.
47. G. Koller. (1930). *Z. Vergleichende Physiol.* **12**: 632–667.
48. B. Hanstrom. (1937). *Kunglica Sven Vetenskapsakademiens Handlingar* **16**: 1–99.
49. B. Hanstrom. (1939). *Hormones in Invertebrates.* Oxford University Press. New York.
50. F.A. Brown. (1940). *Physiol. Zool.* **13**: 343–355.
51. F.A. Brown. (1935). *J. Exp. Zool.* **71**: 1–15.
52. M. Fingerman. (1963). *The control of chromatophores.* Pergamon-MacMillan. New York, 1–84.

53. G. Koller. (1925). *Verh Dtsch Zool. Ges.* **30**, 128–132.
54. L. Hogben, and D. Slome. (1931). *Proc. R Soc. Lond. Biol.* **108:** 10–53.
55. L.H. Kleinholz. (1936). *Biol. Bull.* **70:** 159–184.
56. M. Fingerman, M.E. Lowe,and B.I. Sundararaj. (1959). *Biol. Bull.* **116:** 30–36.
57. C. Ludolph, D. Pagnanelli, and M.I. Mote. (1973). *Biol. Bull.* **145:** 159–170.
58. L.M. Passano. (1951). *Anat. Rec.* **111:** 501–502.
59. D.E. Bliss. (1951). *Anat. Rec.* **111:** 502–503.
60. J.H. Welsh. (1941). *J. Exp. Zool.* **86:** 35–49.
61. I.M. Cooke, and R.E. Sullivan. (1982). *The biology of Crustacea Vol 3, Neurobiology. Structure and function.* Academic Press New York, 205–290.
62. F. Lachaise, A. Le Roux, M. Hubert, and R. Lafon. (1993). *J. Crustacean Biol.* **13:** 198–234.
63. H. Laufer, D. Borst, F.C. Baker, C.C. Reuter, L.W. Tsai et al. (1987). *Science* **235:** 202–5.
64. S. Eastman-Reks, and M. Fingerman. (1984). *Comp. Biochem. Physiol.* **79:** 679–684.
65. A. Sagi, and D. Cohen. (1990). *World Aquacult.* **21:** 87–90.
66. V. Zupo, F. Jüttner, C. Maibam, E. Butera, and J.F. Blom. (2014). *Mar. Drugs* **12:** 547–567.
67. H. Charniaux-Cotton, and G. Payen. (1988). *Endocrinology of selected invertebrate types.* Liss, New York, 279–303.
68. T. Ginsburger-Vogel, and H. Charniaux-Cotton. (1982). *The Biology of Crustacea.* Orlando Fl. Academic press, 257–281.
69. E.A. Santos, and R. Keller. (1993). *Comp. Biochem. Physiol.* **106:** 405–411.
70. F.A. Brown. (1946). *Physiol. Zool.* **19:** 215–217.
71. M. Fingerman. (1966). *Am. Zool.* **6:** 169–179.
72. J. Stangier, C. Hilbich, K. Beyreuther, and R. Keller. (1987). *Proc. Natl. Acad. Sci. USA,* **84:** 575–579.
73. F.I. Kamemoto. (1991). *Zool. Sci.* **8:** 827–33
74. G.K. Kulcarni, and M. Fingerman. (1992). *Biol. Bull.* **182:** 341–347.
75. R. Nagabhushanam, R. Sarojini, P.S. Reddy, and M. Fingerman. (1995). *Curr. Sci.* **69:** 659–671.
76. M. Londershausen, and K.D. Spindler. (1985). *Amer. Zool.* **25:** 187–196.
77. P. Fernlund, and L. Josefsson. (1972). *Science* **177:** 173–175.
78. T. Levy, V. Zupo, M. Mutalipassi, E. Somma, N. Ruocco et al. (2021). *Front. Mar. Sci.* **8:** 745540.
79. D.M. Skinner. (1985). *Amer. Zool.* **25:** 275–284.
80. J.J. Meusy, M.F. Blanchet, and H. Junbra. (1977). *Gen. Comp. Endocrinol.* **33:** 35–40.
81. G. Gäde, and H.G. Marco. (2013). *Handbook of Biologically Active Peptides*, 185–190.
82. M. Iga, T. Nakaoka, Y. Suzuki, and H. Kataoka. (2014). *PLoS One* **9**(7): e103239.
83. C. Nagamine, and A.W. Knight. (1987). *Inv. Reprod. Dev.* **11:** 77–85.
84. V. Zupo, and P. Messina. (2007). *Mar. Biol.* **151:** 907–917.
85. B.M. Carpenter, and R. De Roos. (1970). *Gen. Comp. Endocrinol.* **15:** 143–157.
86. L.H. Kleinholtz, and R. Keller. (1979). *Hormones and Evolution.* Academic PressNew York, 160–213.
87. V. Zupo, P. Messina, A. Carcaterra, E.D. Aflalo, and A. Sagi. (2008). *Invertebr. Reprod. Dev.* **52:** 93–100.
88. W.J. Veith, and S.R. Malecha. (1983). *S. Afr. J. Sci.* **79:** 84–85.
89. J.S. Weis. (1978) *Mar. Biol.* **49:** 119–124.
99. P.S. Reddy, R.V. Katyayani, and M. Fingerman. (1996). *Bull Environ. Contan. Toxicol.* **56:** 425–431.

CHAPTER 2
Sexual Biology and Reproduction

Valerio Zupo[1,]* and *Alan N. Hodgson*[2]

2.1 General considerations

Crustaceans are morphologically, physiologically, and ecologically a highly diverse taxon and correspondingly diverse in their reproductive characteristics [*1*]. Crustaceans evolved various structural adaptations since they first appeared in the Cambrian [*2*] to ensure reproductive success, so providing key study material for comparative morphologists and evolutionary biologists. With such a long evolutionary history, they have had sufficient time to develop a variety of life history specialties [*3*]. These adaptations ensure successful production of offspring up to the first larval molt. Several biological features uniquely characterize crustaceans and this justifies why they are a key taxon in the development of suitable research models to explore new issues and pathways in various fields of physiology and ecology [*4*]. Some special reproductive features, such as the gelatinous fertilization tent and the safety lines of freshwater crayfish, are only found in this large group of invertebrates. In some Crustacea giant spermatozoa (further described in Section 2.6) evolved 100s of millions of years ago, and despite their high energetic cost and mechanical constraints, they still persist [*5*].

Crustaceans exhibit, in various taxonomic groups, unique features in terms of sex determination (ranging from primitive polygenic systems to chromosomal sex determination), cytoplasmic sex factors (exclusive for various arthropods, with the exception of aphids), sex differentiation (with the unique functions of the Androgenic Gland (AG)), sexual systems (according to the stability of AG tissues, as explained in the previous chapter) and intersexuality. Other unique reproductive aspects are found in their mating systems, spermatogenesis, spermatophore production and morphology, sperm transfer mechanisms, functions of the accessory reproductive glands and fertilization (including male or female heterogamety), oogenesis and

---

[1] Stazione Zoologica Anton Dohrn, Department of Ecosustainable of Marine Biotechnology, Villa Dohrn, Punta San Pietro, 80077 Naples, Italy.
[2] Department of Zoology and Entomology, Rhodes University, Grahamstown, South Africa, 6140.
* Corresponding author: vzupo@szn.it

endocrine regulation of vitellogenesis, yolk utilization, embryonic nutrition, mating behaviors [6], sex pheromones [7], reproductive cycles, brood care, larval dispersal systems, and environmental effects on reproduction and development (as outlined in the previous chapter but including epigenetic effects).

A detailed description of all these topics would require a whole book. Nevertheless, we will try to synthesize the most important issues especially useful for reproductive, evolutionary and developmental biologists, as well as for all invertebrate biologists, comparative endocrinologists and aquaculture specialists, by adopting schematic representations to synthesize a huge amount of available information. For example, crustaceans are perfect models to investigate sperm storage and paternity as general biology concepts. Multiple paternity and long-term sperm storage are important factors in population biology and fisheries biology and several mechanisms of paternity assurance are exemplified by crustaceans, such as precopulatory mate guarding [8], frequency of copulation before egg laying [9], and the displacement or removal of sperm of predecessors. In addition, amphipods and isopods are good models for studying mate guarding [10] while decapods may be models for studying sperm displacement, contributing to the key topic of sperm competition [11]. References available at the end of this chapter will help improve the study of these important topics.

Although some aquatic arthropods, including crustaceans, exhibit external fertilization, males and females do not simultaneously broadcast spawn gametes into the water, as commonly observed in many aquatic invertebrates and most fish. In fact, the majority of male crustaceans do not liberate spermatozoa into the water. However, recently the gooseneck barnacle *Pollicipes polymerus* has been shown to release sperm into the water column that are gathered by other distant individuals for fertilization of eggs [12]. This behavior, where only males release gametes into the water column is known as *spermcasting*. Among arthropods, the simplest fertilization mechanism is possibly observed in Xiphosuran crabs (Chelicerata), whose males release sperm directly over eggs spawned in depressions in moist sand, high up on sandy beaches [13]. However, in crustaceans the eggs are normally retained by the female and only released into the water after fertilization. In most cases, part (or, more rarely, all) of the larval development is completed in the egg, when still attached to the body of females. Crustaceans have various modifications to their appendages and body structures for sperm transfer, as compared to insects and arachnids, probably because the latter have fewer appendages (usually only three or four pairs) available for modifications, beyond feeding and locomotion. In addition to evolving appendages for sperm transfer and mechanisms of fertilization, crustaceans have evolved an enormous variety of sperm forms [14].

While sperm deposition and subsequent fertilization are external in a few crustaceans, insemination and fertilization are internal in most crustaceans, with sperm deposited into the female oviducts. Whilst sperm deposition may be internalized, it may not be truly internal, because sperm or packets of sperm ("*spermatophores*") are deposited and protected within cuticular invaginations, the *spermathecae*, described in Section 2.3. In this case, spermatophores may be deposited in the female's body by males using special genital papillae that are extensions of the male gonopores. In other cases, special appendages modified for this purpose from locomotory structures are used to transfer spermatophores that arc further glued to the body of females using a wax-like material produced in accessory sex glands, located at the basis of the first gonopods of males [15]. Various reproductive specialties of crustaceans may also be of interest to conservation ecologists and biotechnologists. Dormant egg and cyst properties may be providential to reconstruct pristine aquatic ecosystems affected by anthropic influences (e.g., pollution). In addition, long-term sperm storage and posthumous paternity of decapods, as well as manipulation of AG endocrinology, may allow a sustainable all-male fishery, still conserving the genetic diversity of exploited stocks.

Crustaceans have a diversity of sexual systems. They may be gonochoristic, hermaphroditic (both simultaneous and sequential, with a prevalence of protandry), androdioecious (hermaphrodites and males coexist within an individual), gynandromorphistic (exhibiting sexual mosaics) or even parthenogenetic, as further detailed in this chapter. However, the evolutionary transition between gonochorism and hermaphroditism, through mixed sexual modalities, may be explained, as described in the previous chapter, based on the tissue stability of their AGs. Given the evidence that the sex of crustaceans is uniquely determined by the presence/absence of an AG, it is evident that the simple persistence of this gland determines gonochoristic sexual determination. In species where the AG is partially ageing during the life cycle, we observe the appearance of complete hermaphroditism, because testes are partially transformed in ovaries (ovotestis). In other species, where the AG is subjected to total ageing and regression, we observe protandric hermaphroditism.

The occurrence of gynandromorphism and parthenogenesis is due to the complex relationships with ovarian hormones, as described in the previous chapter. The picture is made more complex due to interactions of environmental factors, such as photoperiod, temperature, nutrition, and hormones (and symbionts that are transmitted along with germ cells in isopods and amphipods, as intracytoplasmic microbes like proteobacteria and protozoans) influencing life cycles, the reproductive phases, larval development and settlement. For example, the shrimp *Hippolyte inermis* is the only known example of an invertebrate whose sex determination is strongly influenced by the food ingested [16,17] and this same species is sensitive to chemical cues influencing larval settlement [18]. In this case, environmental cues allow the shrimp to exploit the advantages of gonochoristic and hermaphroditic strategies. Similarly, some r-strategy crayfish are capable of facultative parthenogenesis and this tactic explains their ability to invade new environments. For example, females of the freshwater gonochoristic crayfish *Orconectes limosus*, prevented from physical contact with males for ten months, successfully spawned and reared viable offspring having resorted to parthenogenetic reproduction [19]. Facultative parthenogenesis is not a rare phenomenon among invertebrates. However, its occurrence in crustaceans and the observation that the strategy depends on the presence of male pheromones is particularly interesting and promising for novel biotechnologies in aquaculture and conservation techniques. Finally, androdioecious crustaceans may be viewed as a possible adaptive evolution of dwarf males and an example is the sex-specific metamorphosis of cypris larvae of the cirripede (pedunculated thoracican barnacles) *Scalpellum spalpellum* [20]. Thus, at least *S. scalpellum* supports [21] both environmental sex determination (ESD) and genetic sex determination (GSD).

Gamete fertilization methods are quite diverse and both external and internal fertilization occur in crustaceans, also according to their anatomical differences. Various appendages used for reproduction have evolved from other appendages used, for example, for grooming [*13*]. In addition to the issues raised in the previous chapter and in conjunction with reproduction topics, crustaceans have developed effective intraspecific communication through the production of sex pheromones able to attract mates and deter competitors. Two main classifications of sex pheromones may be considered: distance cues and contact cues. In all aquatic animals, including crustaceans, most distance pheromones are polar-water-soluble compounds able to be transmitted in odour water plumes. Female decapod urine contains pheromones that act over long distances and attract male partners. Contact pheromones facilitate mate recognition in caridean shrimps [*22*]. Volatile organic compounds also influence the settlement of decapod larvae, as demonstrated for the shrimp *Hippolyte inermis*, whose zoeae may accelerate their development and settle in response to volatiles emitted by benthic diatoms [*18*]. Similarly, gregarious larval settlement cues demonstrated for the barnacle *Balanus amphitrite* were due to surface bound protein complexes, enabling species recognition and settlement. However, it is worth noting that such sex cues are known for insects and they are chemically characterized as cuticular hydrocarbons. Among marine crustaceans, glycoproteins are proposed as contact pheromones for the harpacticoid copepod *Tigriopus japonicus*. Also, contact cues on the body surface of females of the hermaphroditic shrimp *Lysmata boggessi* are hexane-extractable lipophilic compounds, and they resemble insect cuticular hydrocarbon contact cues. The reproductive cycle and the molting cycle are governed by neuro-hormones produced in the same body organs, so producing a synergism between molting processes and the reproductive cycles. For example, uridine diphosphate is a by-product derived from uridine diphosphate-N-acetyl glucosamine in chitin production during molting of decapods and it acts as a urinary female pheromone, attracting males for copulation in a wide range of crabs. Thus, a single sex pheromone triggers molting of females and prepares the same individual for copulation in a sequential process, in synergy with the activity of sex and molt neuro-hormones. Contemporaneously, as shown in further chapters of this book, crustaceans exhibit a distinct and unique endocrine control of vitellogenesis, including the presence of carotenoids directly linked to the protein chains or esterified to the fatty acids of lipovitellins, and these events are under hormonal control, including most of the bioactive compounds indicated in the previous chapter.

2.2 Anatomy of the reproductive systems of crustaceans

The mechanisms of crustacean sex determination and differentiation are still not totally understood because sex determination and differentiation are genetically determined, but various environmental factors (e.g., daily photoperiod, temperature, abundance and quality of food, maximum irradiance), along with social influences, parasites, and physiological stress [*23*] may dramatically modulate sex determination [*17*]. However, from an anatomical point of view, some important features characterize crustaceans, such as the AG, which is the only endocrine gland specifically related to male sexual functions and whose presence/absence trigger sex determination.

Various anatomical and functional relationships link crustaceans with other arthropods, such as insects. For this reason, some comparisons will appear natural and obvious, given the complexity and the diversity of reproductive systems of insects. In general terms, the evolution of anatomical designs of the reproductive systems of crustaceans and insects, such as the diversity in testes and Sertoli-like cells or sperm maturation, packing and storage of male gametes and the position of *germaria* and genital ducts in females, converged driven by environmental constraints, since similar morpho-functional patterns are found in similar habitats [24]. Crustacean sperm cells are immobile [25] and need to be delivered by the male to the female to fertilize the eggs. Overall, the major reproductive organs of crustaceans exhibit a remarkable diversity of anatomical forms, corresponding to functional variations [26]. Eckelbarger [27] stated that "*The rate of egg production, frequency of breeding, size and energy content of the egg, and the resultant consequences to the larval ecology are strongly influenced by the ovary.*" Thus, the co-evolution of internal organs, mating behaviors and fertilization mechanisms according to phylogenetic constraints is mainly driven by the evolution of female reproductive organs. This is not surprising if we consider that "femaleness" represents the natural physiological trend for several crustaceans, eventually modified by the presence of an AG to produce males. We may conclude that the study of ovarian forms and functions helps in an understanding of the general evolutionary trends of crustacean life-history patterns.

Key structures are the female sperm storage organs, because they are morphologically diverse in crustaceans. They can be absent in various species (in this case these species need to mate at each spawning, involving various mating behaviors) but when present they are an integral part of the mating strategy. This implies adaptation of female mating frequency and control (or further manipulation) of spermatozoa [28]. The physiology of males appears quite simplified, since the AG hormones regulate the differentiation of the primary and secondary sexual characters, including the reproductive behavior and aggression against conspecifics. This is a consistent feature of malacostracan crustaceans and it has been demonstrated in amphipods, isopods, crayfish, and crabs [29].

Adrenal Gland Hormone (AGH) is an insulin-like compound, whose molecular structure is similar in several crustaceans *[30]*. The quantity of AGH produced and the time its production starts or decreases during the life of a crustacean determines the sexual strategy typical for a species. In species that are gonochoristic, AGH is produced throughout the life cycle. In hermaphroditic species AGH levels decrease but are not totally absent during the whole life, whereas in protandric species AGH is only produced during the first phases of life, generally for about 1 year in decapods [*18,31*].

Another a key character of crustacean reproductive systems, besides the possibility of being gonochoristic or hermaphroditic (both sequential and simultaneous), is the production of sperm packages (*spermatophores*) that are transferred to females, and there eventually stored (in *spermathecae*). The huge anatomical diversity and reproductive patterns found across crustacean taxa offer unique opportunities to study the relationships among optimal systems for packaging of spermatozoa, "sperm economy" issues, female selection systems, physiological and morphological adaptations in the male and female reproductive tracts, latency time until fertilization, when sperm are released from spermatophores, metabolic costs of reproduction and strategies of counteracting the risk of sperm competition in each species [*32*]. Although females of a few crustaceans are broadcast-spawners or free-spawners, because they release fertilized eggs into the water, most crustaceans retain and incubate their embryos for some or all of their development in special body structures (described below). In addition, most crustaceans transfer their spermatozoa using copulatory organs. However, this general plan may be valid for large taxa such as decapods, but not for all crustaceans. For example, modifications of appendages for reproduction are rarely observed in Remipedia. They are simultaneous hermaphrodites, exhibiting serially homologous biramous swimming limbs, never modified for insemination: sexual systems are externally distinguished in males and females just by the location of gonopores on different trunk somites [*33*].

An interesting characteristic of some crustaceans is observed in peracarid females that keep their eggs within the brood pouches, and juveniles hatch directly in the pouch as miniature adults [*34*]. This strategy results in offspring adapted for life on the solid substratum (such as seaweeds and hydroids) and therefore juveniles are released onto the best sites for recruitment, avoiding dispersion during larval life. In other species, such as *Hippolyte inermis*, some volatile compounds present on diatoms covering the surface of *Posidonia oceanica* provide a cue for settlement [*18*]] on this seagrass. This demonstrates that chemical cues may be fundamental for larval metamorphosis. In fact, the functional anatomy of reproductive organs is related to the issues of larval dispersal and settlement, according to the physiology and the ecology of each individual species. Laboratory rearing systems are a valuable way to study aspects of the reproductive biology, such as brooding, in amphipods (Peracarida). For example, Nakajama and Takeuchi [*35*] designed a special rearing system for *Caprella mutica* in which they were able to observe and monitor brooding of juveniles on a large screen. Such a system could be replicated for other crustaceans, hopefully increasing our knowledge of this interesting topic.

2.2.1 *Copulatory organs*

The majority of crustaceans transfer sperm by copulatory organs and a variety of structures have evolved for this purpose. In Anostraca, the male gonopods, which are modified thoracic limbs, are located anterior to the abdomen, as observed in brine shrimps [*36*]. Adult males continuously search for receptive females, grasping them with special appendages to reach the brood pouch or genital segments just behind the last pair of appendages [36]. Male tadpole shrimps (Notostraca) exhibit a few modifications to their reproductive appendages, but special male gonopods are observed in their phyllopod appendages. In contrast, phyllopods 1 and 2 of male clam shrimps (conchostracans) terminate in prehensile or sub-chelate claspers, used for grasping the female carapace during pairing and copulation. In water fleas (Cladocera, a diplostracan suborder) the first trunk appendages of males are prehensile or even hooked for grasping their females. In most Ostracoda the copulatory organs are paired and probably evolved from an eighth pair of appendages. The penis or hemipenis can be very large (in relation to body size) and morphologically complex [*37*].

For example, copulatory organs (hemipenes) of podocopans may be incredibly large and intricate. It is thought that such complex reproductive apparatus is needed because of the difficulty of copulating whilst encased in a bivalve carapace.

In barnacles (Thoracica), which are hermaphrodite and ubiquitous filter-feeding sessile organisms, the fertilization organ of "functional males" is a long, remarkably mobile and flexible penis, which introduces a sperm mass into the mantle cavity of another individual serving as a "functional female". The penis departs from the bases of the posterior cirri and it develops from a remnant of the larval abdomen (the terminal body sclerite). It is extended and retracted by modulation of the turgor pressure in the longitudinally running inflatable hemolymph channels. A compact layer of longitudinal musculature under the epidermis enables searching movements. Intertidal barnacles can adapt size and shape of their penises to local hydrodynamic conditions, and to the reproductive seasons [*38*]. For example, on wave-exposed areas they develop a shorter penis with a larger diameter than in wave-protected sites and invest more energy and resources in penis development and functions. In some sites they may have quite a short reproductive season: for this reason, they develop a functional penis only during the reproductive period [*39*]. Consequently, in the waters off New York, where *Semibalanus balanoides* mates between October and November, the penis progressively grows, molt after molt, just in this period and then degenerates [*40*]. A similar long penis is exhibited by Ascothoracida (parasites of cnidarians and echinoderms) and by most acrothoracicans, small cirripedes burrowing in limestones [*41*].

Copepoda are taxonomically and ecologically diverse, with diverse morphology and mating behaviors. In cyclopoid copepods, the copulatory apparatus is simplest, the fifth swimming legs are almost rudimentary and males simply sway their bodies to the correct position, next to females, to transfer spermatophores [*13*]. In calanoids the main reproductive appendages are the male's first legs and last swimming legs in the thorax [*42*]. Calanoid males grasp females with the first legs and then swing their bodies around, so as to seize the female with the right P5 (5th swimming leg), modified for gripping the female urosome. The left P5, in its turn, strokes and examines the female's genital region. A spermatophore is promptly emitted from the male genital pore and attached to the genital somites of females, where sperm are discharged into the female genital opening prior to moving to the spermathecae, for longer storage. In some calanoids the fifth pereiopod of females is also modified to clean off discharged spermatophores using the exopods and coxal serrations.

Quite different are the organs for the sperm transfer of decapods. Decapods are vagile and often large and they possess permanently rigid copulatory organs that are form-invariant [*43*]. Brachyuran crabs and freshwater crayfish have paired copulatory organs composed of the first and second pleopods (transformed in "gonopods") which perform various functions [*44*]. The shape and permanent stiffness of both gonopods is achieved and maintained by means of a thick cuticle. The first gonopod forms a tube where sperm flow from the genital opening, while the second gonopod is inserted into the tube like a syringe needle, producing spermatophores to be transferred into the female's sperm storage site. The size of gonopods changes during the life span of decapods. It increases at each molt during development and, in sex reverting species, decreases progressively when the AG is deteriorating, in parallel with the formation of an ovotestis, up to the complete disappearance when sex reversal is completed. However, given the specificity and the stability of shapes of gonopores of decapods, they probably played a role in the radiation of the 15000 species.

2.3 Sexual dimorphism

Morphological differences between sexes are common in crustaceans and generally derive from secondary sex characteristics related to competition for mates, sperm transfer and brood care. Secondary sex differences can include differences in color, shape, as well as size and arrangement of various appendages. Chelipeds may have a different size (heterochely) in the two sexes in decapods (as exemplified by fiddler-crabs), and body size can be quite different, with larger males or larger females, according to sexual strategies and environmental constraints. These differences may be important in aquacultural practices, because one of the two sexes may reach larger size, grow faster, and show vivid colors that increase the market value.

An extreme form of sexual dimorphism is the tremendous reduction of male size (dwarf males), which has evolved independently in some parasitic isopods and copepods. In the Cirripedia, dwarf males have also evolved independently in the sessile Thoracica and the parasitic Rhizocephala. The dwarf male often holds on to the female for most of its life using special structures. The advantage of dwarf males is the permanent availability of males at a low physiological cost and the conservation of genetic diversity under limited mating opportunities [45]. In the Chondracanthidae (Copepoda), a family that parasitizes marine fishes, the dwarf males become attached to immature females at the second copepodite stage where they complete their development and remain on the female body for their entire life. The dajid isopod *Zonophryxus quinquedens* parasitizes the carapace of the Antarctic shrimp *Nematocarcinus longirostris* [46]. Isopod dwarf males (which reside on females) are typical of the superfamilies Bopyroidea and Cryptoniscoidea, parasitizing other crustaceans and often influencing sexual characters and the reproductive biology of their hosts [47].

Various parasitic crustaceans (and other agents, including protozoans and bacteria) can change the reproductive biology of crustaceans by influencing the sex of their hosts and causing infertility. In the caridean *Hippolyte inermis*, the frequent presence of bophyrids in natural populations leads a shift to female sex with drastic reduction of fertility [48]. Additionally, a well-known intracellular alpha-proteobacterium able to influence crustacean reproduction is *Wolbachia pipientis* [49], detected in several terrestrial, limnic, and estuarine isopods, as well as amphipods and barnacles. *Wolbachia* converts genetic male isopods into functional females and is vertically transmitted via the egg cytoplasm, although horizontal transmission also occurs [50]. In insects *Wolbachia* induces parthenogenesis but this mechanism has not yet been observed in crustaceans. This parasite alters the DNA methylation patterns of genes involved in sex determination, thus disrupting the male genetic imprinting [51] by reprogramming of the host genome.

In contrast, some bacteria such as *Pasteuria ramosa* may produce gigantism in their hosts [52] and parasitic castration. *Pasteuria* is horizontally transmitted by means of spores released from dead hosts and it has a polymorphic life cycle, beginning with cauliflower-shaped rosettes and ending with individual spores. After infection, *Daphnia* continue to produce eggs for some days prior to being castrated, when nutrients and energy are re-directed to the reproduction of *Pasteuria* [*53*]. However, a relevant proportion of the energy that is saved by interrupting the reproduction of *Daphnia* is used for unusual body growth of the host, resulting in gigantism: a larger host produces more spores.

2.4 Ovary and accessory structures

The female reproductive system of crustaceans, consists of paired (or in some groups unpaired) rounded or elongated ovaries that are positioned dorsally although their location may be anterior or posterior in various taxa. Other important variations depend on the mechanism of fertilization and sperm transfer. For example, decapod females store sperm either externally as spermatophore clumps, or internally in specialized seminal receptacles [*43*] and this key difference may produce dramatic modifications of the anatomy in both sexes. The initial and most important part of ovaries is the so-called "*germarium*", i.e., the germinal zone active throughout the reproductive life of the females and easily identifiable in juvenile crustaceans. In several species, oocytes are arranged in the mature ovary in order of size, with the larger oocytes (mature) farther from the *germarium*, often in a ventral position [*54*], and the youngest closer to the *germarium*. As commonly observed in other invertebrates, in addition to oocytes, ovaries also contain various cells that have a trophic role, e.g., follicular cells, accessory cells, nurse cells, and mesodermal *stroma* [*55*]. On the whole, these cells are grouped into the two categories of follicular cells (non-germinative accessory somatic cells involved in the formation of oocytes and egg envelopes) and nurse cells (germinal cells formed by incomplete cytokinesis of oogonial divisions, born to accompany the growing oocytes).

Promiscuity of females and multiple paternity of the offspring is widespread in crustaceans and although it is mainly known for decapods, there are known cases for isopods, cirripeds, and copepods [*56*]. Multiple paternity involves storage of spermatozoa of all mates until egg-laying, which may occur weeks or, in some species, even years after the last mating. For example, in long-lived species like clawed lobsters and some crabs, sperm can be stored by females and utilized for years [*57*]. Females of the crab *Chionoecetes bairdi,* when isolated from males after copulation, produce viable embryos in the following two years [*58*].

Ovaries in Ostracoda, which are in the posterior of the body, have lateral lobules generally fused at the center and they are thus paired. However, ovaries may be in unusual positions in non-malacostracans. For example, in barnacles paired ovaries lie in the mantle tissue of balanomorphs, overlying the basis of the foot, while they lie in the distal half of the peduncle in lepadomorphs. In calanoid copepods a single median ovary is located in the body while paired egg masses lie outside in the posterior part of bopyrids, where they grow enormously in different shapes, with segmentally arranged diverticula [*59*]. However, in most crustaceans, including Malacostraca, elongated paired ovaries are the general condition, with different degrees of fusion in the central part of the body. This produces in various decapods and isopods an H-shaped ovary or eventually a Y-shaped organ. Ovaries are normally surrounded by thin connective tissues (the "ovarian walls") connected to the oviducts [*23*]. Given this ovary structure, the eggs move from lobules, when mature, through the oviducts to reach paired (with a few exceptions) gonopores generally located in the sternite or the coxae of the sixth thoracic somite in malacostracans. This ovary structure is reinforced in some species, as in lobsters, by a thick muscular tunic rich in blood vessels, protecting and irrigating the follicles [*60*] and has a specific function during ovulation, thanks to the production of prostaglandins stimulating the contractions of these tissues. When the tunic is absent, as in amphipods, isopods, crabs and most caridean shrimps, the release of oocytes from the ovary is accompanied by contractions of body muscles during pre-spawning and spawning activities.

According to Ikuta and Makioka [*61*] two types of ovaries are recognized in crustaceans, depending on the position of the growing oocytes along the ovarian lumen: type I (mandibulate) and type II (chelicerate). These two types probably reflect ancient modes of oogenesis of crustacean progenitors, but the principal consequence is that the germ cells are positioned in opposed directions in the *germaria*, and consequently in the ovarian lumen. The main difference, in fact, is in the position of oocytes of various sizes and age, that in type I grow in the inner surface of the ovarian wall, facing the ovarian lumen, while in type II they are attached to the outer surface of the ovary, according to their sizes, with the smaller oocytes closer to the *germarium*. Type I is found in various mandibulate arthropods, including freshwater ostracods and malacostracans, where a compact ovary is observed, because mature oocytes move towards the lumen, filling and compacting it. In contrast, in type II (as observed in myodocopids, branchiurans, pentastomids, and chelicerates) growing oocytes are not stored in the lumen, and secondary oocytes move through the gaps between elongated epithelial cells, forming "stalks" that support them, prior to reaching the oviducts [*62*].

Invertebrates evolved two basic types of oogenesis [*27*]: extra-ovarian oogenesis, that is the most common in invertebrates such as polychaetes, and intra-ovarian oogenesis, as observed in echinoderms and molluscs, when oocytes are retained in the ovary until late development and up to spawning. The latter strategy is found in almost all crustaceans, with the single exception of Cephalocarida, which show late vitellogenesis in the oviducts. During oogenesis there is **an initial proliferative phase** (with mitotic activity) including a premature phase (previtellogenesis, when cytoplasmic organelles start protein synthesis)**, and a growing phase**, when proteins are synthesized in the oocytes and various components are transferred from the nurse cells and from the hepatopancreas. The growing phase also includes primary and secondary vitellogenesis, when the diameter of oocytes increases according to the production of vitellogenin, due to the accumulation of yolk proteins, i.e., materials and energy for the embryos until larvae begin to feed. Yolk, in fact, contains carotenoids and tetraterpenoids fundamental for the development of the embryos, as well as key glycolipoproteins. The growing phase may be followed by means of histological preparations to characterize the size of oocytes, the presence of nucleoli, the consistency of the cytoplasm and the presence of yolk droplets. A specific "cortical differentiation" occurs while oocytes develop, characterized in crustaceans by the differentiation of cytoplasmic organelles known as "cortical granules", i.e., membrane-bound structures that accumulate in the cortex during oogenesis in various animals and after fertilization produce a thick envelope protecting the embryo.

Cortical granules have been identified in copepods, amphipods and decapods, both reptantia and natantia. Furthermore, after vitellogenesis, mature oocytes of crustaceans produce a fibrous glycoprotein coat, adhering to the plasma membrane and play a key role in the sperm-oocyte interaction. The coat is the fertilization membrane (also known as vitelline envelope or vitelline coat) and it is composed of a network of proteins and carbohydrates, already identified in ostracods, branchiurans, copepods, isopods, amphipods and decapods. All these modifications make oocytes stronger, protecting them from external influences and parasites, and trigger the next phases of embryonic development, providing also osmotic protection [63]. From a macroscopic point of view, these maturation processes and histological modifications produce clear changes in the macro-morphology of ovaries that may be used by experts (in aquaculture, or ecological applications) to determine the actual time for reproduction of crustaceans.

Mature ovaries of crustaceans (especially decapods having economic value) can often be observed through the ventral transparent exoskeleton and characterized according to their color. Specific developmental categories have been set to identify the reproductive status of females. To assess the maturation and avoid observer bias, the use of a chromatic Pantone Matching System scale has been introduced by Peixoto et al. [64], and this system may have quite useful applications for various biotechnological purposes. As a general rule, there is a close relationship between body length of females and number of embryos produced and this is demonstrated also by species having multiple reproductive periods at various sizes. In *Hippolyte inermis* two seasons for recruitment are observed. During the first season, in spring, females are large (up to about 20–25 mm length) and may produce up to 250 embryos simultaneously [17,65], while during the second reproductive season females are younger and smaller (less than 12 mm total length) and they produce on average 30–50 embryos.

Crustaceans are among the animals with the largest external egg clutches carried by females. The largest clutches are carried by females of some brachyuran crabs [66]. For example, *Callinectes sapidus*, can brood up to 8-million eggs until the zoeal stage under its pleon, in a special sponge-like structure. This number of embryos brooded simultaneously is, however, surpassed by *Cancer* spp., whose females carry more than 20 million eggs in their clutches. In some crustaceans producing resting stages, as in branchiopods, a thick and dry egg shell is produced by the so called "shell glands" located in the follicular ducts, to ensure the survival of dormant resting eggs. For example, diapausing eggs and cysts are produced by various branchiopods, copepods and ostracods and they may lie dormant for up to 16 years. *Daphnia pulicaria* probably holds the metazoan record of egg viability, since cysts were demonstrated to survive for up to 700 years. Dormant eggs that are produced by parthenogenetic and bisexual species [67] may be tolerant to several environmental stressors, including mechanical damage, drying, freezing, and UV radiation. *Artemia franciscana* is a model for dormancy investigations [67], because their dormant eggs may survive at temperatures ranging from –271°C, up to 100°C. Several limnic and planktonic crustaceans that have a short life cycle as well as species living in ephemeral water bodies may produce dormant eggs that survive adverse environmental conditions up to the next seasons or for several years, as a special adaptation to changing environments. This ability probably evolved in ancient metazoans.

Various crustaceans produce embryos that are prematurely exposed to external influences and need greater protection, as in the case of decapods that incubate their embryos on pleopod sacs, where "tegumental glands" or "cement glands" located on their pleopods produce a thick envelope also involved in the attachment of eggs to the mother's body [68]. All decapod taxa (with the exception of dendrobranchiates) incubate embryos up to the complete development and hatching of larvae in incubation sacs produced by the junction of pleopods under their abdomen. During this period fertilized eggs are aerated by females thanks to pleopodal beats and, in brachyurans, to the abdominal flaps that circulate water and remove wastes. In some species embryos are cleaned using specific grooming appendages [69] and this prevents the intrusion of parasites, bacteria and predators such as nemertean worms that can infest embryonic masses and produce mass mortalities both in nature and in aquaculture. Often, the same structures transporting eggs serve for their fertilization, because female gonopores lead to paired (or unpaired, as in copepods and in groups where a single one is functional) oviducts, i.e., mesodermic or mesoectodermic structures that may be viewed as "genital ducts", because within their lumen mature eggs derived from the ovary meet spermatozoa, often conserved in seminal receptacles creating a spermatheca. Additionally, several crustaceans bear incubation chambers for brooded embryos, also helping egg storage prior to fertilization and release, forming a "brood pouch" (also named "*marsupium*") that represent the fertilization site. According to the presence/absence of receptacles four types of *genital ducts* have been described.

Type I genital ducts do not bear differentiated seminal receptacles. In this case, spermatophores are attached to the female abdomen, because they lack sperm storage structures, reflecting an ephemeral storage [70], as observed in amphipods, isopods, anomurans, and various shrimps and crayfish [71].

Type II genital ducts are paired or unpaired ventral invaginations forming a pocket able to accept spermatophores. In this case gonopores are the natural outlet for gametes, while seminal receptacles are separate and are named differently in various groups. For example, in penaeid prawns and in some crabs, they are called the "*thelycum*", in Cambaridae the "*annulus ventralis*", while in most species they have the generic name of seminal receptacles or spermathecae. In some species a mobile or immobile operculum is soft only during the mating season: it opens the gonopores just for the time required for mating, and further closes the gonopores against environmental threats when the mating season is ended. An operculum has been observed as well in males of various species, as in the case of hermit crabs such as *Clibanarius vittatus*, *Diogenes pugilator*, and in males of various caridean shrimps of the genera *Heptocarpus*, *Lysmata*, and *Corismus*, and it acts as a protection against dehydration during the periods of sub-aerial life in species living at the water interface. In females, the opercula of Brachyura have been extensively studied, while less information is published for opercula of female Anomura. In some crustaceans, such as copepods, spermatophores are attached to the urosomes, and their content passes into a chitin-lined seminal receptacle [72]. Genital ducts may be connected by a thin bridge with the seminal receptacles, or be totally disconnected, as in ostracods, branchiurans, penaeid prawns and other decapods.

Type III genital ducts are characterized by an expanded chamber in the oviducts, where spermatozoa are deposited and fertilization occurs. This type characterizes some isopods and other crustaceans.

Type IV genital ducts are typical of cirripedes, eubrachyuran crabs, terrestrial isopods and anostracans and differ from all the previous from an ontogenetic point of view, because they are not of mesodermic origin but of mixed origin and appear as a mesoectodermic chamber for the fertilization of eggs, after the deposition of spermatozoa by males.

Most crustaceans feature a single type of seminal receptacle, while isopods exhibit a great diversity of strategies because they feature all four designs, so representing an ideal model to study phylogenetic trends in reproductive behavior. In fact, each strategy produces advantages and disadvantages in the possibility of spawning without the need of mating and, simultaneously, limiting or increasing the possibility of sharing genes among different partners during the same reproductive season. In addition, the storage of spermatozoa in the body of females requires various adaptations, such as the need for a secretory epithelium able to nourish the sperm, allow aerobic respiration, stabilize their activities and control bacterial proliferation [70].

2.5 Testes and associated structures

Knowledge of the morphological and ultrastructural features of male reproductive systems, including the position of gonopores and the ability to produce spermatophores, are essential to understand the reproductive practices in different crustaceans. Where paired oval or elongated testes are present, they are positioned dorsolaterally in the cephalothorax, generally located dorsal to the hepatopancreas and midgut, while the paired *vas deferens* are often coiled. In most anomuran decapods, such as *Galathea intermedia*, the testes are also located in the cephalothorax [73]. However, in hermit crabs (Anomura), they are located in the pleon. Generally, testes originate from paired tubes in the anterior region of the body [23]. The tubes are unbranched in amphipods, tanaids and some isopods, and produce paired structures that may be centrally fused or even partially joined. In fact, testes may be partially or totally jointed in the center of the body as observed in copepods, some branchiopods and ostracods. In the latter taxon the testes exhibit the largest relative size because they can be about 1/3 of the body size. The abdominally positioned testes of *Dolops* spp. (Branchiura) are unusual because they are trilobed, and those of Cirripedia are diffused in the connective stroma of the prosoma [74]. In some isopods and cumaceans, the testes consist of lobes extending from each tube [23]. In most crustaceans however testes are paired, distinct, and often elongated, as observed in most Malacostraca and in decapods where testes are V-shaped or H-shaped lobes [75].

The morphological differences characterizing testes in various crustacean taxa correspond to differences in the process of spermatogenesis. For example, in cephalocarids and branchiopods clusters of sperm are observed, while spermatogenic islands are observed in Rhizocephala and follicles are present in the testes of isopods. A variety of tissues characterize the testes of decapods, although two fundamental designs are represented by seminiferous tubules and testicular lobules, including testicular lobules and lobes, acini, cysts, and seminiferous tubules [75], but testicular lobes are present also in cumaceans and isopods [23]. Seminiferous tubules are tubular structures containing various stages of gametogenesis in their lumen. In general, immature stages are located at the periphery, and mature stages in the center, of the lumen, ready to be transported to the outer regions, up to the vas deferens [76]. In contrast, testicular lobules (or testicular acini) perform spermatogenesis in plump pockets made of several lobules where mature sperm are produced and sent to a testicular lumen, finally reaching the vas deferens. Each lobule and each acinus may be in a given stage of spermatogenesis, regardless of the stages occurring in the adjacent lobes and this architecture is evidently different from the seminiferous tubules above mentioned. According to these fundamental differences, Grier [77] defined "restricted testes" as those based on seminiferous tubules and unrestricted testes those based on testicular lobules or *acini*.

Despite the above architectural variations, testes are connected to genital ducts that open into the gonopores of males. For example, in the decapod *Clibanarius erythropus* the gonopores are paired structures, on both the left and right coxa of P5, and are flush with each coxal surface [78]. Each genital duct is made of three parts: a collecting tubule, a vas deferens, and an ejaculatory duct. Ejaculatory ducts may be expanded in some crustaceans to receive sperm masses and spermatophores and in this case they are termed seminal vesicles or terminal *ampullae* [76]. Genital ducts and male accessory glands also exhibit a diversity of designs according to evolutionary and phylogenetic trends. The vas deferens is quite developed in decapods and peracarids, but also in cirripeds and brachyurans. In these structures there is an important secretion of seminal plasma and lubricant agents [79] that may aid the transfer of spermatophores and the activation of spermatozoa. In some species these lubricants may also have antibacterial activity preventing premature degradation of spermatozoa, thus prolonging the life of sperm in the seminal receptacles. The lubricants may also nourish the sperm because they are composed of proteins, carbohydrates and lipids [4]. They may also be involved in sperm activation. The vas deferens has been histologically characterized into distinct regions (from three up to ten, depending on taxon). There are proximal zones that are mainly secretory, and the distal zone that is usually more muscular and involved in the release of spermatozoa during mating activities [80]. However, this differentiation is not observed in several crustaceans. For example, some ostracods and cephalocarids, as well as *Artemia* spp., have a unique differentiation of the genital duct.

As described for females, the position, the structure and the functioning of male reproductive systems of crustaceans, along with the presence and the complexity of copulatory structures (e.g., simple gonopores, elevated papillae, other structures needed to attach spermatophores to the body of females) are quite variable and consequently considered useful taxonomic characters to understand phylogenetic trends ruling the mating activities and crustacean reproduction [81]. In parallel to the diversity of tissues and structures characterizing various taxa, a large diversity of sperm types is observed [*37,14*] and they represent a useful tool to analyse phylogenetic relationships. In a few crustacean taxa (e.g., Cirripedia) the fundamental morphology and organization of the sperm is typical of many other animal taxa consisting of an anterior head containing the nucleus and acrosome, mid-piece with centrioles and mitochondria, and posterior flagellum (tail). However, most crustacean groups have more highly modified sperm that lack a flagellum. For example, ostracods of the suborder Cypridocopina possess giant sperm that range in length from 268 µm up to 11.8 mm, so exceeding the shell length of their producers. Ostracods hold the animal record in aflagellate sperm size and these spermatozoa are unique in their morphology in the animal kingdom. Such sperm have a long evolutionary history as they have been found in fossilized 100-million-year-old ostracods [*82*]. Their aflagellate sperm consist of an elongated nucleus that runs for the entire sperm length and two giant mitochondria. They may move thanks to contractile elements that promote longitudinal rotation and, surprisingly, they penetrate the egg during fertilization, despite their giant size [*5*]. The sperm of decapods are also aflagellate, but they have quite different ultrastructures [*14*]. In the Decapoda the sperm consist of a compact central body containing the nucleus and acrosome, along with some radial arms or spikes that contain microtubules. The sperm penetrate the egg by a unique explosion-like movement [*83*] and therefore in some literature these sperm are referred to as "explosion sperm". The radial arms aiding the attachment to the egg become free when sperm are released from the spermatophores [*84*]. Although they lack a flagellum, the sperm may move briefly thanks to a "jet" mechanism, brought about by an abrupt eversion of the acrosome, which propels the sperm and nucleus forward. For example, in *Homarus americanus* this forward movement is about 18 µm, sufficient to push the nucleus through the egg coat. Explaining the diversity in sperm morphology is one of the challenging questions in evolutionary biology and according to Koch and Lambert [*85*] internal fertilization is probably the ancestral condition in decapods. Research is also urgently needed to understand the mechanisms of spermatozoan recognition and penetration into eggs according to sperm morphologies.

Testes contain various types of cells besides spermatozoa. These include accessory cells, nurse cells, Sertoli or Sertoli-like cells, but also epithelial, follicular, and intercalary cells. Some of these cells are needed for regulating sperm abundance (by phagocytizing excess spermatids) or to remove unneeded cytoplasmic material during spermatogenesis. Some produce cohesion material to keep the sperm in the spermatophores that are widespread in crustaceans and transferred to the body of mates during copulation. For example, during copulation male Paguridae ejaculate spermatophores from the flush gonopores onto the female's gonopores, on the ventral face of the coxae of pereiopod III [*78*]. Spermatophores are structures emitted by the male during copulation, surrounded by a protective and adhesive component produced in the vas deferens, containing spermatozoa [*81*].

The morphological diversity of spermatophores, the origin of their components, and their hardening mechanisms may have ecological importance. This is because different strategies for sperm transfer are related to mating strategies, e.g., external or internal fertilization, time for spermatophore transfer, their dehiscence, and ovulation processes that characterize various crustacean taxa. However, the final strategies applied by each species probably reflect their life strategies and ecological constraints. For this reason, and due to evolutionary convergence, phylogenetically unrelated groups may share similar spermatophore morphology. Thus, given the variability in morphology and methods of spermatophore transfer, this feature does not allow conclusions to be drawn about phylogenetic relationships within crustaceans, but they may help in understanding the various mating systems in crustaceans. Internal fertilization, mediated by spermatophores and copulatory structures, is an evident advantage because the direct transfer of sperm into the body of females increases their fertilization potential and reduces the risk of sperm competition.

2.6 Hermaphroditism as a key study area for crustaceans

Hermaphroditism is a common condition in crustaceans and therefore the group could serve as model organisms for studying the evolution of this reproductive strategy [30]. This condition evolved independently within crustacean taxa, i.e., it is not associated with phylogenetic relationships. For example, in some families of crustaceans (e.g., decapod hippolytids) there are species that are gonochoristic, simultaneous hermaphrodites and also sequential hermaphrodites (mainly protandric). Protogyny has been reported in some crustaceans, e.g., tanaids and some isopods. This fact suggests that all crustaceans have the potential to develop as hermaphrodites, given the single element determining their sexual determination (presence/absence of the AG) and that several taxa took this opportunity to take the advantages connected to this condition and life strategy.

In general, hermaphrodites develop an "ovotestis" during their life, i.e., a gonad made of an anterior half as an ovary and a posterior half as a testis, as observed in caridean shrimps and crayfish. An ovotestis has paired genital ducts for eggs and posteriorly, for spermatozoa, eventually packaged into spermatophores. This may be the original developmental form of their gonad, or be the result of a slow (or fast, as in *Hippolyte inermis* [17]) process of sex reversal. In some groups, as in Cephalocarida, ovary and testis share a common genital duct [86] but most crustaceans have independent genital ducts and gonopores for the two sexes. While in decapods and other crustaceans [31] protandric hermaphroditism (sex change from male to female) is the typical strategy, protogyny (female to male) is rare but is observed in Tanaidacea and in some isopods, e.g., *Gnorimosphaeroma oregonense* [87] and *G. naktongense* [88]. Protogyny is associated with species in which the females have low mobility, a low abundance of males and hence competition amongst males for females. In protogynous isopods the males show mate guarding. An interesting form of protandric hermaphroditism occurs in some caridean shrimp. Individuals first mature as males but as they grow and molt, they become simultaneous hermaphrodites, and can reproduce both as males and females. This is despite the fact that the male portion of the reproductive system is greatly reduced. This type of hermaphroditism was termed protandric simultaneous hermaphroditism by Bauer [89].

In addition to hermaphroditism, several crustaceans exhibit *intersex* individuals, i.e., subjects showing one type of gonad (either ovaries or testes) but with both male and female gonoducts and gonopores, as found in Parastacidae, Hippidae and Diogeneidae. A similar condition is observed in amphipods, isopods, and mysids.

Intersex is also common in some caridean taxa during their transition from male to female. As a general condition, just a small proportion of a crustacean population are intersex (for a review on intersex see [90]).

Besides hermaphrodites (of different kinds) and intersexes, micro-parasites and environmental cues may be involved in processes of sex reversal [91], as observed in some amphipods, where the presence of intersexes is caused by endocrine disruptors [92], or in the caridean decapod *H. inermis*, where the sex change is triggered by the ingestion of marine microalgae [17]. This large diversity of adaptations indicates that male and female reproductive systems of crustaceans coevolved to maximize reproductive efficiency under a range of environmental and population demographic conditions. Further studies on neuro-hormonal pathways will improve the understanding of the action and the impact of various possible endocrine disruptors.

2.7 Reproductive events in crustacean aquaculture

Fertilization and maturation are key features for an understanding of the reproduction of various taxa, to produce the right biotechnological tools in aquaculture or to monitor animal cultures for research purposes. To this end, it is necessary to understand the mechanisms of fertilization used by various crustaceans and recognize the early signs of maturation for both sexes. Various crustaceans show symptoms of maturation which can be used to aid their culture. For example, the glair glands of freshwater crayfish develop some weeks before spawning and are good indicators of forthcoming egg-laying. They appear as creamy-white patches in the last thoracic sternal plates, the sterna and pleura of the pleon, the pleopods, and the uropods, and must be considered as a transient tent-like compartment under the bodies of crayfish, which facilitate egg fertilization and attachment of the zygotes to the pleopods. The fertilization tents seem to be multi-functional, but in freshwater it is likely that they protect the soft and highly labile fresh eggs from osmotic stress. This structure is formed by a secretion from the glair glands and is evident for several hours after fertilization. The female of *Orconectes limosus* first bends the pleon toward the underside of the cephalothorax and then fills this pouch with a gelatinous secretion from the glair glands in a process lasting about 30 minutes. Since the fertilization tent is produced shortly before the onset of spawning, it may be a clear signal for management of their reproduction in aquaculture. Glair glands and fertilization tents are absent in marine crustaceans and they should be seen as a special adaptation to reproduction in freshwater, to protect eggs and facilitate fertilization, keeping both sperm and eggs in a physically and chemically controlled milieu in species characterized by external fertilization. Freshwater crabs, for example, have internal fertilization and do not require such a protective mechanism.

The simplest pattern of reproduction observed in a few marine crustaceans is represented by the "broadcasting strategy", i.e., the release of eggs and sperm into the water and larval development in the plankton, but this strategy is quite rare since most crustaceans evolved structures to brood their embryos, in some case even up to the first juvenile stages. Therefore, in many species a brood compartment is more common for developing embryos. Brood care is obligatory in peracarids, and morphologically specific. Thus, it may be used as a character for phylogenetic analyses [*93*], and as a clear symptom of the pending release of larvae. The simplest type of brood care is carrying of the eggs with the mother until hatching of a *nauplius*. This mechanism is used by copepods and euphasiaceans. Dorsal brood chambers, where eggs are laid and embryos develop until a miniature adult is released, characterize some entomostracan groups such as cladocerans and ostracods. It is worth noting that some cladocerans secrete a nutritive fluid in this chamber, able to nourish embryos up to hatching. A further evolution is represented by the marsupium of peracarids, present in both marine crustaceans and terrestrial isopods, since in this case the original water environment must be assured for developing embryos. It also contains nutrients to assist development. A simple but efficient marsupium made by junction of pleopods is also present in all decapods, evolved from a common ancestor of Pleocyemata 430 million years ago [*94*]), absent only in some Dendrobranchiata, but present in all Pleocyemata. The length of the brood care is quite variable and ranges from a few weeks in various shrimps up to 16 months in *Homarus americanus*. However, the period may be prolonged or shortened according to environmental constraints, to assure the release of larvae in the ideal period for their correct development [*65,95*]. Finally, crayfish and other freshwater crustaceans (mainly Cambaridae and Parastacidae) evolved some "safety lines", unique in the animal kingdom, represented by telson thread and anal thread, to protect larvae during the immobile and helpless phases of hatching up to the first molt.

2.8 Sex chromosomes and sex determination of crustaceans

Crustaceans may take advantage of both genotypic sex determination (GSD) and environmental sex determination (ESD). In the case of GSD, genetics are the switch to specify whether given individuals are females or males, or even hermaphrodites. In ESD external stimuli such as temperature, photoperiod, food quantity and quality and even social factors control sex determination and, in some crustaceans, they take the role of epigenetic factors, leading to a variety of sexualisation systems. The line between GSD and ESD is vague in several crustaceans, with environmental triggers altering the genetically determined sex of target species. In fact, ESD may be a primary cause of intersexuality when coupled to GSD. Intersex conditions may arise during the period of sex switching in calanoid copepods and in some decapods, e.g., *Hippolyte inermis* [*17*]. However, GSD may evolve and the chromosome pair that determines sex may change, especially when the ancestral sex chromosome exhibits little genetic differentiation, since WW or YY combinations are then less likely to be lethal. This evolution may thus lead to the transitions within and between different XY (male heterogamety) and ZW (female heterogamety) sex chromosome systems, according to the crustacean groups observed.

In conclusion, sex determination is quite a complex process in crustaceans, often mixing GSD and ESD and involving a network of interactions among genes, influenced by environmental triggers [96]. Cytoplasmic sex determination is still another system found exclusively in crustaceans, with the possible exception of a few species of aphids. However, strong chromosomal determination is observed in most crustaceans but several of them are complicated by ESD. Whilst genetic determination of sex characterizes all Crustacea, it largely varies, ranging from the most primitive and weak polygenic systems, to strong chromosomal sex determination. For example, an archaic system of polygenic sex determination was demonstrated in various crustaceans, such as the copepod *Tigriopus californicus* [97] but, as a general rule, sex determination of copepods is under strong environmental control. The genes involved in gonad differentiation pathway are predominantly transcription factors, e.g., DMRT-1, DSX-1, and SOX-9.

We should consider in this context that the human X and Y sexual chromosomes lead to the impression that sex determination mechanisms are a conserved ancient feature of life on earth [98]. However, each eukaryote taxon independently evolved primary mechanisms of sex determination, besides further complications above reported about sex reverting species, parthenogenesis and hermaphroditism. It is likely that homomorphic chromosomes (morphologically similar) are evolutionarily young, and it is likely that they too will degenerate in future evolutions [99].

Heteromorphic chromosomes evolve from initially identical autosomes. However, during their evolution they stop recombining and differentiate. Further evolution of anisogamy (small male and large female gametes) led to the evolution of male and female functions. Separate sexes evolved many times independently (both in plants and animals). This evidence suggests that hermaphroditism, often observed in crustaceans, has evolutionary costs and constraints. The sex of eukaryotes is determined either by male (XX/XY) or female heterogamety (ZZ/ZW). In crayfish, for example, female heterogamety is normally observed. Consequently, males have a ZZ chromosomal assemblage, while in females there is a ZW assemblage. This is quite different from other crustaceans, where male heterogamety is observed (XY males vs. XX females). In addition, several crustaceans express in their developing gonads regulatory genes having a largely conserved role in the molecular pathways leading to male or female gonad development, such as the *doublesex-mab3* (abbreviated as "DM") family genes [100].

The number of chromosomes of crustaceans is variable, and often it is quite high. For example, the crayfish *Pacifastacus leniusculus* is considered to be a record holder in the animal kingdom for its chromosome number, since it bears a diploid set of 376 chromosomes, corresponding to 188 chromosomes in its gametes [101]. Another crayfish, *Procambarus virginalis*, has a triploid set of 276 chromosomes [102]; however, it reproduces by apomictic parthenogenesis, and consequently its unfertilized oocytes have 276 chromosomes as well [103]. The large number of chromosomes characterizing some crayfish is probably due to genome duplication events [104] or to chromosome fragmentation events.

In conclusion, crustaceans are a valuable group to investigate all possible factors ruling determination and maturation of sexes and they also represent key models to investigate chromosomal assemblages and sex determination systems.

References

1. F.R. Schram. (2013). *The Natural History of the Crustacea, Vol.1: Functional Morphology and Diversity*. New York, NY, Oxford Univ Press, 1–33.
2. J.J. Sepkoski Jr. (2000). *Cont. Zoo.* **69:** 213–222.
3. G. Vogt. (2016). *BioRxiv*, doi.org/10.1101/047654.

4. T. Subramoniam. (2017). *Sexual Biology and Reproduction in Crustaceans.* Academic Press Print Book ISBN: 978-0-12-809337-5: 526 pp.

5. R. Matzke-Karasz. (2005). *J. Exp. Zool. B Mol. Dev. Evol.* **304B:** 129–149.

6. F. Gherardi. (2002). *Biology of Freshwater Crayfish.* Oxford: Blackwell, 258–290.

7. D. Zhang, J.A. Terschak, M.A. Harley, J. Lin, and J.D. Hardege. (2011). *PlosOne* **6**(4): e17720.

8. D.O. Elias, S. Sivalinghem, A.C. Mason, M.C.B. Andrade, and M.M. Kasumovic. (2014). *Anim. Behav.* **97:** 25–33.

9. S.A. Crowe, O. Kleven, K.E. Delmore, T. Laskemoen, J.J. Nocera et al. (2009). *Anim. Behav.* **77:** 183–187.

10. V. Jormalainen. (1998). *Q. Rev. Biol.* **73:** 275–304.

11. S. Wigby, and T. Chapman. (2004). *Curr. Biol.* **14:** R100–R103.

12. M. Barazandeh, C.S. Davis, C.J. Neufeld, D.W. Coltman, and A.R. Palmer. (2013). *Proc. R Soc. Lond. B.* **280:** 20122919.

13. R.T. Bauer. (2013). *Functional Morphology and Diversity. Oxf. Mag.* **1:** 337–375.

14. B.G.M. Jamieson. (1991). *Mem. Queensl. Mus.* **31:** 109–142.

15. D. Brandis, V. Storch, and M. Türkay. (1999). *J. Morphol.* **239:** 157–166.

16. V. Zupo, and P. Messina. (2007). *Mar. Biol.* **151:** 907–917.

17. V. Zupo. (2000). *Mar. Ecol. Prog. Ser.* 201: 251–259.

18. V. Zupo, M. Mutalipassi, F. Glaviano, A.C. Buono, A. Cannavacciuolo et al. (2019). *Sci. Rep.* **9:** 12336.

19. M. Buřič, M. Hulák, A. Kouba, A. Petrusek, and P. Kozák. (2011). *PLoS One* **6**(5): e20281.

20. N. Dreyer, J. Olesen, R.B. Dahl, B.K. Kan Chan, and J.T. Høeg. (2018). *PLoS one* **13**(2): e0191963.

21. J.T. Høeg, Y. Yusa, and N. Dreyer. (2016). *Biol. J. Linn. Soc. Lond.* **118:** 359–368.

22. R.T. Bauer. (2011). *In: Chemical Communication in Crustaceans,* 277–296.

23. W.S. Johnson, M. Stevens, and L. Watling. (2001). *Adv. Mar. Biol.* **39:** 105–260.

24. L.S. López Greco. (2013). *Functional Morphology and Diversity.* Oxford Univ Press, 413–450.

25. J. Pochon-Masson. (1994). *J. Zoo.* **7**(1): 727–783.

26. G.T. Miller, and S. Pitnick. (2002). *Science* **298:** 1230–1233.

27. K.J. Eckelbarger. (1994). *Proc. Biol. Soc Wash* **107:** 193–218.

28. D.M. Neubaum, and M.F. Wolfner. (1999). *Dev. Biol.* **41:** 67–97.

29. G. Martin, R.M. Laulier, and P. Juchault. (1996). *Crustaceana* **69:** 349–358.

30. T. Levy, V. Zupo, M. Mutalipassi, E. Somma, N. Ruocco et al. (2021). *Front. Mar. Sci.* **8:** 745540.

31. R.T. Bauer. (2000). *J. Crustacean Biol.* **20:** 116–128.

32. A. Rondeau, and B. Sainte-Marie. (2001). *Biol. Bull.* **201:** 204–217.

33. J. Yager. (1991). *In: Crustacean Sexual Biology.* Columbia Univ Press, New York, 271–289.

34. I. Takeuchi, and R. Hirano. (1991). *Mar. Biol.* **110:** 391–397.

35. K. Nakajama, and I. Takeuchi. (2008). *J. Crust. Biol.* **28**(1): 171–174.

36. D.C. Rogers, B.V. Timms, M. Jocquè, and L. Brendonck. (2007). *Zootaxa* **1551:** 49–59.

37. A.C. Cohen, and J.G. Morin. (1990). *J. Crust. Biol.* **10:** 184–211.

38. J.M. Hoch. (2009). *Evol.* **63:** 1946–1953.

39. W. Klepal, C. Rentenberger, V. Zheden, S. Adam, and D. Gruber. (2010). *J. Exp. Mar. Biol. Ecol.* **392:** 228–233.

40. W. Klepal. (1990). *Oceanogr Ma.r Biol.* **28:** 353–379.

41. M.J. Grygier. (1996). *J. Zool.* **7**(2): 722–726.

42. P.I. Blades, and M.J. Youngbluth. (1979). *Mar. Biol.* **51:** 339–355.

43. C.L. McLay, and C. Becker. (2015) *Treatise on Zoology – Anatomy, Taxonomy, Biology. The Crustacea. Vol. 9, Part C1: Decapoda: Brachyura (Part 1).* Leiden: Brill, 165–184.

44. C. Ewers-Saucedo, S. Hayer, and D. Brandis. (2015). *J. Morph.* **276:** 77–89.

45. Y. Yusa, M. Yoshikawa, J. Kitaura, M. Kawane, Y. Ozaki et al. (2012). *Proc. R Soc. Lond. B* **279:** 959–966.

46. A. Brandt, and H.H. Janssen. (1994). *Polar Biol.* **14:** 343–350.

47. G. Reverberi, and M. Pitotti. (1942). *Pubbl. Stazione Zoologica Napoli* **19:** 111–184.

48. G. Reverberi. (1942). *Pubbl Stazione Zoologica Napoli* **19:** 225–316.

49. D. Bouchon, T. Rigaud, and P. Juchault. (1998). *R. Soc. Lond. B* **265:** 1081–1090.

50. T. Rigaud, P.S. Pennings, and P. Juchault. (2001). *J. Invertebr. Pathol.* **77:** 251–257.

51. I. Negri, A. Franchini, E. Gonella, D. Daffonchio, P.J. Mazzoglio ct al. (2009). *Proc. R. Soc. Lond. B* **276:** 2485–2491.
52. C.E. Cressler, W.A. Nelson, T. Day, and E. McCauley. (2014). *Proc. R. Soc. Lond. B* **281:** 20141087.
53. D. Ebert, P. Rainey, T.M. Embley, and D. Scholz. (1996). *Philos. Trans. R Soc. Lond. B Biol. Sci.* **351:** 1689–1701.
54. K. Ikuta, and T. Makioka. (1997). *J. Morph.* **231:** 29–39.
55. K.J. Eckelbarger, and A.N. Hodgson. (2021). *Invertebr. Reprod. Dev.* **65:** 71–140.
56. S. Dennenmoser, and M. Thiel. (2015). *Cryptic Female Choice in Arthropods.* Heidelberg: Springer, 203–237.
57. L.M. Pardo, M. Riveros, J.P. Fuentes, and L. Lopez-Greco. (2013). *Invertebr. Biol.* **132:** 386–393.
58. AJ. Paul. (1984). *J. Crustacean Biol.* **4**(3): 375–381.
59. J.W. Wägele. (1992). *Crustacean sexual biology.* Columbia Univ Press, New York, 529–617.
60. A. Elorza, and E. Dupré. (1999). *Investig. Mar.* **28:** 175–194.
61. K. Ikuta, and T. Makioka. (1999). *Crustaceans and the biodiversity crisis.* Brill, Leiden, 91–100.
62. L. Soranzo, R. Stockmann, N. Lautie, and C. Fayet. (2000). *Proceedings of the 19th European Colloquium of Arachnology. European Arachnology 2000.* Aarhus Univ Press, 91–96.
63. G. Charmantier, and M. Charmantier-Daures. (2001). *Am. Zool.* **41:** 1078–1089.
64. S. Peixoto, G. Coman, S. Arnold, P. Crocos, and N. Preston. (2005). *Aquac. Res.* **36:** 666–673.
65. V. Zupo. (1994). *J. Exp. Mar. Biol. Ecol.* **178**(1): 131–145.
66. A.H. Hines. (1991). *Can. J. Fish. Aquat. Sci.* **48:** 267–275.
67. J. Radzikowski. (2013). *J. Plankton Res.* **35:** 707–723.
68. P. Talbot. (1991). *Crustacean egg production.* Balkema, Rotterdam, 9–17.
69. R.T. Bauer. (1981). *J. Crustacean. Biol.* **1:** 153–173.
70. B. Sainte-Marie. (2007). *In: Evolutionary Ecology of Social and Sexual Systems: Crustaceans as Model Systems.* Oxford Univ Press, 191–210.
71. J.L. Wortham-Neal. (2002). *J. Crustacean Biol.* **22:** 728–741.
72. M.G. Corni, V. Vigoni, and F. Scanabissi. (2000). *Crustaceana* **73:** 433–445.
73. K. Kronenberger, D. Brandis, M. Türkay, and V. Storch. (2004). *J. Morph.* **262:** 500–516.
74. G. Walker. (1992). *In: Crustacean Sexual Biology.* Columbia Univ Press, New York, 249–311.
75. H. Hobbs Jr., C. Harvey, and H. Hobbs. (2007). *Smithson Contrib. Zool.* **624:** 223–226.
76. R.M. Krol, W.E. Hawkins, and R.M. Overstreet. (1992). *Crustacea* **10:** 295–343.
77. H. Grier. (1993). *The Sertoli cells.* Cache River Press, Clearwater, FL, 703–739.
78. T. Tirelli, E. Campantico, D. Pessani, and C. Tudge. (2007). *J. Crust. Biol.* **27**(3): 404–410.
79. R.R. Hessler, R. Elofsson, and A.Y. Hessler. (1995). *J. Crust. Biol.* **15:** 493–522.
80. A. Avenant-Oldewage, and J.H. Swanepoel. (2005). *J. Morphol.* **215:** 51–63.
81. R.T. Bauer. (1991). *Crustacean sexual biology.* Columbia Univ Press, 183–207.
82. H. Wang, R. Matzke-Karasz, D.J. Horne, X. Zhao, M. Cao et al. (2020). *Proc. R. Soc. Lond. B* **287:** 1661.
83. P. Galeotti, G. Bernini, L. Locatello, R. Sacchi, M. Fasola et al. (2012). *PLoS One* **7:** e43771.
84. H. Niksirat, A. Kouba, and P. Kozak. (2014). *Anim. Reprod. Sci.* **149:** 325–334.
85. R.A. Koch, and C.C. Lambert. (1990). *J. Electron Microsc. Tech.* **16**(2): 115–154.
86. R.R. Hessler, and R. Elofsson. (1992). *Microscopic anatomy of invertebrates.* Crustacea, Vol. 9. Wiley-Liss, New York, 9–24.
87. H.J. Brook, T.A. Rawlings, and R.W. Davies. (1994). *Biol. Bull.* **187**(1): 99–111.
88. M. Abe, and H. Fukuhara. (1996). *Zool. Sci.* **13**(2): 325–329.
89. RT. Bauer. (2002). *J. Crustacean Biol.* **22**(4): 742–749.
90. T.F. Grilo, and R. Rosa. (2017). *Sci. Total Environ.* **592:** 714–728.
91. D.G. McCurdy, M.R. Forbes, S.P. Logan, M.T. Kopec, and S.I. Mautner. (2004). *J. Crustacean Biol.* **24:** 261–265.
92. A.T. Ford, C. Sambles, and P. Kille. (2008). *Mar. Environ. Res.* **66:** 146–148.
93. S. Richter, and G. Scholtz. (2001). *J. Zoolog. Syst. Evol. Res.* **39:** 113–116.
94. M.L. Porter, M. Perez-Losada, and K.A. Crandall. (2005). *Mol. Phylogenet. Evol.* **37:** 355–369.
95. G. Vogt. (2013). *Biol. Rev.* **88:** 81–116.
96. J.J. Legrand, E. Legrand-Harmelin, and P. Jachault. (2008). *Biol. Rev.* **62**(4): 439–470.

97. H.J. Alexander, J.M. Richardson, S. Edmands, and B.R. Anholt. (2015). *J. Evol. Biol.* **28**(12): 2196–2207.

98. M.T. Ghiselin. (1974). *The Economy of Nature and the Evolution of Sex.* Berkeley, CA: Univ California Press, 346 pp.

99. D. Bachtrog, J.E. Mank, C.L. Peichel, M. Kirkpatrick, S.P. Otto et al. (2014). *PLoS Biol.* **12**(7): e1001899.

100. E.S. Haag, and A.V. Doty. (2005). *PLoS Biol.* **3**: e21.

101. H. Niiyama. (1962). *Annot. Zool. Jpn* **35**: 229–233.

102. P. Martin, S. Thonagel, and G. Scholtz. (2016). *J. Zoolog. Syst. Evol. Res.* **54**: 13–21.

103. G. Vogt, C. Falckenhayn, A. Schrimpf, K. Schmid, K. Hanna et al. (2015). *Biol. Open* **4**: 1583–1594.

104. S.P. Otto, and J. Whitton. (2000). *Ann. Rev. Genet.* **34**: 401–437.

CHAPTER 3
Crustacean Yolk Proteins
Structure, Function and Diversity

Ulrich Hoeger[1],* and *Sven Schenk*[2]

3.1 Introduction

As in other oviparous animals, the oocytes of crustacea accumulate large amounts of yolk proteins as a source of nutrients and energy for the developing embryo. Yolk proteins are synthesized as precursor and converted into the final yolk protein by partial proteolysis for which the term "vitellin" is used. The term was coined in 1967 when Wallace et al. [1] described the major yolk proteins isolated from crustacean eggs as "lipo-vitellins". Based on this the term "vitellogenin" has been introduced by Pan et al. [2] to designate the precursor form in insects and this term has been applied to the yolk precursor proteins of other animals. As pointed out by Pan et al. [2], the term vitellogenin was intended "to designate a function and to imply nothing about molecular structure". In fact, while the most vitellogenins/vitellins belong to the family of large lipid transfer proteins (see below), some exceptions exist: In higher dipteran insects, like, e.g., Drosophila (the cycloraphan flies), the large lipid-transfer protein-like vitellogenins are replaced by proteins derived from lipoprotein lipases [3]. In the sea urchin, the major yolk protein is an iron-binding, transferrin-like protein synthesized by the adult gut [4]. In general, vitellogenins are considered sex-specific but surprisingly, there is one example of a male-specific vitellogenin found exclusively in the spermatozoa of the mud crab [5].

In this chapter, we will review the present findings on the structure and function of the yolk proteins in Crustacea. Previous reviews on vitellogenin (now termed apolipocrustaceins in the case of decapod crustacea; see below) have been published with focus on prawns by Wilder et al. [6], on vitellogenesis and yolk proteins in

[1] Institut für Molekulare Physiologie, Johannes Gutenberg-Universität, D-55099 Mainz, Germany.
[2] Max F. Perutz Laboratories; Vienna Biocenter (VBC), Dr. Bohr-Gasse 9/4, A-1030 Vienna, Austria.
  Email: Sven.Schenk@univie.ac.at
* Corresponding author: uhoeger@uni-mainz.de

Crustaceans and Molluscs [7], on mechanisms and control of vitellogenesis [8] and on vitellogenin genes and genomes [9]. While the term "Crustacea" will be maintained in line with earlier work, it should be noted that the Crustacea are no longer considered as monophyletic group as a result of recent phylogenetic analyses. Crustacea are now considered paraphyletic and part of Pancrustacea/Tetraconata that include both "Crustacea" and Hexapoda [10].

3.2 The superfamily of large lipid transfer proteins

The known lipid transporting proteins of crustaceans, including yolk proteins, belong to the superfamily of the large lipid transfer proteins (LLTP's). Exceptions represent the high-density lipoproteins/β-glucan binding proteins (see [11] for a recent overview) and the intracellular fatty acid binding proteins of crustaceans [12] and other animals, which belong to different protein families. The subfamilies of large lipid transfer proteins include the vitellogenins (vitellogenin-like LLTP's, VTG's), the apoB type LLTP's and the extracellular cytosolic large subunit of the microsomal triglyceride transfer proteins (MTP's). The vitellogenin branch includes the VTG's of vertebrates, molluscs, nematodes, insects, the clotting proteins of decapod crustaceans and the melanin-engaging proteins. Melanin engaging proteins are involved in the phenoloxidase system enhancing melanin synthesis. First described in the flour beetle *Tenebrio*, related proteins were found recently in the copepod *Tigriopus kingsejongensis* (Order Hexanauplia, [13]). The apoB family (apoB-like LLTP's) includes vertebrate apoB LLTP's such as human apoB 100 (the major apolipoprotein of VLDL and LDL), the apolipocrustaceins (apoCr; see Section 3.7 below) of decapod crustaceans and insect apolipophorins II/I (apoLp II/I) [14]. MTP's are essential for the lipidation of intracellular lipoprotein particles in apoB, apoLp II/I and VTG. MTP forms a dimer with protein disulfide isomerase which assists in the folding of the nascent lipoprotein particle and interacts with the lipid binding module (see below) of LLTP's (reviewed in [27]). For crustaceans, the presence of three isoforms of MTP has recently been demonstrated for the first time in the ectoparasitic copepod *Lepeophtheirus salmonis* [15].

In recent years, new members of LLTP's have been discovered. For instance, a vitellogenin-related protein, Crossveinless, was found to be required for BMP (bone morphogenetic protein) signaling in wing development of *Drosophila* [16]. The apolipophorin-II/I-related protein (ARP, [17]) is also an apoB type LLTP. This protein (accession nr. BAN58736) has now been identified as the insect lipid transfer particle (Ltp) based on its sequence. Ltp is responsible for the lipidation of the major lipid transport protein in the insect hemolymph.

Finally, the large subunit of the large discoidal lipoprotein found in *Astacus leptodactylus* [18] is also a new member of the LLTP family.

3.3 Evolution and origin of vitellogenin-like large lipid transfer proteins

It may be speculated that the LLTP's could have evolved in multicellular organisms when the necessity for exchange of lipids between cells and tissues arose (cf. [*17*]). The transcriptome of the simplest eumetazoan known to date, *Trichoplax adhaerens*, contains several putative lipoproteins (see below) displaying the same lipoprotein N-terminal domain (LPD_N) as other invertebrates and vertebrates. In contrast, a tblastn search among the unicellular organisms (NCBI database) using the lipid binding domains of *Trichoplax* as query did not reveal the presence of this characteristic domain in unicellular organisms (unpublished observations).

By molecular phylogenetic analysis, it has been demonstrated in several studies that the family of large lipid transfer proteins evolved from a common ancestor [*14,19–21*]. The evolution of MTP and apoB from an ancestral VTG likely arose from gene duplications and the presence of both VTG type and apoB type LLTP's in protostomes, e.g., VTG and apoLp in insects or the clotting proteins as derived VTG's and apolipocrustaceins in decapod crustaceans (see Section 3.7), suggested that this step arose at an early level of bilaterian evolution [*19*]. After the evolution of Bilateria, MTP was suggested to have evolved by another event of gene duplication leading to a truncation of its sequence as adaptation to its major function in the biogenesis of lipoprotein particles [*20*].

Due to its wide evolutionary distribution, VTG was initially suggested as the ancestral member of the LLTP gene family from which other LLTP's, MTP and apoB, arose via gene duplication [*20*]. In contrast, MTP should be considered as the oldest LLTP family member. This view was based on the finding that secretion of VTG in *Caeonorhabditis elegans* is dependent on the dsc-4 protein, an orthologue of MTP [*22*]. Likewise, Sellers et al. [*23*] have demonstrated that secretion of the VTG of *Xenopus* expressed in COS cells was enhanced five-fold by coexpression with MTP. This dependence of VTG secretion from MTP was seen as support for the ancient position of MTP arguing against the ancestral position of VTG.

More recently, Hayward et al. [*24*] noted the absence of MTP-like lipoproteins in the transcriptomes of Cnidaria while the presence of VTG's in Cnidarians has been established and was described first in the coral, *Galaxea fascicularis* (accession nr. BAD74020.2). At present, more than 100 cnidaria species are listed in the transcriptome shotgun assemblies (TSA) database of the NCBI. In a survey of this database, we found no hits for putative MTP-like proteins using the MTP's of the crustacean *Lepeophteirus salmonis* (accession nr. ASA47097.1) or *Drosophila melanogaster* (accession nr. AAF53946.2) as query. In contrast, in the sequences of several other Cnidarian VTGs are annotated in the NCBI nr protein database besides that of *Galaxea*. If this view is correct, however, the absence of MTP-like lipoproteins in the transcriptomes of Cnidaria would also imply that lipid acquisition during biogenesis of VTG's at least in Cnidaria is independent on the activity of MTP and the dependence of MTP might have evolved during VTG evolution in metazoans.

A more recent genome study of *Trichoplax* (NCBI Bioproject PRJNA393433) has again changed this picture. In this study, the authors have obtained by conceptual translation two VTG-like sequences (accession nr. RDD43034.1, RDD46142.1) and in addition, a MTP-like lipoprotein (RDD38051.1) and even an apoB-related protein (RDD42209.1). This would place the origin of MTP and apoB-like proteins much earlier than previously thought. *Trichoplax* reproduces sexually and the presence of oocytes containing yolk bodies has been demonstrated in electron micrographs [*25*]. However, the putative *Trichoplax* VTG and the other putative LLTP's have not yet been biochemically and functionally characterized. If these findings can be corroborated in future studies, this would call for a reanalysis of the evolution of LLTP family at the beginning of eumetazoan evolution. Obviously, the apparent absence of MTP proteins in the Cnidaria and its presence in *Trichoplax* needs to be explained. In any way, these findings highlight the ancient origin of the VTG's, MTP's and apoB proteins and show that the complex structure of the lipid binding domain (see below) evolved very early in (eu)metazoan evolution.

3.4 Domain structures of vitellogenins and apolipocrustaceins

LLTP's are characterized by the presence of a N-terminal lipid binding region, the LLT module. This module encompasses the lipoprotein binding domain (LPD_N), the DUF1943 domain, and—in case of apoB-type LLTP's—the DUF1081 domain. The LLT module is followed by a C-terminal region which contains various lipid binding elements (see Figure 2 below). The C-terminal region shows considerable differences in the length depending on the type of LLTP. MTP's contain a LPD_N domain and no other domains and its sequence is relatively short with lengths of less than 900 amino acids in all known MTP's from both vertebrates and invertebrates. The apoB type LLTP's represent the largest LLTP's with apoB100 as largest known LLTP, with a C-terminal region of more than 3500 amino acids. In crustacean apoCr's, which belong to the apoB branch (see Figure 3), C-terminal regions range around 1500 amino acids. The VTG's of both vertebrates and invertebrates, show an intermediate length of the C-terminal lipid binding region ranging between 500 and 1000 amino acids. The length of the C-terminal portion is important for the lipid carrying capacity of the lipoprotein particle as will be explained in more detail below. At the end of the C-terminal region, there is a von Willebrand factor type D motif in VTG's and apoCr's which is lacking in MTP's and vertebrate apoB-100 (see Figure 3). Instead, apoB-100 contains a unique C-terminal domain (apoB100_C).

3.4.1 N-terminal lipid binding domain, DUF1943 and DUF1081 domain

The N-terminal lipid binding domain (LPD_N) is well conserved within all known members of the large lipid transfer proteins including vitellogenin [*17*]. Even in the putative lipoproteins of the simplest multicellular animal known, the placozoan *Trichoplax*, this domain is very similar to the LPD_N of lamprey lipovitellin, the first LLT protein which has been crystallized [*26*]. The lamprey lipovitellin (protein data bank, pdb 1LSH) thus has served as reference for modelling this domain in other lipoproteins and provided the structural background to elucidate lipid binding and transfer in the LLT protein family. Only recently, the crystal structure of a second LLTP, the human MTP, has been published in the protein data bank (pdb 6I7S). In mammals, MTP is involved in the transfer of lipids for the assembly and secretion of very low density lipoproteins and likewise in the lipophorins of insects, working in concert with protein disulphide isomerase to form a functional complex [*27*].

In crustaceans, the role of MTP has been investigated for the first time in a parasitic copepod, the salmon louse *Lepeophteirus salmonis* [*15*]. There it was found that the offspring of RNAi mediated MTP knock down females had a very low survival rate (~ 10% for offspring from newly emerged preadult female II), and that the surviving nauplii had little to no lipid droplets. Furthermore, the total lipid content of these nauplii was 83% lower than that of control animals. The authors further showed that the incorporation of lipid into the eggs is directly dependent from the amount of MTP transcripts. In a second series of RNAi mediated knock down experiments the authors used older females (newly molted young adult) than in the first experiments which already had transcribed MTP. Offspring of these females with a very low MTP transcript level, showed survival rates of ~ 20%. This together with the virtual absence of lipids in the offspring clearly demonstrated the importance for MTP-mediated oocyte lipid supply, most likely via efficient VTG lipidation.

A model of the LLT module of the apolipocrustacein of *Macrobrachium rosenbergii* (GenBank accession nr. BAB69831.1) is shown in Figure 1. The LLT module contains several characteristic structure elements: an N-terminal barrel-like β-sheet, a helical region containing a series of α-helices, a C-sheet and an A-sheet (named according to [26]). The N-sheet does not bind lipids; however, it contains one major region serving as docking site for the binding to the apoCr-receptor [28] (marked in red in Figure 1). The A- and C-sheets and part of the α-helical region containing amphipatic elements form a lipid binding cavity.

This lipid binding pocket, which can accommodate seven molecules of phospholipids and 43 lipid hydrocarbon chains could be modelled into the lipid binding cavity of a lamprey VTN monomer [26]. These authors showed that the LPD_N domain forms a funnel like structure with two major openings and a lipid accessible area of 60,000Å. In the native lamprey vitellin, two adjacent LPD_N domains form a lipid binding pocket by dimerization of the two protein subunits. Based on these data, this pocket has a lipid binding surface of 120,000Å and can complex 100 lipid molecules. Within the lipid binding pocket, the hydrocarbon chains of lipids are primarily associated with β-sheets running roughly parallel to the hydrocarbon chain [26]. In the contrast to the lamprey lipovitellin model, the A-sheet in *Macrobrachium* apoCr is truncated at the C-terminus leaving the base of the lipid cavity open as also true for other apoB lipoproteins (human apoB 100, insect lipophorin). The open cavity allows for further binding of lipids mediated by the DUF1081 domain (see below). With regards to the designation of the domain boundaries of the LLT module (see Figure 3), the N-terminal lipid binding domain in *M. rosenbergii* apoCr (amino acids 41–589) covers the N-sheet and parts of the α-helical region, while the DUF1943 domain (amino acids 621–920) covers the C-sheet and parts of the A-sheet. The DUF1081 domain (amino acids 940–1050 in the complete sequence) covers a second part of the A-sheet. This domain, however, is only partially represented in the model since a DUF1081 domain is not present in the lamprey lipovitellin as a member of the VTG branch of the LLTP's.

Segrest et al. [29] found that the first 1200 residues of human apoB-100 show sequence and amphipathic motif homologies to the lipid-binding pocket of lamprey lipovitellin and found evidence for the presence of these motives in the vitellogenins of four other vertebrates and also of *C. elegans*. In a later study, Richardson et al. [30] showed a very similar architecture for the structure of the nascent apolipoprotein B.

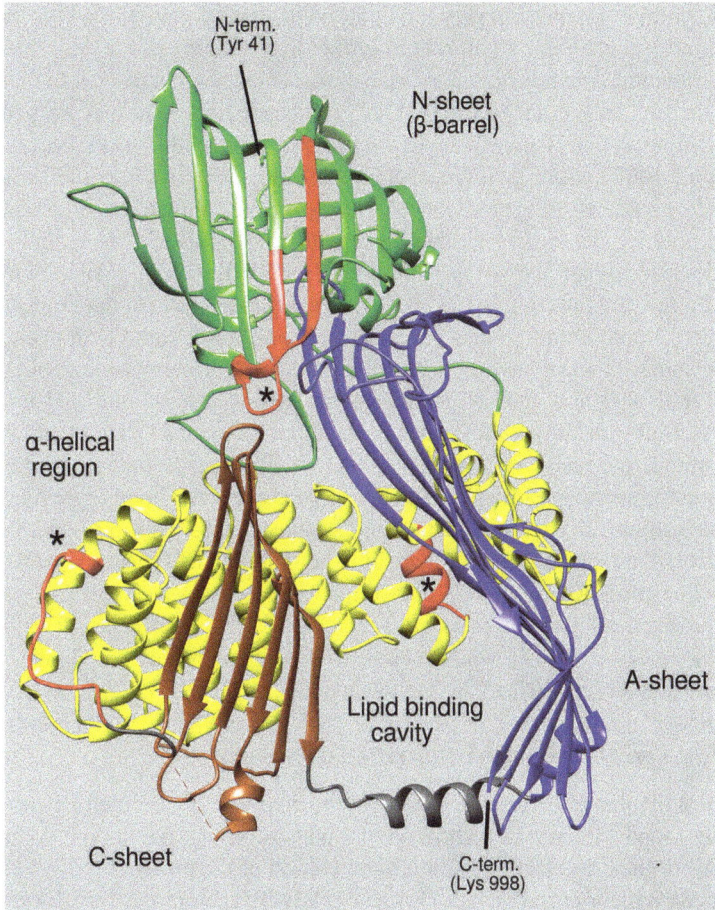

Figure 1. Ribbon diagram of the LLT module of the apolipocrustacein of *Macrobrachium rosenbergii* generated using the Phyre2 web server (http://www.sbg.bio.ic.ac.uk/phyre2/html/page.cgi) and the lamprey lipovitellin (pdb 1LSH) as template. The model was visualized using Chimera v. 1.3.1 (http://www.rbvi.ucsf.edu/chimera). The region covering amino acid residues Tyr 41 (N-term.) to Lys 998 (C-term.) was modeled with a model coverage of 95% and a confidence of 100%. The different structural elements: N-sheet (amino acid residues 41–296, green); A-sheet (residues 796–998, blue); C-sheet, (residues 621–757, brown); α-helical region (residues 297–607, yellow) are named according to Thompson and Banaszak [*26*]. The A- and C-sheets and part of the α-helical region form a lipid binding cavity. The base of the lipid cavity is open. The grey helix at the bottom is not consistent with the lamprey lipovitellin model. Asterisks denote three regions interacting with the VTG/apoCr-receptor (marked in red, data taken from [*28*]).

In this study the authors were able to model 48 phospholipid (POPC) molecules into the lipid binding pocket [*30*], thus highlighting not only structural but also functional conservation of the LPD_N domain.

As mentioned above, the DUF 1081 domain is found in apoB type lipoproteins only and therefore in the apoCrs of Crustacea which are members of this family (see Section 3.3 above). Located directly on the C-terminal of the LPD_N and DUF 1943 domains, crucial for the initiation of lipoprotein lipidation [*21, 30, 31*], the DUF 1081 seems to play an important role in further lipid loading of the lipoprotein. This was experimentally demonstrated by Manchekar and co-workers [*31*]. In their work, the authors used transformed McA-RH7777 (rat hepatoma) cells to express truncated human apolipoprotein B, consisting of the first 800 (apoB:800), 931 (apoB:931), and 1000 (apoB:1000) amino acids, respectively. Analysis of the secreted lipoprotein particles by non-denaturing gradient gel electrophoresis followed by autoradiography and immunoblot demonstrated that apoB:800 and apoB:931 were only poorly lipidated with buoyant densities > 1.23 g/mL; apoB:1000 was lipidated to a density of < 1,23 g/mL and thus within the HDL_3 range. Lipid analysis of the lipids bound to the different lipoprotein particles revealed that apoB:1000 had 50–56 phospholipid molecule and 8–11 triacylglycerols as well as up to 7 cholesterols bound per particle, whereas apoB:800 and apoB:931 only bound some phospholipids and no neutral lipids [*31*]. Since (in human apoB-containing lipoproteins) the LPD_N and DUF1943 occupy amino acid residues 21–918, while DUF 1081 starts at amino acid 933, this clearly highlights the importance of DUF 1081 for extended lipoprotein lipidation.

Malacostracan lipo-vitellogenins (apolipocrustaceins, apoCr's) are members of the apoB-type of lipoproteins and thus contain (at least partly) a DUF 1081 domain. The presence of this domain might serve as a structural explanation of why Malacostracan yolk proteins are rather lipid-rich lipoproteins (lipid contents ~ 28–35%, reviewed in [*11*]) while "classical" vitellogenins are rather lipid poor with lipids contributing ~ 3–20% to the particle mass [*11, 26, 32, 33*].

3.4.2 The Von Willebrand factor type-D domain

The von Willebrand factor is a glycoprotein involved in the clotting cascade of vertebrate blood. It contains 4 different domains of which the D-type domains are important for the dimerization and multimerization of the protein.

The von Willebrand factor type D-domain (VWD) is commonly found in vertebrate and invertebrate vitellogenins and it is therefore also found in Malacostracan apoCr's (for a recent review see, e.g., [*11*]). A characteristic feature of the VWD domain in the apoCr's of Decapod crustaceans is the GLLG motive which is orthologue to the GL/ICG domain present in other animal's vitellogenins [*34–37*]. Even though its role has not been established, its presence in the VWD of many LLTP's and its functions in the von Willebrand factor led to the assumption of its participation in invertebrate VTG's and apoCr's multimerization. However, multimerization of the human von Willebrand factor requires the presence of two vicinal cysteines in the D-domain which are separated by glycine and leucine forming the motif C-G-L-C-G and multimerization was lost when this motif was experimentally manipulated [*38*]. For several decapod species, we have found no cysteines near the decapod GLLG motif within the VWD (unpublished observation). This renders a similar function for the VWD at least in decapod crustaceans rather doubtful. In isopods and amphipods, a VWD domain was even found absent based on the SMART domain search (see Table 1). While the function of the VWD in crustaceans remains to be elucidated, the present data suggest that a VWD is not an absolute requirement for the apoCr's.

3.4.3 Polyserine stretches

In vitellogenins of most vertebrate and insect species polyserine stretches (the so-called phosvitin domain) have been detected. The role of polyserine stretches seems not to be clear, however, serines are good substrates for phoshorylation by various kinases. And indeed, vertebrate and *Aedes aegypti* VTG's have been found to be heavily phosphorylated [39–41] and this may serve for enhancing their solubility or their ability to complex essential ions like, e.g., Ca^{2+} [42]. Furthermore, VTG uptake into the oocyte has been shown to be dependent on phosphorylation [42]. In most insects (with the exception of Apocrita) the polyserine stretch harbours a consensus cleavage site (R/K)XX(R/K) for subtilisin-like endoproteases [41]. Further studies [43] have shown that the polyserine stretches represent low complexity polypeptides which are presumably disordered and show a random coil structure as suggested by NMR studies (see [43]). This is thought to facilitate the access for proteases.

In contrast, no polyserine stretches have been found in the crustaceans characterized so far [35,36,44–48], and it has been suggested that receptor binding is mediated by other mechanisms. And indeed, in *Macrobrachium rosenbergii* it was demonstrated that the receptor binding domain lies in the β-sheet domain of the N-terminus [28] (see Figure 1 and below, Chapter 3.12; [36,44,45]).

It could be argued that the replacement of VTG by apoCr's in malacostracans has led to the absence of polyserines in the yolk proteins of this group; however, a survey of the VTG's in copepods and cirripeds (see Table 1) likewise revealed the absence of polyserine stretches. Patches of mostly di-serines have been found in the apoCr of *Callinectes sapidus* [47] and this is also true for the apoCr's listed in Table 1 (unpublished data). These di-serines are believed to be phosphorylated by caseine kinase II rendering VTG/apoCr a higher affinity to the oocyte VTG receptor [49,50]. The effect of phosphorylation of these serines on the receptor mediated endocytosis remains to determined in crustacean VTG's and apoCr's.

3.5 The C-terminal region of the LLT proteins and the formation of lipoprotein particles

The lipoprotein assembly of LLT apoproteins depends on the N-terminal LLT module and its interaction with the MTP/protein disulfide isomerase (PDI) complex to achieve the initial lipidation and folding of the nascent lipoprotein particle during translation of the lipoprotein. The second determinant is the lipid binding capacity of the C-terminal region which is size dependent and shows widely differing lengths depending on the lipoprotein. While the N-terminal LLT module comprises about 900–1000 amino acids in most LLTP's, the C-terminal part can be very short in the case of the MTP's (< 400 aa) or more than 3500 aa long as in the case of human apoB 100. In crustaceans the C-terminal parts range between 800-900 aa and about 1700 aa for VTG's and apoCr's, respectively.

Considering the lipid binding properties of the C-terminal region in LLTP's, most of the studies have been carried out on human apoB-100. ApoB-100 is the largest known lipoprotein consisting of 4536 amino acid residues. It is required for the formation of triglyceride-rich lipoproteins in the liver and intestine and is the major

Table 1. Domain structure and distribution of vitellogenins and apolipocrustaceins in different crustacean groups. The presence and absence of domains is indicated. Sequences marked with an asterisk were retrieved by BLAST searches using annotated LLTPs as query. The open reading frames and the domain boundaries were determined using the online tools Translate (http://web.expasy.org/translate/) and SMART (http://smart.embl-heidelberg.de/), respectively. SOD, superoxide dismutase domain; LPD, lipoprotein N-terminal domain; DUF1943, DUF1081, domains of unknown function, VWD, von Willebrand factor type D domain.

	VTG/ApoCr	SOD	LPD	DUF 1943	DUF 1081	VWD	Sequence length (aa)	Nr. of species tested[1]	Genbank Acc. nr.	Representative species
Remipedia[2]	VTG	-	+	+	-	+	1700	1	*GCBC01036284.1	*Xibalbanus tulumensis*
	VTG	-	+	+	-	+	1735		*GCBC01048359.1	*Xibalbanus tulumensis*
	VTG	-	+	+	-	+	1712		*GCBC01048358.1	*Xibalbanus tulumensis*
Branchiopoda	VTG	+	+	+	-	+	2001	4	BAD05137.1	*Daphnia magna*
	VTG	+	+	+	-	+	1991		QJE49262	*Diaphanosoma celebensis*
	VTG	+	+	+	-	+	2219		HQ647327.1	*Artemia parthenogenetica*
Malacostraca										
Phyllocarida	ApoCr	-	+	+	+	+	2552	1	*GCDB01029281	*Nebalia bipes*
Eucarida	ApoCr	-	+	+	+	3	> 2200[3]	1	*GETT01201529.1	*Meganyctiphanes norwegica*
Stomatopoda	ApoCr	-	+	+	+	+	2506	1	ALI16501.1	*Oratosquilla oratoria*
Decapoda	ApoCr	-	+	+	+	+	2537	>10	BAB69831.1	*Macrobrachium rosenb.*
		-	+	+	+	+	2587		AAP76571.2	*Litopenaeus vannamei*
		-	+	+	+	+	2583		ABO09863.1	*Homarus americanus*
		-	+	+	+[4]	+	2563		AEI59132.1	*Callinectes sapidus*
		-	+	+	-[4]	+	2562		AGM75775.1	*Eriocheir sinensis*
		-	+	+	[4]	+	2561		ACO36035.1	*Scylla paramosain*
Isopoda	ApoCr	-	+	+	-	+	2512	6	*HAFH01019191.1	*Proasellus racovitzai*
		-	+	+	-	+	2514		*HAEM01113487.1	*Bragasellus molinai*
Amphipoda	ApoCr	-	+	+	-	-	2212	5	*GHCP01099668.1	*Gammarus pulex*
		-	+	+	-	-	2234		*GHHW01012450.1	*Eulimnogammarus cyaneus*
		-	+	+	-	-	2212		*GEPA01030211.1	*Oxyacanthus_curtus*
		-	+	+	-	-	2211		*GEQP01092177.	*Micruropus wahlii*

Copepoda	VTG	-	+	+	-	+	1769	>10	AGH68974	*Eurytemora affinis*
		-	+	+	-	+	1905		ADD73551	*Paracyclopina nana*
		-	+	+	-	+	1842		ABZ91537	*Tigriopus japonicus*
		-	+	+	-	+	1964		ABU41134	*Lepeophteirus salmonis*
Cirripedia	VTG	-	+	+	-	+	1603	7	*GIJW01050126.1	*Octolasmis warwickii*
		-	+	+	-	+	1601		*GIJX01041139.1	*Glyptelasma gigas*
		-	+	+	-	+	1597		*GGJM01156104.1	*Lepas anatifera*
		-	+	+	-	+	1602		*GGJN01034354.1	*Pollicipes pollicipes*

1   Total nr. of species analyzed within the same taxonomic group displaying identical domain structures and very similar sequence lengths (data not shown).

2   The systematic position of the Remipedia within the taxon "Tetraconata" (comprising Crustacea and Hexapoda) is uncertain.

3   Sequence incomplete

4   In some brachyuran crabs (*Callinectes sapidus* and *C. toxotes* (not shown), the DUF1081 domain was detected by the SMART domain search, while in other crabs, this domain was – unexpectedly - not found.

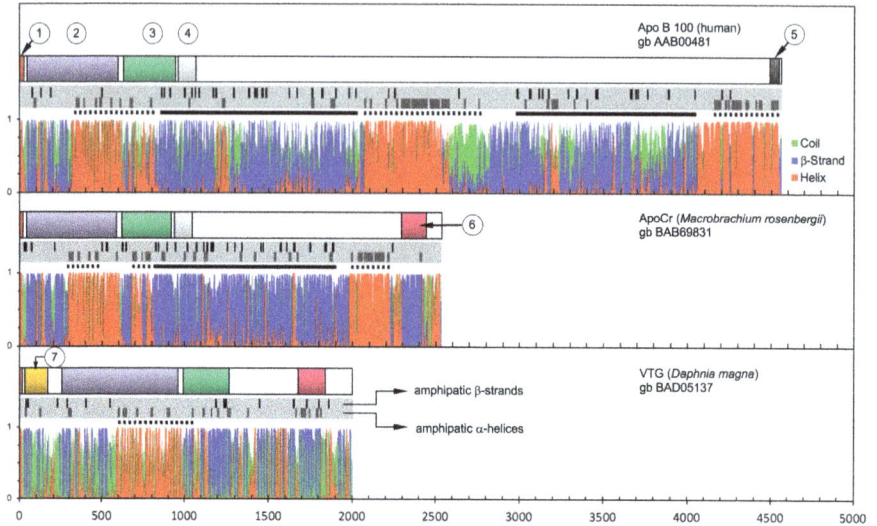

Figure 2. Comparison of the domain organization (horizontal bars), the secondary structure predictions and the presence of amphipatic α-helices and amphipatic β-strands in the sequences of human Apolipoprotein B100, apolipocrustacein (apoCr) of *Macrobrachium rosenbergii* and vitellogenin (VTG) of Daphnia magna. Circled numbers indicate signal sequence (1), N-terminal lipid binding domain (2), DUF1943 domain (3), DUF 1081 domain (4), apoB100_C domain (5), von Willebrand factor type D domain (6) and superoxide dismutase domain (7). The domain boundaries were determined using the online tool SMART (https://smart.embl-heidelberg.de/smart). The secondary structures (green: unassigned (coil) regions, blue: β-strands, red: α-helix) were predicted using PSIPRED 4.0 (http://bioinf.cs.ucl.ac.uk/psipred). The predictive significance is between 0 and 1. The presence of α-helices and β-strands exhibiting amphipatic properties are indicated by black and grey rectangles, respectively, below the domains. Dotted and solid black lines indicate the clusters enriched in either amphipathic α-helices or amphipathic β-strands, respectively. The data for amphipatic properties have been modified from [17]. The x-axis indicates the amino acid positions.

protein component of the low-density lipoproteins. Segrest et al. [29] developed the program LOCATE to identify amphipathic and amphipatic α-helices as the main lipid binding motifs in LLTP's [29]. For human apoB 100, two clusters of putative lipid-associating amphipathic helices were located in the middle and at the C-terminal end of the protein (see Figure 2). Several clusters of amphipatic β-sheets were identified in the regions dominated by β-sheets corresponding to amino acid positions 2103 through 2560 and 4061 through 4338, respectively. Smolenaars et al. [17] also used the program LOCATE to identify amphipatic β-sheets and amphipatic helices in a variety of LLTP's from vertebrates and invertebrates including several crustaceans and found similar patterns in the distribution of amphipatic helices and β-sheets in the C-terminal region of VTG's and apoCr's (Figure 2). It can therefore be assumed that the basal lipidation mechanisms as found for apoB-100 are also true for VTG's and apoCr's crustaceans.

3.5.1 Role of the microsomal triglyceride transfer protein

MTP is not involved in the transport of lipids to other tissues. It is present in intracellular compartments of the secretory pathway and serves for the assembly of apoB and VTG lipoproteins [27]. During assembly of lipoproteins, MTP acquires lipids from the ER and transfers them to the nascent polypeptide chain of the apolipoprotein. Within the LPD_N, the N-terminal β-barrel domain mediates interaction with the N-terminus of apoB proteins, the α-helical region is involved in the binding of both PDI and apoB and the C-terminal β-sheet has both lipid binding and lipid transfer activities [27].

MTP's from different sources have different lipid transfer properties and can thus determine the composition of the lipid moiety of the lipoprotein particle. Rava et al. [51] compared human and *Drosophila* MTP with regards to their lipid transfer specificity and to assemble apoB lipoproteins. The transfer of phospholipids was similar in both orthologues, however, unlike human MTP, the *Drosophila* orthologue was unable to transfer neutral lipids which resulted in the formation of a phospholipid-rich, high-density apoB lipoprotein. This demonstrated the successful activity of MTP in a heterologous system and also showed the MTP as a determinant of the lipid composition of the lipoprotein particle. In a further study, Rava et al. [52] compared the triglyceride transfer activity of MTP's in different vertebrates and found that fish, amphibians, and birds displayed 27%, 40%, and 100% of the triglyceride transfer activity, respectively, compared to mammals. The authors concluded that the phospholipid transfer activity of MTP is an early function present in invertebrates. It is sufficient for the assembly and secretion of primordial apoB lipoproteins, however, the evolution of triglyceride transfer activities was a process unique to vertebrate lipid transport systems. The observation that phospholipids are the dominant lipids in VTG's and apoCr's of crustaceans (see Section 3.6 below) and other invertebrates are in line with these results. Studies on the interaction of MTP with either VTG's or apoCr's and its role in determining their lipid composition are currently lacking in crustaceans. But the requirement for MTP in the vitellogenesis and in the deposition of lipidated vitellogenins in the eggs of the copepod *Lepeophteirus salmonis* has been clearly demonstrated ([15], see Section 3.5.1 above).

3.5.2 *Influence of the C-terminal region on the lipidation of The LLTP particle*

In continuation of their earlier studies [31], Manchekar et al. [53] have investigated the influence of the residue length of the C-terminal region on the lipidation of the apoB-100 lipoprotein particle using truncated variants of apoB-100 with a length between 1000 (apoB:1000) and 1700 amino acids (apoB:1700) expressed in a murine cell line. Variants up to apoB:1400 produced lipoprotein particles containing phospholipids as the main lipid moiety and less triglycerides while apoB:1400 and apoB:1500 showed increasing content of triglycerides. apoB:1700 lipoprotein particles showed equal content of phospholipids and triglycerides and the number of lipid molecules per particle increased twofold. This showed the effect of chain length on the lipid content of the lipoprotein particle. The results also suggested that apoB:1700 marks the threshold for the formation of a TAG-rich particle. Additional results using a MTP-specific inhibitor showed that MTP was required only for assembly and secretion of triglyceride-rich particles, apoB:1700 and the full-length apoB-100, suggesting that MTP mediates in the transition from phospholipid-rich to triglyceride-rich lipoprotein particles. The initiation of lipoprotein assembly thus proceeds independent of MTP and is likely to be mediated by another protein, the phospholipid transfer protein [*54*].

With respect to the Crustacean apoCr's and VTG's, the properties of the MTP would allow the formation of predominantly phospholipid-rich lipoprotein particles. The size of Malacostracan apoCrs (> 2500 aa; see Table 1) exceeds this critical length suggesting that MTP would be required for apoCr assembly and secretion. In contrast, the size of copepod VTG's (1700–1900 aa; see Table 1) lies within the critical size length. This raises the question if the assembly of copepod VTG's also requires the activity of MTP and/or that of the phospholipid transfer protein. At least for the VTG assembly in *Lepeophteirus salmonis* with a size of 1900 amino acids, which is beyond the critical size of apoB:1700, MTP seems required [*15*] (see Section 3.5.1 above).

3.6 Lipid content of apolipocrustacein

Most invertebrate lipoproteins belong to the high-density lipoproteins with buoyant densities between 1.064 and 1.210 g/ml [*55*] and this is also true for crustacean VTG/apoCr's [*56*] as well as for other hemolymph lipoproteins (see Yepiz-Plascencia et al. 1995 and cited references). In three decapod VTGs/apoCr's, the total lipid content ranged between 45 and 50% [*57–59*]; a lower figure was measured in the mole crab, *Emerita* (28%, [*60*]) and for the putative VTG in the branchiopod *Triops* (19%, Hoeger and Schilz, unpublished). Phospholipids are the dominant lipid fraction in invertebrate lipoproteins. In the VTG/apoCr of *Penaeus semisulcatus*, Lubzens et al. [*59*] found phospholipids (70%), cholesterol (18%) and only low amounts (4.5%) of triglycerides and a similar situation was found in the putative VTG/apoCr of *Cancer antennarius* (80% phospholipids, 4% cholesterol, 8% triglycerides, [*58*]). In addition to neutral- and phospholipids, crustacean lipoproteins contain various amounts of carotenoids, mainly astaxanthin, canthaxanthine and β-carotene (for an overview, see [*60*]). In comparison, the fully lipidated apoB-100 particle has a protein content of 22% and a lipid content of 78% consisting of 22% phospholipids, 8% cholesterol, 42% cholesteryl esters, and 6% triglycerides [*61*].

3.7 Vitellogenins and apolipocrustaceins in different crustacean groups

Vitellogenins and yolk proteins in crustacea were characterized first in decapod crustaceans only [*1, 60*]. Avarre et al. [*14*] showed that the decapod "VTG's" were more closely related to the apoB-type LLT proteins and not to the classical VTG's of other invertebrates such as insects, mollusks and vertebrates. Therefore, the designation apolipocrustacein (apoCr) was suggested for the former decapod VTG's to avoid confusion [*14*]. To date, more than 30 decapod apolipocrustaceins are currently annotated in the NCBI non redundant protein database and a recent list is given by Jimenez-Gutierrez et al. ([*62*]; see Table A-III ibid.). They show a narrow sequence length range (2500–2600 amino acids) and a characteristic domain structure including the N-terminal lipid binding domain (LPD_N), two domains of unknown function (DUF1943 and DUF1081) and a von Willebrand factor type

D-domain (Table 1). The gene organization is also very similar in many decapods showing a structure of 15 exons and 14 introns [*62*]. Apolipocrustaceins have native molecular masses ranging between 325 and 700 KDa. An overview is given by Lee [*60*] (see Table 5 ibid.).

In contrast to apolipocrustacein, another group of crustacean LLTP's, the clotting proteins, was found to be related to the classical VTG group of LLT proteins [*14,17,24*] (see above). Clotting proteins are involved in defensive functions [*63*] but also in the differentiation of hematopoetic stem cells [*64*]. Their lipid content is much lower than that of VTG's and apoCr's ranging between 5 and 15% (see Table 2 in Hoeger and Schenk [*11*]). As pointed out above, apolipocrustaceins (formerly called vitellogenins) have been studied only in decapod crustaceans. Given the presence of the "classical" vitellogenins in other taxa such as the copepods and branchiopods (see below), it is of interest to compare the VTG's of basal taxa which could shed some light on the evolution of Crustacean VTG's and apolipocrustaceins. In addition, investigating the VTG/apoCr sequences of Malacostracan Crustacea in taxonomic groups other than those of the well-studied decapods could further elucidate at which taxonomic level the transition towards the use of the novel apolipocrustacein arose as the major yolk protein in the evolution of the Malacostraca. To elucidate this aspect, we searched the transcriptomes (TSA) of *Anaspides tasmaniae*, a Malacostracan which retains ancient traits within this group, and *Nebalia bipes* belonging to the Leptostraca considered a sister group to the Eumalacostraca.

In *Nebalia*, an apoCr-like protein was found (Acc. Nr. GCDB01029281) in addition to four LLTP's (GCDB01031228.1, GCDB01032583.1, GCDB01032579, GCDB01032580) which clustered with the clotting proteins in the phylogenetic analysis (Figure 3). In *Anaspides*, we did not find an apoCr (presumably due to a lack of transcriptome coverage), however, we found a LLTP (Acc. Nr. GCBI01016509) with a sequence length of about 1800 aa which clustered with the clotting proteins (Figure 3). This suggests for *Nebalia* and *Anaspides* a situation as found in other Malacostracans and indicates that apoCr arose early in the evolution of Malacostraca.

The phylogenetic relationship of the decapod clotting proteins with the VTG's is also supported by the absence of a domain of unknown function (DUF 1081) in both the clotting proteins and the classic VTG's, while this domain is present in the decapod apolipocrustaceins and in other apoB proteins such as the human apoB 100. Unexpectedly, we found that the DUF1081 domain was lacking in the apoCr's of two Malacostracan groups (isopods and amphipods; see Section 3.7.3 below) and in the Brachyuran branch of the decapods.

3.7.1 Branchiopoda

In contrast to the apolipocrustaceins, the VTG's of crustaceans other than decapods were characterized much later. The VTG of the branchiopod *Daphnia* was described in 2004 by Kato et al. [*65*]. Unusually, the *Daphnia* VTG contained a superoxide dismutase (SOD)-like domain at its N-terminus. This SOD-like domain is also present in the VTG's of two other Branchiopods, *Artemia parthenogenetica* (Acc. Nr. HQ647327.1, [*66*]) and the same domain structure is present in *Diaphanosoma celebensis* (Acc. Nr. QJE49262). The SOD domain seems to be a unique feature of the

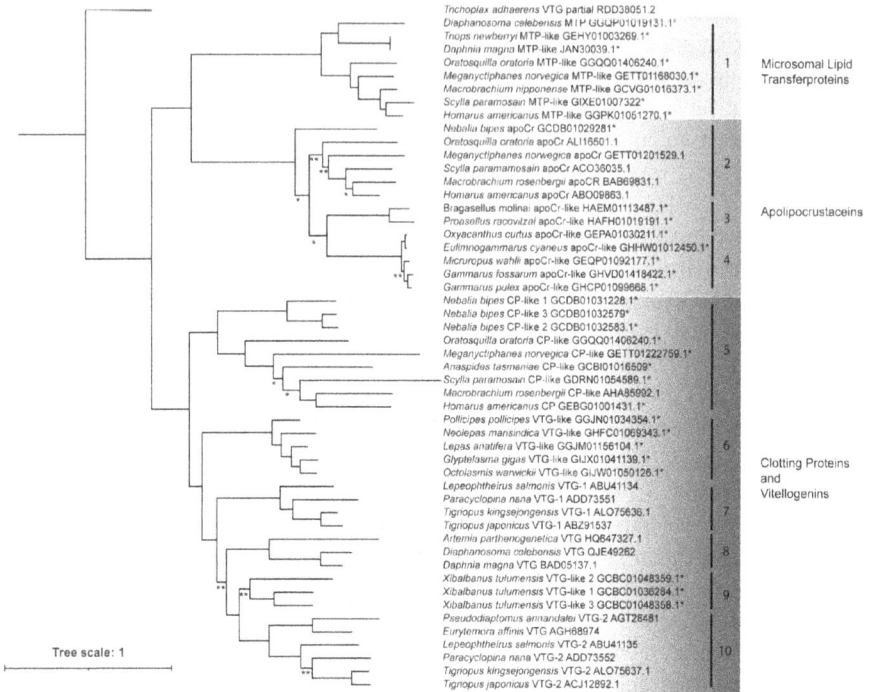

Figure 3. Phylogenetic relationships of known and newly predicted microsomal triglyceride transfer proteins (MTP), clotting proteins (CP), apolipocrustaceins (ApoCr) and vitellogenins (VTG) within different crustacean groups. The phylogenetic tree was constructed using the alignments of the conserved LLT module. Phylogeny was calculated using the IQ-tree web server (http://iqtree.cibiv.univie.ac.at). Numbers show the percentages of support based on 1000 bootstraps steps. The scale bar equals 1 substitutions per site. The VTG of *Trichoplax adhaerens* (Accession Nr. RDD38051.1) was used as outgroup.

Black bars and numbers designate the following branches:

1. MTP Branchiopoda and Malacostraca; 2. ApoCr Malacostraca; 3. ApoCr Isopoda; 4. ApoCr Amphipoda; 5. CP Malacostraca; 6. VTG Cirripedia; 7. VTG-1 Copepoda; 8. VTG Branchiopoda; 9. VTG Remipedia 10. VTG-2 Copepoda.

Bootstrap values of all individual branches are > 90% except when indicated: * 66–75%; ** 81–89%. Sequences marked with an asterisk were retrieved by BLAST searches from the NCBI TSA database (https://www.ncbi.nlm.nih.gov/Traces/wgs/?page=1&view=tsa&search=crustacea) using annotated LLTPs. The open reading frames and the domain boundaries were determined using the online tools Translate (http://web.expasy.org/translate/) and SMART (http://smart.embl-heidelberg.de/), respectively. The terms "ApoCr-like", "VTG-like", etc., are used to indicate the preliminary assignment to the corresponding protein group.

VTG in this group. A phylogenetic analysis by Kato et al. revealed that the *Daphnia* VTG is more closely related to insect VTG's than those of decapod crustaceans. This indicates that the branchiopods are members of the "classic" VTG's. Chen et al. [66] further investigated the SOD-containing VTG of *Artemia parthenogenetica* and suggested that the SOD-domain could serve as protection against oxygen toxicity during the diapause stage of the embryo. In addition, SOD may have a protective function against hemoglobin used as oxygen carrier in branchiopods (e.g., *Artemia* and *Daphnia)* since hemoglobin may generate reactive oxygen species due to autoxidation.

3.7.2 Copepoda

Copepods are among the most abundant taxa in marine invertebrates and the dominating organisms in the planktonic food chains transferring energy from phytoplankton to higher trophic levels. For instance, *Calanus finmarchicus* constitutes a major part of the mesozooplankton biomass in areas of the northeast Atlantic during summer plankton bloom. Copepods are thus another important crustacean group. In addition, the induction of VTG expression was found after exposure of *Paracyclopina nana* to $HgCl_2$ suggesting VTG as an indicator for heavy metal stress in copepods [67]. The VTG's of several copepods, *Eurytemora affinis, Tigriopus kingsejongensis, T. japonicus, Paracyclopina nana* and *P. annandalei* were sequenced more recently (see [48] and cited refs). The authors performed a phylogenetic analysis and found that the copepod VTG also clustered with the classic VTG-type LLTP's as found for the *Daphnia* VTG [65] (see Figure 1). The SOD-like domain is lacking in copepod VTG's. The annotated VTG sequences of copepod VTG's are shorter than the apolipocrustaceins with sequence lengths between 1700 and 1900 amino acids (Table 1; see also Figure 2 in [9]) and lack the DUF1081 domain found in apoB-type LLTP's.

3.7.3 Vitellogenin- and apolipocrustacein-related proteins in isopoda, amphipoda and cirripedia

Besides Branchiopoda and Copepoda, transcriptomes from taxa other than decapods have been added to the NCBI database in recent years, especially of isopods, amphipods, and several cirripeds. Although VTG sequences have been rarely annotated for these groups, prediction of "new" vitellogenins can be made relatively easy from the transcriptome shotgun database (TSA; https://www.ncbi.nlm.nih.gov/genbank/tsa/). This allows for the prediction of VTG's which differ from the "standard" VTG's/apoCr's which mostly refer to those of decapod crustacea. However, further biochemical characterization and/or the demonstration of expression changes during reproduction and oogenesis is also necessary to unequivocally demonstrate function of predicted VTG's as proteins involved in oogenesis and reproduction and this is also true for the yet putative VTG's and apoCr's presented in this chapter. With regards to non-decapod species, VTG gene expression studies related to reproduction have been performed in some copepods ([13], [97], [48] and cited refs). To obtain a more complete picture about the distribution of apoCr within the crustacea we have searched the NCBI TSA database for related sequences within groups other than decapods. Using the apoCr of *Macrobrachium rosenbergii* as query we found ApoCr related proteins in isopods and amphipods and in *Nebalia*, a representative of the leptostracans as the most primitive subclass within the Malacostraca. As malacostracans, amphipods and isopods were expected to possess apoB type LLTP's with a domain structure close to those of decapods including the DUF1081 domain typical of apoB-like LLTP's. However, in the isopods investigated, the DUF1081 domain was found absent and surprisingly in the amphipod species analyzed, even the VWD domain was apparently lacking in this group and this was consistently found in several species (Table 1). On the other hand, the N-terminal lipid-binding domain of both groups used as the basis for the phylogenetic comparison clearly clustered with those of the other apoCr's suggesting that they belong to the apoB branch. Together, with the presence of the DUF1081 domain in the basal Malacostracan *Nebalia*, these findings indicate a secondary loss of the DUF1081 and VWD domains in isopods and amphipods, respectively. Another enigmatic finding is the apparent loss of the DUF1081 domain in some members of brachyuran crabs. It was present in two *Callinectes* species but absent in several other crabs (Table 1). It should be noted that the apparent "absence" is relative and depends on the domain search criteria of the SMART program used here. It is likely that these domains have been modified in these groups so they are below the confidence limit set by the program—in contrast to the lipid-binding and the DUF1943 domains which are well represented in isopods and amphipods. However, it clearly shows an extensive modification of apoCr's in these groups.

For the transcriptomes of cirripeds, we have searched for VTG-like proteins in this group using the *Eurytemora affinis* VTG (GenBank AGH68974) as query. In 7 species, we found putative VTG's similar to those of copepods (Table 1) and in the phylogenetic analysis, these putative VTG's formed a distinct group within the VTG clade (Figure 1). In *Xibalbanus tulumensis*, a representative of the Remipedia whose systematic position within the tetraconata (i.e., crustacea and insecta) is uncertain, a similar putative VTG was found (Table 1).

3.8 Glycosylation of apolipocrustacein

Like other arthropod apolipoproteins, apoCr's are glycosylated. Khalaila et al. [*68*] characterized the N-glycan moieties of the apoCr of *Cherax quadricarinatus* by exoglycosidase sequencing and mass spectrometry. The authors determined 10 putative N-glycosylation sites with the consensus motive Asn-X-Ser/Thr. Three of these sites were found to be glycosylated and were located in three different subunits of apoCr. The N-glycans were of the oligomannose type with five to nine mannose residues linked to two basal N-acetylglucosamine residues ($Man_5HexNAc_2$ to $Man_9HexNAc_2$), the latter being the dominant N-glycan. Unusually, a glucose-capped N-glycan ($Glc_1Man_9HexNAc_2$) was also found. In a similar study, the presence of these high-mannose type N-glycans was confirmed in the apoCr of *Macrobrachium rosenbergii* by Roth et al. [*69*]. The presence of high mannose-type N-glycans appears to be a typical pattern of arthropod apolipoproteins [*70–72*]. Roth et al. [*69*] have compared the putative N-glycosylation sites in two subgroups of decapod crustaceans, the Pleocyemata and the Dendrobranchiata. In eight species of the Pleocyemata, apoCr was found to contain multiple N-glycosylation sites while in eight members of the Dendrobrachiata, only O-glycosylation sites were identified and N-glycosylation sites were missing. The reason for this clear-cut difference between both groups is not yet clear. The authors [*69*] speculate that this could have to do with the lecitotrophic Nauplius stage in the Dendrobrachiata, while the Pleocyemata brood their eggs until completion of larval development.

While N-glycosylation in general is involved in diverse biological functions and essential for organismal development, body growth, and organ formation [*73*], the specific role of the glycosylation of apoCr and VTG is not clear at the moment. The hydrophilic glycosylation could reduce or at least modify lipophilic properties and thus lipid binding of apoCr. Soulages et al. [*74*] studied the influence of glycosylation on lipid binding in an exchangeable insect lipoprotein, apolipophorin-III. Their results suggested that glycosylation inhibits the lipid binding activity by preventing the exposure of hydrophobic domains and/or decreasing the conformational flexibility of the protein. On the other hand, glycosylation of VTG (apoCr) may be necessary in folding and subunit assembly and to keep the lipoprotein soluble in the hemolymph. In the N-terminal lipid binding domain at least, the two glycosylation sites determined in *Cherax quadricarinatus* (Asn^{152} and Asn^{160}) and three for *Macrobrachium rosenbergii* (Asn^{152}, Asn^{160} and Asn^{168}) are exposed to the outer environment and do not interfere with the lipid binding pocket of the lipid binding domain [*69*].

3.9 Multiple forms of apolipocrustacein and vitellogenin

In 2007, [*14*] Avarre et al. introduced the term apolipopcrustacein (apoCr) to highlight the affiliation of the decapod yolk proteins with the apoB LLTP's (see Section 3.6 above). Multiple forms have been described for both the apoCrs of decapoda and the VTG's of copepoda and in all cases, two copies of VTG or ApoCr have been found. As in earlier and recent studies, the decapod apoCr's are still designated "vitellogenins", the term "VTG/apoCr" will be used here when referring to the earlier "VTG" literature. Presently, the specific function of the different isoforms is unknown and there are large differences in the similarity of the sequences depending on the species (see below). In decapoda, multiple copies of *vtg/apoCr* genes have been found previously in the suborder Dendrobranchiata (such as the penaeid shrimps and prawns) (*Metapenaeus ensis* and *M. japonicus*, [*75, 76*]), but two copies of VTG have also been found in a representative of the Pleocyemata (caridean shrimps and most other decapod crustaceans) in *Pandalopsis japonica* [*35*] and in the astacidean *Homarus americanus* [*77*]. In the former study, the differences in the open reading frames and in the 5' and 3'-UTR's suggested the presence of two distinct genes [*35*]. In *Metapenaeus ensis*, the second *vtg/apoCr* gene (termed MeVg2) contained less exons and introns (13 and 12, respectively) compared to the 15/14 exons/introns usually present in malacostracans (see Table A-II in [*62*], for an overview) and MeVg1; a similar situation seems to exist in *Portunus trituberculatus* [*78*]. In *M. ensis*, the MeVg2 gene had a 54% identity to the MeVg1 gene and was suggested to have evolved from the MeVg1 gene [*76*]. In addition, the MeVg2 contained a higher number of potential subunit cleavage sites indicating differences in subunit processing. In contrast, the two *vtg/apoCr* genes found in *Penaeus merguiensis* were almost identical and the same finding was obtained in *Callinectes sapidus* [*79*]. In *Pandalopsis japonica*, two different *vtg/apoCr* genes were found in the hepatopancreas, while the ovaries expressed only one of these genes [*35*]. In the crayfish, *Procambarus clarkii*, two *vtg/apoCr* transcript groups were found and their expression was either hepatopancreas- or ovary-specific. Neither of the transcripts were found expressed in both tissues [*80*].

In copepods, (at least) two isoforms of VTG are also present in several species [*13, 48, 97*]. As already found by these authors, the two VTGs of each species appear in separate clusters in the phylogenetic analysis (Figure 1) suggesting an early event of gene duplication during copepod evolution as suggested for decapod crustaceans [*35*].

However, multiple VTGs/apoCrs may also arise from alternative splicing, a situation which has also been suggested for some species. This possibility has been inferred from clustering of different vitellogenins from different organs and species according to the species, rather than the organ which would have been expected if the split in organ specific VTG's would have occurred before speciation [*9, 76, 78*]. Chan et al. [*81*] demonstrated the expression of full-size and smaller transcripts of *vtg/apocr* mRNA in the hepatopancreas of *Charybdis feriatus* and showed that their sequences contain alternative splicing products of the full-length sequence. However, on the protein level, neither the full-length *vtg/apoCr* nor the splice products could be detected in the hepatopancreas.

3.10 Tissue expression of vitellogenin and apolipocrustacein

In the well studied decapods, the hepatopancreas and the ovary are the main sites of apoCr synthesis. Differences were previously thought to exist in the two major decapod groups, the Dendrobranchiata (such as the penaeid shrimps and prawns) and the Pleocyemata (caridean shrimps and most other decapod crustaceans). In the Dendrobranchiata, *vtg/apocr* expression has been found in both the hepatopancreas and the ovaries [*14,46,75,82–85*]. In *Penaeus merguiensis*, expression of *vtg* mRNA was much higher in the ovary compared to the hepatopancreas at all stages of ovarian development indicating that the ovary is the main site of VTG/apoCr synthesis [*75*]. Tiu et al. [*86,87*] determined that the ovary constituted 50% of the total *vtg*-mRNA in *Metapenaeus ensi*s. Tsutsui et al. [*82*] reported the follicle cells as the site of VTG/apoCr synthesis. In contrast in the Pleocyemata, VTG/apoCr synthesis was previously found in the hepatopancreas only in caridean shrimps [*14,35,45,88,89*], in the crayfish *Cherax quadricarinatus* [*44*] and in the brachyuran crab *Oziotelphusa senex* [*90*]. However, expression of *vtg/apocr* in both tissues has recently been found in other species of Pleocyemata as well such as in the blue crab *Callinectes sapidus* [*47*] *Upogebia major* [*34*], *Carcinus maenas* [*91*] and *Scylla paramosain* [*36*], indicating that this taxon-specific difference is not as clear-cut as thought although the hepatopancreas appears to be the most important synthetic site in the Pleocyemata. In *Callinectes sapidus*, the hepatopancreas produced more than 99% of the total *vtg/apocr* mRNA [*79*] and in the same species, the relative expression levels were 300-fold higher in the hepatopancreas compared to the ovaries [*47*]. Interestingly, Jeon et al. [*35*] found the expression of the VTG-1 variant also in the male hepatopancreas of *Penaeus japonica*. The authors suggested that this was related to the proterandric hermaphroditism of this species, in which initially functional males develop into functional females. Based on the same finding in *Pandalus hypsinotus* [*92*], the authors suggested VTG/apoCr as a marker for sex change.

The gonochoristic mud crab *Upogebia major* seems to be a special case; in this species, both males and females express VTG/apoCr [*34*]. Males have feminized endocrine status from their immature stages and the male testis has a posterior part in which oocytes develop although the oocytes are not ovulated due to lack of gonoducts.

An exotic finding is the finding of a male-specific VTG/apoCr found exclusively in the spermatozoa of the mud crab (designated VTG2, [5]). Sequence analysis indicated that the male-specific isoform arises from alternative splicing. The expression of VTG2 could be induced by the injection of lipopolysaccharide and lipoteichoic acid (LTA) injection at both transcriptional and translational levels suggesting an immune-relevant function [5].

3.10.1 *Expression of VTG and apolipocrustacein during ovary development*

Ovary growth depends on the supply of VTG/apoCr from the hemolymph and the increase in oocyte volume due to yolk protein incorporation is accompanied by an increase in the synthesis in hepatopancreas and ovary tissue. The rates of VTG/apoCr synthesis and mRNA expression are in phase with the molting cycle of the animal; they increase during the intermolt stage, remain high during premolt with a decrease prior to oviposition and ecdysis as found in *Macrobrachium rosenbergii* and several penaeid shrimps ([93,94] and cited references). For instance, Jia et al. [36] measured relative expression levels in different ovarian stages of *Scylla paramosain* by real-time PCR and found an about 30–35-fold increase from the youngest (proliferation) stage to the late vitellogenic stages, respectively, while the increase in the hepatopancreas was less (16 to 18-fold, respectively). In *Macrobrachium rosenbergii*, the VTG/apoCr transcript levels in the hepatopancreas increased 500-fold until the vitellogenic phase with a strong decline thereafter. The increase was only about 30-fold in the ovary tissue with a maximum at the maturation stage [37]. A similar trend was found in *Penaeus monodon* [85], however, the differences in the relative expression levels measured between different stages of ovary development were much less. In contrast in *Penaeus merguiensis*, ovary expression of *vtg/apocr* mRNA increased nine-fold and was maximal in the early vitellogenic stage compared to previtellogenic ovaries with a nine-fold decline towards late vitellogenic stages, while in the hepatopancreas, the overall expression was about ten-fold lower reaching the maximum at the main vitellogenic stage [46]. A difficulty in comparing the results from different authors and across species is the different methods applied for the measurement of relative expression rates (Northern blot with luminescent detection, real-time PCR) and different calculation methods (standard curve, $\Delta\Delta$Ct-method).

The expression of VTG/apoCr does not cease after the vitellogenic period but is also found during development [37]. In *Macrobrachium nipponense*, expression levels were low but a sharp increase was found in zoea and first day larvae with a subsequent decline in later stages. A second, strong increase was found in 30-day postlarvae.

The increase in the rate of VTG/apoCr synthesis during oogenesis leads to corresponding increase in the hemolymph concentration with a subsequent uptake by the oocytes (see Section 3.12 below). In *Pandalus hypsinotus*, low concentrations (< 0.02 mg/ml) were found in previtellogenic oocytes and increased to variable concentrations averaging 3 mg/ml [94]. In *Macrobrachium rosenbergii*, Jayasankar et al. [93] and cited references) reported low hemolymph VTG/apoCr concentrations at the beginning (1 mg/ml) and high (5 mg/ml) at the end of the reproductive cycle covering a period of 24 days. Zmora et al. [*47*] found similar relative changes during the reproductive molting cycle of *Callinectes sapidus* following the changes in the hemolymph levels over a period of 10 weeks after the final molt. In this species, the highest absolute concentrations did not exceed 0.2 mg/ml hemolymph. This could be caused either by slower rates of VTG/apoCr synthesis and release into the hemolymph due to the longer time of sexual maturation in *Callinectes* or by a rapid uptake by the gonads. Eye stalk ablation results to the removal of the gonad inhibiting hormone and leads to an extreme increase in the VTG/apoCr levels and acceleration of gonad development [*84,93*]. The endocrine control and regulation of VTG-synthesis will not be the focus of this review and the reader is directed to appropriate reviews [*8,95*].

Information on VTG/apoCr expression in groups other than decapods is scarce. In the copepod *Labidocera aestiva*, both autosynthetic and heterosynthetic synthesis of VTG have been discussed [*96*]. The follicle cells surrounding the oocyte have been suggested to participate in the supply of VTG but evidence was based on electron microscopical observations only. Only recently, a more specific study on the VTG synthesis in copepods was carried out in the salmon louse *Lepeophtheirus salmonis* [*97*]. In this parasitic copepod, the synthesis of two types of VTGs has been localized to a specific cell type of the subcuticular cells in the cephalothorax and the presence of VTGs in the haemocoel suggested its transport to the ovary indicating again a heterosynthetic origin. After uptake by the oocytes, the VTG-2 form was processed into 2 smaller fragments, whereas VTG-1 was not cleaved. In *Lepeophtheirus*, the subcuticular cells synthesize and release another female-specific protein [*97*] which was found to associate with the yolk protein complex of the oocyte. The protein (GenBank accession nr. ABU41111) contained a Fasciclin I domain, a domain believed to act in cell attachment and adhesion. Gene silencing by RNAi was found to disrupt embryonic development [*97*]. No other reports exist on the occurrence of similar proteins in other crustaceans. With regards to the synthesis of *vtg*-RNA, a situation similar as shown for the decapod crustaceans has been observed in two copepods, *Eurytemora affinis* and *Tigriopus kingsejongensis*, with a several fold increase in mature and ovigerous females compared with immature females [*13*].

While hepatopancreas and ovaries are clearly the most important sites of VTG/apoCr synthesis, other tissues have also reported to synthesize VTG/apoCr. In the shrimp *Palaemon serratus*, *Parapenaeus longirostris* and in the mud crab *Scylla*, the subepidermal tissue has also been reported as a site of VTG/apoCr synthesis [*98–100*], but the relative contribution for the overall synthesis has not been evaluated. In *Pandalopsis japonica*, Jeon et al. [*35*] did not find the expression of VTG/apoCr in the epidermis. Bai et al. [*37*] compared the relative expression rates of different tissues in *Macrobrachium rosenbergii* by real-time PCR and found the hemocytes as an additional cell type expressing VTG/apoCr; the expression rates were comparable to those found in the ovaries and thus much lower than in the hepatopancreas as decribed above.

With regards to other malacostracan groups, the fat body has been reported as the site of VTG/apoCr synthesis in the isopod *Armadillidium* and in the amphipod

Orchestia, respectively [*101,102*], but further information on these and crustacean groups other than decapods is scare or lacking.

3.11 Processing of VTG/apocr

In egg laying animals, most vitellogenins as well as apolipocrustaceins are partially cleaved after uptake into the oocyte which proceeds by receptor mediated endocytosis (see Section 3.12 below). In crustaceans, the situation is more complex since the VTG/apoCr precursor is cleaved already at the time of release into the hemolymph and inside the oocyte, further cleavage occurs. This results in several cleavage products which can be isolated from the oocyte as found mostly in decapod crustaceans as the economically most important group. Three to 5 cleavage products of VTG/apoCr have been found by different authors (see Table 2 in Wilder et al. [*6*]). For instance, Raviv et al. [*84*] isolated five polypeptides of masses 179, 113, 78, 61, and 42 kDa from the ovaries of *Litopenaeus vanamei* and showed that these derived from one *vtg/apocr* gene. In *Macrobrachium rosenbergii*, detailed studies by Okuno et al. [*102*] revealed a first cleavage site after amino acid positions 707–710 (RQRR) representing a general furin-type consensus motif R-X-K/R-R where X represents any amino acid. This first cleavage occurs before the release into the hemolymph. A second furin-type cleavage site was found to occur within the hemolymph after amino acid positions 1747–1750 (RDRR). The corresponding extracelluar protease has yet to be identified, however. Avarre et al. [*14*] also found a primary cleavage site at amino acid position 710 and a second cleavage site in the ovary in *Penaeus semisulcatus*. In a comparative study, Jeon et al. [*35*] aligned the furin cleavage sites in 20 decapods. For all species investigated, a highly conserved region was found within the amino acid positions 725–730. The primary cleavage site lies within the C-sheet of the N-terminal lipid binding region which is part of the lipid pocket of the LLT module serving for the formation of the nascent lipoprotein particle (see Section 3.4.1 above). The cleavage in the C-sheet does not necessarily impair the function of the LLT module, since it might be stabilized by interaction with the lipid moiety or by the α-helix region surrounding the lipid pocket (see Figure 1). Likewise in insects, furin cleavage of the apolipophorin II/I takes place within the N-terminal lipid binding domain (at position 720, [*17, 103,*]). The former authors showed that inhibition of cleavage does not prevent lipidation and secretion of the lipophorin particle.

In addition to the second furin-like cleavage site found in *Macrobrachium rosenbergii*, a putative cleavage site of the subtilisin-type was also found as a second site. In contrast to the more stringent substrate specificity of furins requiring the R-X-K/R-R motif, the subtilisin-like convertases involved in similar functions are active on more variable motifs which usually contain the consensus sequence R/K-Xn-R/K, where X indicates any amino acid residue and n represents the number of spacer amino acid residues which can be 0, 2, 4 or 6 [*104*]. The second cleavage site was found in some, but not all decapods studied, in *Macrobrachium rosenbergii*, *Homarus americanus*, *Cherax quadricarinatus* and *Pandalopsis japonica* [*35*]. In the latter species, the second site was present in only one of the VTG/apoCr isoforms (termed Pj-Vg2) found in this species. In brachyuran and penaeid VTG/apoCr's, no second site was found.

The situation as shown for decapods does not represent a general picture for apoCr's of other malacostracan groups and for crustacean VTG's. In the basal malacostracan *Nebalia*, a furin cleavage site (RSKR, residues 740-743) was present which aligned well with the consensus site found for other decapod species (see above). In amphipods and isopods, however, we did not find furin-specific cleavage sites in the species listed in Table 1 using the ProP prediction server (https://services. healthtech.dtu.dk/service.php?ProP-1.0). Although putative subtilisin-convertase-like cleavage sites were detected, no consensus regions were found. In the *Artemia parthenogenetica* VTG, one furin-type cleavage site was found in addition to five putative subtilisin-like protein endopeptidase motifs [66], but we could not detect similar sites in the VTG sequences of *Daphnia* and *Diaphanosoma*. Likewise, no clear picture emerged in our survey of the copepods and cirriped species listed in Table 1. This suggests a high variability in the processing of VTG of taxonomic groups other than malacostracans but too few studies have been carried out to have a clear picture. This is in contrast to insect VTG's, in which the cleavage sites and the corresponding furin- and subtilisin-like convertases are much better documented with a much larger body of existing literature [41,42].

Following uptake and proteolytic processing to the final storage form of VTG and apoCr, the lipid content decreases significantly. Being relatively high in VTG's/ apoCr's of decapods (up to 50%; see Section 3.6 above), the lipid content of the storage form decreases to values between 28 and 35%. This has been measured in seven different decapod species [1, 105]. The structural background for this is has not yet been investigated, but removal of lipids from the storage protein may allow for a denser packaging and thus increase the energy content of the protein moiety during storage in the egg. In the tadpole shrimp *Triops longicaudatus*, the lipid content of vitellin is even lower (3%; Schilz and Hoeger, unpublished data). Here, the packaging of vitellin seems extremely dense and normal physiological buffers fail to solubilize the protein when extracting the oocytes. This is only achieved by high salt concentrations which apparently dissociate the protein (Hoeger, unpublished observation) and the same has been applied for the isolation of *Artemia* vitellin [106]. For ephemeral species like *Triops* and *Artemia* this trait is likely to be related to the tolerance of the eggs for dessication and freezing.

Lipids are important energy reserves and there is a dramatic increase in the oocyte lipid content during ovary development [107] and cited references). In *Penaeus semisulcatus*, the relative proportion of phospholipids decreased from 66% to 49% during the course of vitellogenesis, while that of triglycerides increased from 20% to 47%. There was a gradual decrease in the relative abundance of polyunsaturated fatty acids (which indicated their exogenous uptake) and an increase in the saturated fatty acids indicating their autosynthetic origin and use for ovarian lipid synthesis. This situation seems to be a general trait of oocytes and has also been found in polychaetes and insects, where oocytes carry out *de novo* synthesis of triglycerides during development [108-110].

3.12 Uptake of VTG/apoCr into the oocyte

Incorporation of the VTG/apoCr by the oocytes is achieved by receptor-mediated endocytosis. This is true for oocytes of all egg laying animals and has first been demonstrated in the pioneering study on mosquito oocytes by Roth and Porter [*111*], which first observed and described coated pits and vesicles. VTG/apoCr interacts with the receptor at the oocyte surface under the formation of a clathrin-coated vesicle. Following incorporation of the receptor-ligand complex, the vesicles become uncoated, fuse with other uncoated vesicles to form primary or early endosomes, which are acidified. In the acidic environment of the endosomes, receptor and ligand dissociate and the receptor returns to the cell surface. The endosomes further fuse to the final yolk bodies. Still maintaining an acidic pH, they were termed "sleepy" lysosomes [*112*], in which the yolk protein is stored and degraded only at the onset of embryonic development.

The LDL receptor gene family represents the main class of endocytic lipoprotein receptors, which are expressed not only in oocytes but also in many tissues in organisms as distantly related as nematodes, insects, frog, chicken and mammals [*113–115*]. LDL receptors exhibit similar structural features consisting of a single transmembrane domain, a short cytoplasmic tail, and an amino terminal portion extending into the extracellular space. The amino terminal regions contain five modular domains (Figure 4): (1) A ligand binding domain (LBD) consisting of several cysteine-rich repeats (type A repeats) requiring calcium for efficient ligand binding, (2) epidermal growth factor (EGF)-like domain (type B repeats) which is required for the dissociation of the lipoprotein from the ligand-binding region [*116*], (3) repeats containing a YWTD consensus motif which form a β-propeller structure which is also essential for the displacement of the ligand in the acidic endosome, (4) a transmembrane domain, and (5) the cytoplasmic region containing a variable sequence motif acting as an internalization signal targeting the receptor to coated pits [*114*]. In vertebrate LDL-receptors and insect lipophorin receptors only a single LBD region is present while two LBDs are present insect and crustacean VTG-receptors [*114,117*].

So far, VTG/apoCr receptors have been identified and studied in decapods only. A biochemical study was carried out in *Scylla serrata* [*118*] and in further molecular studies, sequences were obtained for *Penaeus monodon* (GenBank acc. nr. ABW79798.1, [*85*]), *Marsupenaeus japonicus* (AB304798.1, [*119*]), *Macrobrachium rosenbergii*, (ADK55596.1, [*120*]), *Pandalus japonicus* (AHL26192.1, [*121*]), *Palaemon carinicauda* (AHB12420.1, [unpublished]) and recently in *Litopenaeus vanamei* (MN807241, [*117*]). The known crustacean VTG/apoCr receptor sequences are 1890–1940 aa long resulting in theoretical molecular masses between 208–213 kDa. A shorter sequence (1120 aa) was published for *Marsupenaeus japonicus* by Mekuchi et al. [*119*], however, in a recent transcriptome of this species (NCBI Bioproject PRJDB11152), we found by a BLAST search a contig (GenBank acc. nr. ICRK01014319.1) with an ORF which N-terminally extends the former sequence (AB304798.1) to a length of 1932 amino acid residues which is in line with the results above. Specific for decapod VTG/apoCr-receptors, two EGF repeats are present in the C-terminal side adjacent to the transmembrane domain; this differs

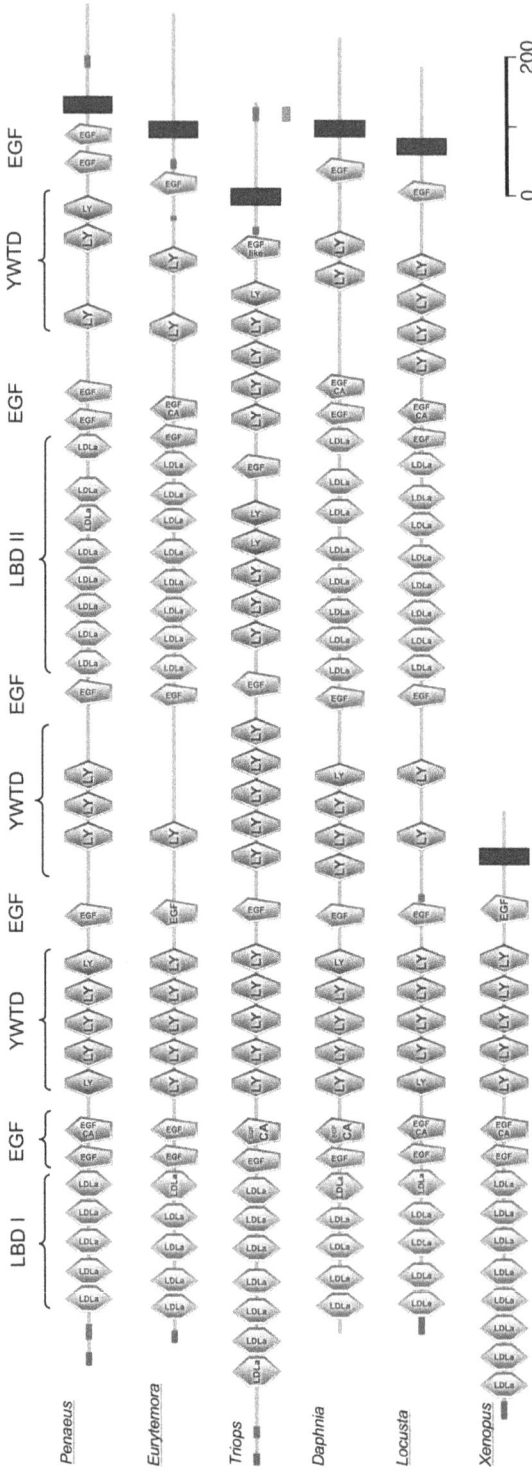

Figure 4. Domain structures of the VTG/ApoCr-receptor (GenBank accession nrs in parentheses) of *Penaeus monodon* (ABW79798.1), the predicted VTG-receptors from the copepod *Eurytemora affinis* (GBGO01104473.1 TSA), and the branchiopods *Triops newberryi* (GEHY01004018.1) and *Daphnia magna* (GDIP01034982.1). The VTG-receptors of *Locusta migratoria* (QCX35737.1) and *Xenopus laevis* (QCX35737.1) are shown for comparison to demonstrate the conserved structure of the VTG-receptors and the duplication of the LBD and YWTD-regions in invertebrate VTG-receptors. For the sequences of *Eurytemora*, *Triops* and *Daphnia*, the open reading frames were determined using the online tool Translate (http://web.expasy.org/translate/). The domain boundaries and their graphical representations were obtained using SMART (http:// smart.embl-heidelberg.de/), respectively.

LBD: ligand-binding region containing cysteine-rich repeats of the low-density lipoprotein receptor class A domain; LDLa: individual repeats; EGF: Epidermal growth factor-like domain; EGF CA: Calcium-binding EGF-domain; YWTD: region with repeats containing the YWTD consensus motive (low density lipoprotein receptor class B repeats); LY: individual repeats. The N-terminal signal sequences and the C-terminal transmembrane domains are shown in small and large rectangles, respectively. The cytoplasmic region is on the C-terminal side of the transmembrane domain. Scale bar indicates a sequence length of 200 amino acids. For further explanations, see text.

from the vitellogenin-receptors of insects and vertebrates with only a single EGF repeat ([*120*], see Figure 4). The same was found for the *Pandalopsis japonica* VTG/apoCr-receptor [*121*].

The domain structures of different VTG/apoCr-receptors are shown in Figure 4. Together with the known sequence of the *Penaeus monodon* VTG/apoCr-receptor, an insect and a vertebrate VTG-receptor, and three sequences predicted from the transcriptomes of two Branchiopods (*Triops*, *Daphnia*) and the copepod *Eurytemora affinis* have been included for comparison. The sequences are similar in length and show a similar domain organization in the first half of the sequence. The domain pattern of *Triops* is unusual and shows a third region of YWTD repeats instead of a second ligand binding region.

The receptor has been located exclusively in the ovary membrane in *Penaeus monodon* [*85*], *Macrobrachium rosenbergii* [*120*] and *Litopenaeus vanamei* [*117*] but not in the follicle cells as also true in insects [*42*]. In a comparative study, Lee et al. [*121*] found 3 additional lipoprotein receptors in *Pandalopsis japonica* including a VTG (apoCr)-receptor which was expressed almost exclusively in the ovaries. The other lipoprotein receptors were smaller (864–867 amino acid residues) with a domain structure reminiscent of the vertebrate type VTG-receptors (see Figure 4) and although one of these receptors showed an ovary-specific expression, the relative expression rates were much lower than that of the VTG/apoCr-receptor. However, this suggests alternative "vitellogenin" (whether apoCr's or clotting proteins) uptake mechanisms or even the contribution of yet unknown lipoproteins for the crustacean oocyte.

The VTG/apoCr-receptor is expressed throughout ovary development and blocking expression of the receptor by RNAi demonstrated a greatly impaired gonad development [*117,122*]. The relative changes in the receptor expression rates measured in different stages vary considerably in different studies. In *Macrobrachium nipponense*, Bai et al. [*122*] found a more than 60-fold increase in the receptor mRNA of the oil globule stage (i.e., before the onset of exogenous VTG/apoCr incorporation) compared to early (proliferating) oocytes. A strong decline was found in the later stages of oocyte growth and maturation. Tiu et al. [*85*] and Roth and Khalaila [*120*] also found an expression maximum before the onset of VTG/apoCr incorporation with a subsequent decline in later stages in *Penaeus monodon* and *Macrobrachium rosenbergii*, respectively, however, the relative expression rates varied only two- to three-fold fold between different stages. In contrast, Mekuchi et al. [*119*] found the highest expression of receptor in the earliest stage of ovary development in *Marsupenaeus japonicus* with a strong and continuous decline during further ovary growth.

This situation is similar to that in insects, where the findings suggest that the gene involved in VTG uptake is particularly expressed at high levels long before vitellogenesis is initiated. In cockroaches, a high VTGR mRNA level is found during previtellogenesis followed by increased protein levels during the vitellogenic phase [*41*]. Considering the decline in the receptor mRNA observed in the aforementioned studies, it should be considered that in the measurement of the relative expression rates by semiquantitative RT-PCR reference is usually made to total RNA rather than to the expression of a house keeping gene or determination of total copy number.

However, an increase in the pool of RNA other than receptor-specific mRNA during oocyte growth would result in its relative decrease while the absolute amount of receptor-specific mRNA could remain unchanged. The observed changes in the above studies thus do not necessarily reflect a lower capacity for receptor synthesis. In fact, Ruan et al. [*117*] calculated the expression rates of receptor mRNA in relation to that of a house-keeping gene, GAPDH, and found in *Litopenaeus vanamei* that the relative receptor expression rates do not change between previtellogenic stage and mature stage.

The expression of the VTG/apoCr-receptor continues during embryonic development with a gradual decline to the zoea stages in *Litopenaeus vanamei* and *Macrobrachium nipponense* [*37,117*]. The latter authors suggested that this reflected the continuous consumption of yolk proteins during development until the larvae start feeding. The receptor expression showed a strong increase from 20 to 30 days post larval which is consistent with the expression pattern of VTG/apoCr [*37*]. The authors reported that ovary development can already be observed in 20 day postlarvae.

With regards to VTG/apoCr, the receptor binding region has been characterized in *Macrobrachium rosenbergii* and a major region was found located in the β-sheet domain of the outer core of the N-terminal region (residues 237–260, [*69*], see Figure 1). In addition, at least two additional regions were identified where interaction with VTG could take place (see Figure 1). A similar receptor binding region was also found in the VTG of the Tilapia *Oreochromis* [*123*]. The interactions of VTG/apoCr with the receptor might require adjustments in its conformation and we have suggested before [*11*] that the conserved cleavage site of VTG/apoCr within the lipid binding region (see above) could facilitate its interaction with the VTG/apoCr receptor. In contrast, the receptor interacting site of human apoB-100 which is also internalized by a LDL-receptor, is located towards the C-terminal end in a relatively small stretch encompassing the residues 3,359–3,369 [*124*]. Similarly in insect and vertebrate vitellogenins the receptor binding domain has been suggested to be located in polyserine stretches located in the C-terminal half of the protein [*39,40,42*].

3.13 Additional functions of VTG and apolipocrustacein

In addition to its "classic" function as a storage protein providing nutrient reserves for the developing embryo, earlier and recent progress in the field has revealed novel functions of vitellogenin. Although most of these discoveries have been made on vertebrates and insects, they will be mentioned here since they provide new ideas for further research on VTG's and apoCr's in crustaceans. One aspect concerns the lipid fraction of VTG/apoCr, which contains various amounts of carotenoids, mainly astaxanthin, canthaxanthine and β-carotene addition to neutral- and phospholipids (for an overview, see [*60*]). VTG-bound carotenoids may protect from excess radiation ([*125*] and cited references). Carotenoids may also serve for the pigmentation of the early larvae, since the carotenoids are not synthesized by the animal and need to be acquired by food. Babin et al. [*19*] found that dietary carotenoids enhanced phenoloxidase activity and resistance to bacterial infection in Gammarid amphipods. In the clam *Chlamys nobilis*, the upregulation of VTG was found after challenge with the pathogen, *Vibrio anguillarum* [*126*]. Interestingly, a correlation between VTG transcript levels and the carotenoid content of the ovary was observed, indicating the role of carotenoids in supporting the immune reaction. Antibacterial properties of VTG's have been found before in several fishes (overview in [*127*]) and upregulation of VTG was found after challenge with bacterial lipopolysaccharide (LPS) and lipoteichoic acid (LTA, [*128*]). More specifically, [*129*] found that recombinant expressed DUF1943, DUF1944 and VWD domains from zebrafish VTG2 interacted with both Gram-positive and Gram-negative bacteria as well as their signature components LTA and LPS. The recombinant DUF1943 and DUF1944 domains promoted the phagocytosis of *E. coli* and *S. aureus* by carp macrophages suggesting that the DUF1943 and DUF1944 domains of VTG have a function as opsonins.

Similarly in honey bees, Salmela et al. [*130*] have also shown that vitellogenin is a carrier of immune elicitors. The authors found that VTG binds to pathogen-associated molecular patterns, LPS, peptidoglycan and zymosan, and further showed that these components are transported into eggs via VTG uptake. VTG thus acts as a carrier of immune-priming signals. Furthermore, honey bee VTG has cell and membrane binding activity and binds preferentially to dead and damaged cells via the alpha-helical region of the LLT module and accumulates on damaged cell surface. It thus has a putative anti-inflammatory role in the honeybee. VTG also binds to live cells, improving cell oxidative stress tolerance [*131*], a function suggested before for the VTG of *C. elegans* [*132*]. These beneficial properties of VTG have important consequences in the social organization of the honey bee. Although functionally sterile, worker bees synthesize VTG. Nurse bees inside the hive synthesize large amounts of VTG which is used for the production of larval food. When nurse bees develop into foragers, their juvenile hormone titers increase, and this causes down regulation of their vitellogenin production leading to a reduced life span [*133*]. The queen in contrast maintains constant high VTG hemolymph levels and has a life expectancy of several years [*134*].

At first, the role of VTG in eusocial insects seem far from the known biology of VTG/apoCr in crustaceans. However, eusociality in crustaceans has independently evolved multiple times within the genus *Synalpheus*, sponge-dwelling shrimps exhibiting a range of social systems, from pair-living to social groups with varying numbers of queens and workers [*135*]. In *Synalpheus regalis,* shrimp colonies may have > 300 individuals, but have only one reproductive female and most larger colony individuals apparently never bred. It is interesting to speculate if VTG/apoCr could have acquired similar functions in these eusocial shrimps in the light of the plasticity of the VTG/apoCr expression system in *Upogebia* where the gene is expressed in the hepatopancreas of both females and males, the ovary, and the ovarian part of testis [*34*].

References

1. R.A. Wallace, S.L. Walker, and P.V. Hauschka. (1967). *Biochemistry* **6**: 1582–1590.
2. M.L. Pan, W.J. Bell, and W.H. Telfer. (1969). *Science* **165**: 393–394.
3. T.W. Sappington. (2002). *J. Mol. Evol.* **55**: 470–475.
4. J.M. Brooks, and G.M. Wessel. (2002). *Dev. Biol.* **245**: 1–12.
5. Y. Yang, B. Zheng, C. Bao, H. Huang, and H. Ye. (2016). *Reproduction* **152**: 235–243.
6. M.N. Wilder, T. Okumura, and N. Tsutsui. (2010). *Aqua-Bio. Sci. Monogr.* **3**: 73–110.

7. M.N. Wilder, K. Bong Jung, and H. Junya. (2018). pp. 290–296. *In:* Skinner, M.K. [ed.]. Encyclopedia of Reproduction 2nd ed. Elsevier.
8. T. Subramoniam. (2011). *Fisheries Sci.* **77:** 1–21.
9. S. Jimenez-Gutierrez, C.E. Cadena-Caballero, C. Barrios-Hernandez, R. Perez-Gonzalez, F. Martinez-Perez et al. (2019). *Crustaceana* **92:** 1169–120544.
10. Wolff, C., and Gerberding, M. 2015. "Crustacea": Comparative aspects of early development. pp. 39–61. *In:* Wanninger, A. [ed.]. Evolutionary Developmental Biology of Invertebrates 4. Springer, Vienna.
11. U. Hoeger, and S. Schenk. (2020). *Subcell. Biochem.* **94:** 35–62.
12. I. Soderhall, A. Tangprasittipap, H. Liu, K. Sritunyalucksana, P. Prasertsan et al. (2006). *FEBS J.* **273:** 2902–2912.
13. S.R. Lee, J.H. Lee, A.R. Kim, S. Kim, H. Park et al. (2016). *Comp. Biochem. Physiol. B* **192:** 38–48.
14. J.C. Avarre, E. Lubzens, P.J. Babin. (2007). *BMC Evol. Biol.* **7:** 3.
15. M.T. Khan, S. Dalvin, F. Nilsen, and R. Male. (2017). *J. Lipid Res.* **58:** 1613–1623.
16. J. Chen, S.M. Honeyager, J. Schleede, A. Avanesov, A. Laughon et al. (2012). *Development* **139:** 2170–217.
17. M.M.W. Smolenaars, O. Madsen, K.W. Rodenburg, D.J. Van der Horst. (2007). *J. Lipid Res.* **48:** 489–502.
18. S. Stieb, Z. Roth, C. Dal Magro, S. Fischer, E. Butz et al. (2014). *Biochim. Biophys. Acta* **1841:** 1700–1708.
19. A. Babin, C. Biard, and Y. Moret. (2010). *Am. Nat.* **176:** 234–241.
20. C.J. Mann, T.A. Anderson, J. Read, S.A. Chester, G.B. Harrison et al. (1999). *J. Mol. Biol.* **285:** 391–408.
21. P.J. Babin, and G.F. Gibbons. (2009). *Prog. Lipid Res.* **48:** 73–91.
22. Y. Shibata, R. Branicky, I.O. Landaverde, and S. Hekimi. (2003). *Science* **302:** 1779–1782.
23. J.A. Sellers, L. Hou, D.R. Schoenberg, S.R. Batistuzzo de Medeiros, W. Wahli et al. (2005). *J. Biol. Chem.* **280:** 13902–13905.
24. A. Hayward, T. Takahashi, W.G. Bendena, S.S. Tobe, and J.H.L. Hui. (2010). *FEBS Lett.* **584:** 1273–1278.
25. M. Eitel, L. Guidi, H. Hadrys, M. Balsamo, and B. Schierwater. (2011). *PLoS One* **6:** e19639.
26. J.R. Thompson, and L.J. Banaszak. (2002). *Biochemistry* **41:** 9398–9409.
27. M.M. Hussain, P. Rava, M. Walsh, and M. Rana. (2012). *J. Iqbal Nutr Metab. (Lond.)* **9:** 14.
28. Z. Roth, S. Weil, E.D. Aflalo, M. Manor, A Sagi et al. (2013). *Chembiochem* **14:** 1116–1122.
29. J.P. Segrest, M.K. Jones, and N. Dashti. (1999). *J. Lipid Res.* **40:** 1401–1416.
30. P.E. Richardson, M. Manchekar, N. Dashti, M.K. Jones, A. Beigneux et al. (2005). *Biophys. J.* **88:** 2789–2800.
31. M. Manchekar, P.E. Richardson, T.M. Forte, G. Datta, J.P. Segrest et al. (2004). *J. Biol. Chem.* **279:** 39757–39766.
32. J.L. Baert, P. Sautiere, and M. Porchet. (1984). *Eur. J. Biochem.* **142:** 527–532.
33. S. Schenk, C. Krauditsch, P. Fruhauf, C. Gerner, and F. Raible. (2016). *Elife* **5:** e17126.
34. B.J. Kang, T. Nanri, J.M. Lee, J.H. Saito, C.H. Han et al. (2008). *Comp. Biochem. Physiol. B* **149:** 589–598.
35. J.M. Jeon, S.O. Lee, K.S. Kim, H.J. Baek, S. Kim et al. (2010). *Comp. Biochem. Physiol. B* **157:** 102–112.
36. X. Jia, Y. Chen, Z. Zou, P. Lin, Y. Wang et al. (2013). *Gene* **520:** 119–130.
37. H. Bai, H. Qiao, F. Li, H. Fu, S. Sun et al. (2015). *Gene* **562:** 22–31.
38. T.N. Mayadas, and D.D. Wagner. (1992). *Proc. Natl. Acad. Sci. U S A* **89:** 3531–3535.
39. B.M. Byrne, M. Gruber, and G. Ab. (1989). *Prog. Biophys. Mol. Biol.* **53:** 33–69.
40. T.S. Dhadialla, A.S. Raikhel. (1990). *J. Biol. Chem.* **265:** 9924–9933.
41. M. Tufail, and M. Takeda. (2008). *J. Insect. Physiol.* **54:** 1447–1458.
42. T.W. Sappington, and A.S. Raikhel. (1998). *Insect Biochem. Mol. Biol.* **28:** 277–300.
43. H. Havukainen, J. Underhaug, F. Wolschin, G. Amdam, and O. Halskau. (2012). *J. Exp. Biol.* **215:** 1837–1846.
44. U. Abdu, C. Davis, I. Khalaila, and A. Sagi. (2002). *Gen. Comp. Endocrinol.* **127:** 263–272.

45. A. Okuno, W.J. Yang, V. Jayasankar, H. Saido-Sakanaka, D.T. Huong et al. (2002). *J. Exp. Zool.* **292**: 417–429.
46. P. Phiriyangkul, P. Puengyam, I.B. Jakobsen, and P. Utarabhand. (2007). *Mol. Reprod. Dev.* **74**: 1198–1207.
47. N. Zmora, J. Trant, S.M. Chan, and J.S. Chung. (2007). *Biol. Reprod.* **77**: 138–146.
48. C. Boulange-Lecomte, B. Xuereb, G. Tremolet, A. Duflot, N. Giusti et al. (2017). *Comp. Biochem. Physiol. C* **201**: 66–75.
49. F. Meggio, and L.A. Pinna. (1988). *Biochim. Biophys. Acta* **971**: 227–231.
50. W. Wahli. (1988). *Trends Genet.* **4**: 227–232.
51. P. Rava, G.K. Ojakian, G.S. Shelness, and M.M. Hussain. (2006). *J. Biol. Chem.* **281**: 11019–110272006.
52. P. Rava, and M.M. Hussain. (2007). *Biochemistry* **46**: 12263–122742007.
53. M. Manchekar, R. Kapil, Z. Sun, J.P. Segrest, and N. Dashti. (2017). *Biochemistry* **56**: 4084–4094.
54. M. Manchekar, Y. Liu, Z. Sun, P.E. Richardson, and N. Dashti. (2015). *J. Biol. Chem.* **290**: 8196–8205.
55. M.J. Chapman. (1980). *J. Lipid Res.* **21**: 789–853.
56. M. Komatsu, S. Ando, and S.I. Teshima. (1993). *J. Exp. Zool.* **266**: 257–265.
57. E. Spaziani, R.J. Havel, R.L. Hamilton, D.A. Hardman, J.B. Stoudemire et al. (1986). *Comp. Biochem. Physiol. B* **85**: 307–314.
58. R.F. Lee, and D.L. Puppione. (1988). *J. Exp. Zool.* **248**: 278–289.
59. E. Lubzens, T. Ravid, T.M. Khayat, N. Daube, and A. Tietz. (1997). *J. Exp. Zool.* **278**: 339–348.
60. RF. Lee. (1991). pp. 187–207. *In*: Gilles, R. [ed.]. *Adv. Comp. Environ. Physiol. Springer Verlag, Berlin*.
61. E.V. Orlova, M.B. Sherman, W. Chiu, H. Mowri, L.C. Smith et al. (1999). *Proc. Natl. Acad. Sci. USA* **96**: 8420–8425.
62. S. Jimenez-Gutierrez, C. Cadena-Caballero, C. Barrios-Hernandez, Raul Perez-Gonzalez, and F. Martinez-Perez. (2019). *Crustaceana* **92**(10): 1169–1205.
63. L. Cerenius, and K. Soderhall. (2011). *J. Innate Immun.* **3**: 3–8.
64. K. Junkunlo, K. Soderhall, and I. Soderhall. (2018). *Dev. Comp. Immunol.* **78**: 132–140.
65. Y. Kato, S. Tokishita, T. Ohta, and H. Yamagata. (2004). *Gene* **334**: 157–165.
66. S. Chen, D.F. Chen, F. Yang, H. Nagasawa, and W.J. Yang. (2011). *Biol. Reprod.* **85**: 31–41.
67. D.S. Hwang, K.W. Lee, J. Han, H.G. Park, J. Lee et al. (2010). *Comp. Biochem. Physiol. C* **151**: 360–368.
68. I. Khalaila, J. Peter-Katalinic, C. Tsang, C.M. Radcliffe, E.D. Aflalo et al. (2004). *Glycobiology* **14**: 767–774.
69. Z. Roth, S. Parnes, S. Wiel, A. Sagi, N. Zmora et al. (2010). *Glycoconj. J.* **27**: 159–169.
70. J.H. Nordin, C.H. Gochoco, D.M. Wojchowski, J.G. Kunkel. (1984). *Comp. Biochem. Physiol. B* **79**: 379–390.
71. N.H. Haunerland, and W.S. Bowers. (1987). *Comp. Biochem. Physiol. B* **86**: 571–5741987.
72. P.M. Weers, W.J. Van Marrewijk, A.M. Beenakkers, and D.J. Van der Horst. (1993). *J. Biol. Chem.* **268**: 4300–4303.
73. D.F. Zielinska, F. Gnad, K. Schropp, J.R. Wisniewski, and M. Mann. (2012). *Mol. Cell.* **46**: 542–548.
74. J.L. Soulages, J. Pennington, O. Bendavid, and M.A. Wells. (1998). *Biochem. Biophys. Res. Commun.* **243**: 372–376.
75. W.S. Tsang, L.S. Quackenbush, B.K.C. Chow, S.H.K. Tiu, J.G. He et al. (2003). *Gene* **303**: 99–109.
76. S.Y. Kung, S.M. Chan, J.H. Hui, W.S. Tsang, A. Mak et al. (2004). *Biol. Reprod.* **71**: 863–870.
77. S.H. Tiu, H.L. Hui, B. Tsukimura, S.S. Tobe, J.G. He et al. (2009). *Gen. Comp. Endocrinol.* **160**: 36–46.
78. Y. Yang, J. Wang, T. Han, T. Liu, C. Wang et al. (2015). *PLoS One* **10**: e0138862.
79. W. Thongda, J.S. Chung, N. Tsutsui, N. Zmora, and A. Katenta. (2015). *Comp. Biochem. Physiol. A* **179**: 35–43.
80. H. Shen, Y. Hu, Y. Ma, X. Zhou, Z. Xu et al. (2014). *PLoS One* **9**: e110548.
81. S.M. Chan, A.S. Mak, C.L. Choi, T.H. Ma, J.H. Hui, and S.H. Tiu. (2005). *Ann. N Y Acad. Sci.* **1040**: 74–79.
82. N. Tsutsui, I. Kawazoe, T. Ohira, S. Jasmani, W.J. Yang et al. (2000). *Zool. Sci.* **17**: 651–660.

83. D.Y. Tseng, Y.N. Chen, K.F. Liu, G.H. Kou, C.F. Lo et al. (2002). *Invert. Reprod. Dev.* **42:** 137–143.
84. S. Raviv, S. Parnes, C. Segall, C. Davis, and A. Sagi. (2006). *Gen. Comp. Endocrinol.* **145:** 39–50.
85. S.H. Tiu, J. Benzie, and S.M. Chan. (2008). *Biol. Reprod.* **79:** 66–74.
86. S.H. Tiu, J.H. Hui, J.G. He, S.S. Tobe, and S.M. Chan. (2006a). *Mol. Reprod. Dev.* **73:** 424–436.
87. S.H.K. Tiu, J.H.L. Hui, A.S.C. Mak, J.G. He, and S.M. Chan. (2006b). *Aquaculture* **254:** 666–674.
88. W.J. Yang, T. Ohira, N. Tsutsui, T. Subramoniam, D.T.T. Huong et al. (2000). *J. Exp. Zool.* **287:** 413–422
89. N. Tsutsui, H. Saido-Sakanaka, W.J. Yang, V. Jayasankar, S. Jasmani et al. (2004). *J. Exp. Zool.* **301:** 802–814.
90. B.P. Girish, C. Swetha, and P.S. Reddy. (2014). *Biochem. Biophys. Res. Commun.* **447:** 323–327.
91. X. Ding, G.P.C. Nagaraju, D. Novotney, D.L. Lovett, and D.W. Borst. (2010). *Aquaculture* **298:** 325–331.
92. T. Okumura, H. Nikaido, K. Yoshida, M. Kotaniguchi, Y. Tsuno et al. (2005). *Mar. Biol.* **148:** 347–361.
93. V. Jayasankar, N. Tsutsui, S. Jasmani, H. Saido-Sakanaka, W.J Yang et al. (2002). *J. Exp. Zool.* **293:** 675–682.
94. T. Okumura, K. Yamano, and K. Sakiyama. (2007). *Comp. Biochem. Physiol. A* **147:** 1028–1037.
95. G.P. Nagaraju. (2011). *J. Exp. Biol.* **214:** 3–16.
96. P.I. Blades-Eckelbarger, and M.J. Youngbluth. (1984). *J. Morphol.* **179:** 33–46.
97. S. Dalvin, P. Frost, E. Biering, L.A. Hamre, C. Eichner et al. (2009). *Int. J. Parasitol.* **39:** 1407–1415.
98. J.J. Meusy, H. Junera, P. Cledon, and M. Martin. (1983). *Reprod. Nutr. Dev.* **23:** 625–640.
99. M. Tom, M. Goren, and M. Ovadia. (1987). *Int. J. Invert. Reprod. Dev.* **12:** 1–12.
100. K. Rani, and T. Subramoniam. (1997). *J. Crustacean Biol.* **17:** 659–665.
101. H. Junera, and Y. Croisille. (1980). *Cr. Acad. Sci. D. Nat.* **290:** 703–706.
102. A. Okuno, H. Katayama, and H. Nagasawa. (2000). *Comp. Biochem. Physiol. B* **126:** 397–407.
103. M.M. Smolenaars, M.A. Kasperaitis, P.E. Richardson, K.W. Rodenburg, and D.J. Van der Horst. (2005). *J. Lipid Res.* **46:** 412–421.
104. P. Duckert, S. Brunak, and N. Blom. (2004). *Protein Eng. Des. Sel.* **17:** 107–112.
105. W.E Fyffe, J.D. O'Connor. (1974). *Comp. Biochem. Physiol. B* **47:** 851–867.
106. D. de Chaffoy de Courcelles, and M. Kondo. (1980). *J. Biol. Chem.* **255:** 6727–6733.
107. T. Ravid, A. Tietz, M. Khayat, E. Boehm, R. Michelis et al. (1999). *J. Exp. Biol.* **202:** 1819–1829.
108. F. Fontaine, M.H Gevaert, and M. Porchet. (1984a). *Comp. Biochem. Physiol. B* **78:** 581–584.
109. F. Fontaine, M.H. Gevaert, M. Porchet. (1984b). *Comp. Biochem. Physiol. A* **77:** 45–50.
110. R. Ziegler, and R. Van Antwerpen. (2006). *Insect Biochem. Mol. Biol.* **36:** 264–272.
111. T.F. Roth, and K.R. Porter. (1964). *J. Cell Biol.* **20:** 313–332.
112. F. Fagotto. (1995). *J. Cell Sci.* **108:** 3645–3647.
113. T.E. Willnow, A. Hammes, and S. Eaton. (2007). *Development* **134:** 3239–3249.
114. M. Tufail, and M. Takeda. (2009). *J. Insect Physiol.* **55:** 87–103.
115. M. Tufail, and M. Takeda. (2018). pp. 285–289. *In*: Skinner, M.K. [ed.]. *Encyclopedia of Reproduction*. Elsevier, Amsterdam.
116. M.S. Brown, J. Herz, and J.L. Goldstein. (1997). *Nature* **388:** 629–630.
117. Y. Ruan, N.K. Wong, X. Zhang, C. Zhu, X. Wu et al. (2020). *Front. Physiol.* **11:** 485.
118. S.A. Warrier, and T. Subramoniam. (2002). *Mol. Reprod. Dev.* **61:** 536–548.
119. M. Mekuchi, T. Ohira, I. Kawazoe, S. Jasmani, K. Suitoh et al. (2008). *Zool. Sci.* **25:** 428–437.
120. Z. Roth, and I. Khalaila. (2012). *Mol. Reprod. Dev.* **79:** 478–487.
121. J.H. Lee, B.K. Kim, Y.I. Seo, J.H. Choi, S.W. Kang et al. (2014). *Comp. Biochem. Physiol. B* **169:** 51–62.
123. A. Li, M. Sadasivam, and J.L. Ding. (2003). *J. Biol. Chem.* **278:** 2799–2806.
124. R. Prassl, and P. Laggner. (2009). *Eur. Biophys J.* **38:** 145–158.
125. A. Sagi, M. Rise, I. Khalaila, and S.M. Arad. (1995). *Comp. Biochem. Physiol. B* **112:** 309–313.
126. Q. Zhang, Y. Lu, H. Zheng, H. Liu, and S. Li. (2016). *Fish Shellfish Immunol.* **50:** 11–15.
127. S. Zhang, S. Wang, H. Li, and L. Li. (2011). *Int. J. Biochem. Cell Biol.* **43:** 303–305.
128. Z. Tong, L. Li, R. Pawar, and S. Zhang. (2010). *Immunobiology* **215:** 898–902.
129. C. Sun, L. Hu, S. Liu, Z. Gao and S. Zhang. (2013). *Dev. Comp. Immunol.* **41:** 469–476.
130. H. Salmela, G.V. Amdam, and D. Freitak. (2015). *PloS Pathog.* **11:** e1005015.

131. H. Havukainen, D. Munch, A. Baumann, S. Zhong, O. Halskau et al. (2013). *J. Biol. Chem.* **288:** 28369–28381.

132. A. Nakamura, K. Yasuda, H. Adachi, Y. Sakurai, N. Ishii et al. (1999). *Biochem. Biophys. Res. Commun.* **264:** 580–583.

133. G.V. Amdam, Z.L. Simoes, A. Hagen, K. Norberg, K. Schroder et al. (2004). *Exp. Gerontol.* **39:** 767–773.

134. H. Havukainen, O. Halskau, and G.V. Amdam. (2011). *Mol. Ecol.* **20:** 5111–5113.

135. K. Hultgren, J.E. Duffy, and D.R. Rubenstein. (2017). pp. 224–249. *In*: Rubenstein, D.R., and P. Abbot [eds.]. Comparative Social Evolution. Cambridge University Press, Cambridge.

CHAPTER 4

Infochemicals Recognized by Crustaceans

Joerg D. Hardege, * *Nicky Fletcher* and *Thomas Breithaupt*

4.1 Introduction

Arthropods are undoubtedly the cornerstone of the natural world; they make up approximately 75% of all animals on Earth and are vitally important in maintaining balance within the ecosystems they reside. They assist with the transfer of nutrients through the terrestrial environment, aiding with mineralisation of elements through the soil and influence the amount of living and decaying organic matter [1]. Arthropods successfully inhabit all environments and are excellent communicators. The 'language of life' for many organisms, involves chemical communication, Arthropods are amazingly adept at synthesising natural compounds in their endocrine glands and utilising these in a seemingly simple but yet complex communication system.

There is extensive evidence of effective chemical communication systems in insects [2]. Insects have the ability to communicate using visual, tactile, acoustic and chemical cues, the latter being able to be dispersed by air or water, left on food or plants or through trophallaxis. The chemicals can be used to elicit a diverse range of responses in the receiver's physiology or behaviour of a sympatric species or be used to attract or repel across different species and have been shown to be particularly well developed in social insects [3]. Infochemicals are divided into two categories; those that pass information between species are known as 'allelochemicals' and those that communicate information within a species are "*pheromones*". Despite this wealth of information in insects of the terrestrial world little in comparison is known about the Arthropods of the oceans; the crustaceans.

School of Biological, Biomedical, Environmental Sciences, University of Hull, Hull, UK.
* Corresponding author: j.d.hardege@hull.ac.uk

Chemical signals mediate elements of behaviour throughout all life stages of aquatic crustaceans and are fundamental in the success of the species [*4*]. Despite this, much is still to be uncovered on the characteristics of pheromone compositions and mechanisms used by different species to communicate. In many crustaceans, pheromones are released through a controlled directional spray of urine, these can provide information on the sex, dominance status, health, moult state and individual identity to conspecifics [*4*]. Crustaceans are used as bioindicators for habitat conditions and are utilized in biomonitoring for environmental stressors, and they are one of the most used organisms in determining environmental quality in European waters [*5*]. Their economic importance in world fisheries and aquaculture, their positive contributions to food security and their link to understanding global health in accordance to the One Health approach linking crustacean, environmental and human health together [*6*] make understanding the communication mechanisms of these valuable species of vital importance.

4.2 Infochemicals used in behaviour

There has been considerable focus paid to both vertebrate and insect sensory ecology, with numerous studies investigating odorant-binding proteins and chemosensory proteins used in communication [*7*]. In contrast, in aquatic crustaceans few specific chemicals have been identified and the mechanisms, roles and physiology of these are comparatively little understood.

Many aquatic organisms use chemosensory cues to determine risk levels particularly to predation, early detection of a predator is literally life or death and many prey species have adapted enhanced detection and evasion skills for survival. With prior knowledge of an imminent predator, prey animals can evade through crypsis, seeking refuge or activate fight responses [*8*]. Chemical cues can be used to provide information on habitat selection [*8*] or cost-benefit analysis [*9*] in relation to predation risk factors. Cues can be released by the predator (kairomones), by a conspecific in response to a predator (active alarm pheromones) or by caught or injured animals (passive alarm cues), the receiver can then use this information to determine their behaviour. Numerous organisms elicit avoidance or escape responses [*8*] whilst others can change their morphological characteristics in an attempt to minimize their risk of predation [*8*]. These predation-driven plasticity changes within organisms allows for adaptation to habitat and seasonal changes in exposure to predators. Whilst many studies have been done looking at the behavioural responses to predator cues using conditioned seawater, relatively few have identified specific chemicals [10]. Exceptions to this are the identification of the fatty acids conjugated to the glutamine amino group that triggers a defense response in the freshwater crustacean *Daphnia* [*11*]. Mud crabs have also shown decreased foraging in the presence of homarine and trigonelline, two urinary metabolites found in blue crab urine, a known predator of mud crabs. This response was elicited by naturally occurring concentrations of either chemical or as a mixture of both [*12*].

Crustaceans use chemical cues to navigate towards potential food sources, many using low molecular weight metabolites like amino acids or nucleotides and nucleosides released by injured animal flesh [*13*]. These are mainly multiple compounds that elicit feeding responses, as they provide more information in an already complex chemical environment. These responses to different cues are highly specialized and vary between species and across trophic levels. Different trophic levels often show responses to different types of feeding cues with carnivores being more sensitive to nitrogen-containing compounds that are found in decomposing prey, whereas, herbivores and omnivores are more reactive to sugars found in plant materials [14]. Extensive research has been focussed on crustacean chemical feeding stimulants with amino acids showing feeding responses in many crustacean species [13]. Crustaceans can identify different chemical compounds within mixtures and even different quantities of compounds to ensure the correct behaviour is triggered. This is achieved through sensitive receptor neurons located on the distal end of the antennule of many crustaceans, these complex processors allow for decoding of blend ratios and quality of food stimuli [13].

4.3 Conspecifics

It is widely accepted that crustaceans use self-propelled chemical communication signals to elicit specific behaviours including sex, dominance and receptivity to name a few. These intraspecific cues allow animals to communicate in a complex environment and ensure optimal responses. Many decapod crustaceans use these cues released in the urine to establish dominance hierarchies and social status (Figure 1).

Chemorecognition is a key component of establishing dominance hierarchies which play an important part in population distribution and genetic structure as well as habitat use and energetics [16]. Lobsters are perhaps the most studied crustacean in relation to dominance status and recognition of dominance hierarchies, from the initial discovery of dominant and subordinate individuals this has been a much-favored

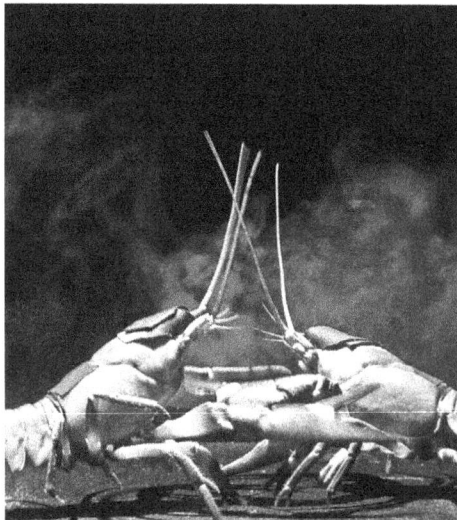

Figure 1. A visual display of crayfish urine release from the anterior nephropore (taken from Berry and Breithaupt [15].

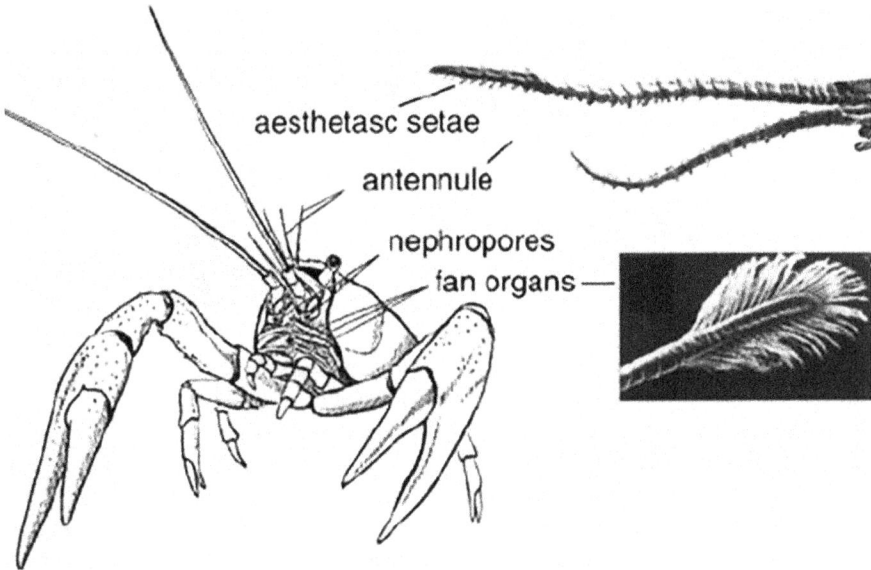

Figure 2. Location of the primary chemoreceptor on the crayfish, showing the location of the aesthetasc setae and the fan organs (taken from Breithaupt 2011).

area of research [*4*]. Intentional urine release is used as an early fight intimidation tactic and can be used to recognise social status. By reducing the quantity of fights through recognising dominance after an initial encounter, energy is conserved by both parties, with animals shown to remember and recognise previous competitors at least in the short term [*4*]. Crayfish develop a linear dominance hierarchy through regular agonistic encounters, this dominance gives priority over shelter, food and mating partners [*17*]. Chemical signals are a crucial element in the establishment and maintenance of these hierarchies with chemicals providing a wealth of information on the combatants [*17*]. Urine in crayfish is released through the nephropores on the anterior end of the crayfish through controlled release by a sphincter muscle ([*18*], Figure 3). This allows the crayfish to time the release of its pheromones for maximum efficiency. These pheromones are then dispersed through gill currents or currents generated by the fan organs transporting the pheromone to the conspecific, the receiver also uses fan generated currents to guide the molecules to the antennules, the major chemoreceptors, to detect and elicit an appropriate response ([*18*], Figure 2).

Hermit crabs show frequent agonistic displays over shell ownership, these shell fights use a range of visual, chemical and tactile cues to provide information of resource potential [*19*].

4.4 Sex pheromones

Sex pheromone's structural identification in many insects has long been established and is now successfully utilised in insect control. Despite significant end user demand for control of crustaceans for example invasive *Carcinus maeanas* (green crab), relatively few chemicals have been identified primarily due to the complex physico-chemical nature of the medium water, especially seawater limiting successful purification methodologies for aquatic environments [20]. Marine crustaceans release their sex pheromones through their urine and primarily by the female of the species [15]. Many attempts have been made at identification but to date only in a few major inroads have been made to characterise these. Novel ceramides are pheromone candidate compounds in *Erimacrus isenbeckii* but did not verify their bioactivity in bioassays [11], and a protein [22] is the mate-recognition in the copepod *Tigriopus japonicus*, and N-acetylglucosamino-1,5-lactone (NAGL), is an indicator of sexually active females in blue crabs, *Callinectes sapidus* [23]. Whilst initial studies on the moulting hormone crustecdysone suggested a function as female sex pheromone [24] this compound is more likely a feeding deterrent for shore crabs, *Carcinus maenas* preventing female cannibalism [25].

The best understood pheromone system is that in the shore crab, *Carcinus maenas* where the nucleotide Uridine-di-phosphate [26] attracts males and induces the mating stance. This nucleotide and its tripeptide form UTP also forms part of the female pheromone complex in *Lysmata* shrimps [27]. Sex pheromones come in two categories; distance pheromones and contact pheromones and are often accompanied by series of pre-mating behaviours. Distance sex pheromones are used to attract a mate and are generally polar, water-soluble compounds that are capable of transmission in odour plumes. Contact pheromones like that used by multiple shrimp species [28] need to be relatively insoluble in water to ensure they remain on the exoskeleton with 9-octadecene-aminde (Oleamide) inducing mating in *Lysmata* shrimps [28]. Interestingly Oleamide is also used as slipping agent in plastic and when leaching into the environment can impact hermit crabs, *Pagurus bernhardus* olfactory behaviours inducing food search and feeding behaviour [29]. As many crustaceans have a short window for mating, that is usually around the time of the female moult, distinct, timed signals need to be utilized; these may range from complex species specific bouquets like that in insects [7] or simple byproducts of physiological state [26] that portray information to the receiver on the condition and molt stage of the sender. Both, UDP in shore crabs and NAGL in the blue crab urine are compounds related to the crustacean moult when the female soft carapace is aiding the mating process. UDP is released during chitin production (mainly post-molt) and NAGL is an oxidized form of N-acetylglucosamine important in chitin degradation aiding the molting process. As such these sex pheromones may have evolved as sensory traps indicating the presence of a molting individual that is a potential mating partner but lacking sex specificity as for example in shore crabs in the field mature males are frequently found cradling freshly molted small males. Fitting this concept, there is a distinct lack of species specificity in this type of crustacean sex pheromones [30] with species isolation relying on spacial or time separation.

As aquatic crustaceans live in an environment containing a constant and ever-changing deluge of complex chemical stimulants, they need to be able to not only differentiate between the cues/signals from background noise but then activate the appropriate behavioral response to the communication. Hardege and Terschak [20] reviewed both the complexity of identification of sex pheromones in crustacea as well as mechanisms used by different organisms to successfully reproduce.

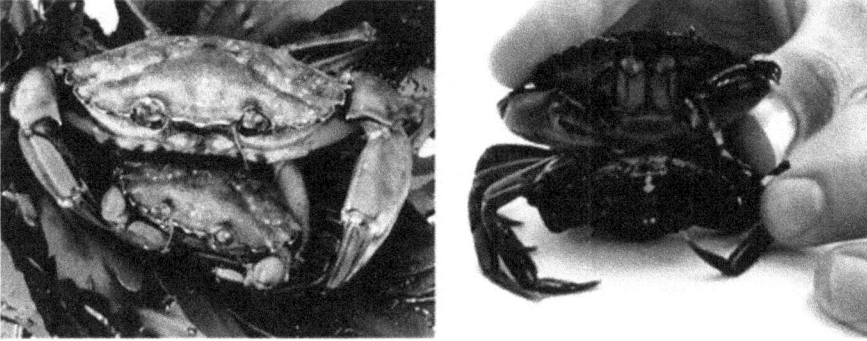

Figure 3. Male mate guarding and copulation in the shore crab Carcinus maenas. Photos provided by Drs. N. Fletcher and R. Bublitz.

Male sex pheromones have been established and well documented in insects [2,7] but less so in crustaceans. There has been some evidence of male cues stimulating female choice and evaluation in lobsters [31] and the potential for tactile and chemical cues from the male, impacting ovarian growth in juvenile caridean shrimp [32] but to date very little is known on male chemical communication in these taxa. Males have well established mating behaviors that generally involves cradling and mate guarding in many decapod crustaceans (Figure 3).

Due to the complex nature of the aquatic environment and the presence of an exoskeleton, crustaceans have sensory sensilla over most of their body that traverse the chemical signals from the exterior world to the interior processes providing both chemical and mechanical information [33]. The primary chemoreceptor in many crustaceans is the aesthetasc sensillum located on the flagella of the antennule ([*14*], Figure 2). These specialized receptors allow for detection of the female released sex pheromone [*34*]. In comparison to these previously discussed uses for chemical communication little is known about brood pheromones, although brood recognition and care has been observed in some crustacean species [*35*].

4.5 Olfactory disruption—the disturbance of the sensory system and chemical communication

In recent years focus has been turned to what impact the increasing presence of aquatic pollutants and ocean acidification has on these communication mechanisms. There are a wide variety of potential influencers on the different processes involved in chemical communication [*10*] from pheromone release, to transmission of the chemical through the complex aquatic environment, to the ability of the receiver to not only receive but also decode the message to elicit the appropriate behavior. At each stage these signals could be changed, interrupted or misread and can impact on effective communication between aquatic organisms (Figure 4).

The reduced ability to detect or respond to predation cues through increased plastics and copper pollution has been observed in *Daphnia* [*36,37*] resulting in morphological implications of reduced size and growth rate to maturity suggesting long term impacts on not only a species level but also a community one [*37*]. Copper

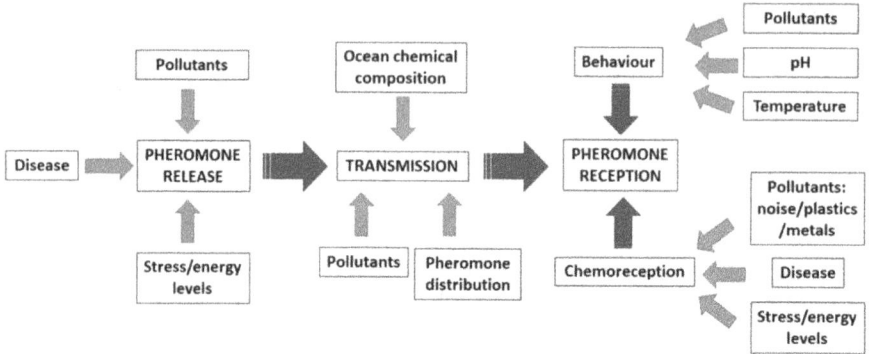

Figure 4. Impacts of external factors throughout the different processes involved in pheromone communications.

pollution has also been shown to have an effect on pheromone detection in the shore crab [*38*]. The impacts of microplastics on olfactory communication in oceanic organisms is a more recent subject focus but has indicated what a devastating impact this microscopic polluter can have on organisms' behavior.

Plastics that mimic prey scent through being characteristically similar to prey odour are particularly harmful acting as an attractant to animals who then ingest them. Seabirds and larger animals can visually misinterpret plastics for prey but more recently evidence has been seen that odours from marine plastics elicit keystone infochemicals to marine organisms [*39*] and Dimethyl Sulphide (DMS) induces feeding behavior. Similarly, Greenshields et al. [*29*] showed how Oleamide may be perceived as a feeding cue in hermit crabs. This ingestion of microplastics particularly at the primary trophic levels will result in bioaccumulation through the food web, an example can be seen in the shore crab whose primary food source is the blue mussel, *Mytilus edulis* who are known to be susceptible to the ingestion of microplastics [*40*]. Through both the diet and gill transfer of the shore crab microplastics are taken into the gut and retained for up to 21 days depending on the uptake method, this allows a sufficient time period to pass on to the next trophic level [*41*].

Abiotic factors including those brought about by Ocean Acidification through the increased uptake of CO_2 resulting in the decrease of pH levels in our oceans as well as increased temperatures are proving to have a profound effect on our aquatic wildlife. The higher CO_2 and subsequent pH reduction in the oceans are changing ocean chemistry altering the impacts of those environmental factors shown in Figure 4. Research on chemosensory capabilities of animals under the duress of future Oceanic conditions is still very much underway with little being completely understood also on their long-term ability to adapt. For more in-depth information see chapter 10 by Hardege and Fletcher. The chemosensory ability of the spiny lobster is shown to be impacted by increased temperature and decreased pH, though they still elicited shelter searching behaviors the antennule flicking rate (smelling) of the lobster decreased, aggregation and suitable shelter choice was lost suggesting a partial loss of chemosensory function [*6*]. This reduced ability to detect odor sources could link to protonation of signalling molecules due to changing pH altering molecule structure or charge that can alter the olfactory response of a receiver through a possible change in the mechanistic pathway of communication [42]. The mechanisms that are most impacted by climate change will affect proteins and peptide communication the most significantly, and with proteins and associated amino acids being some of the most used chemicals in predator-prey communication [10] this could drastically impact organisms and community structure.

4.6 The application of crustacean info chemicals: Aquaculture and Fisheries

To understand the mechanisms underpinning communication and identify chemicals involved would be infinitely valuable in aquaculture, an industry that between 1990 and 2018 had increased by 527% globally and was standing at a yield of 82.1 million tonnes of live weight aquatic animals by 2018 and a projected 109 million tonnes by 2030 [43]. With the need for fish-based products ever increasing and the recognition that regional and global food security rely on this protein source to feed their nations [43] (FAO 2020) it is ever more important to maximise this potential through understanding chemical processes, attractants and deterrents that can be used in aquaculture. To date, with few chemical cues having been identified in crustaceans only speculations and naive observations have been made as to the uses of chemical cues in an aquaculture system that could be beneficial. Some examples are the regulation of reproductive cycle phases, synchronizing female mating, timing of larval release, minimizing social control of growth and feeding attractants to reduce costs and maximize growth rate [44].

4.7 Applications in pest management

Chemical cues have long been established and effectively utilised in pest control management of a variety of insect species [45]. In more recent years the switch from the more toxic insecticides to more environmentally friendly use of semiochemicals has been well received. Semiochemicals in particular sex phermones are the most intensively used in insect management; they are potent, species specific, relatively non-toxic to the environment and target species appear less likely to develop resistance to them [45].

Pest management in aquatic environments is still largely illusive, with the lack of knowledge on the specific chemicals utilised by many crustaceans it is difficult to identify how successful this initiative would be. Work has been undertaken on invasive crayfish to see if sex pheromones could be effective for controlling this invader, however, although the pheromone baited traps attracted more males of the species, they were not more attractive than food baited traps [46]. This would suggest the need for a multidisciplinary approach, sex pheromone traps could be improved with the purification of the sex pheromone and despite the small window of trapping opportunity, breeding season only, this has been shown to be effective in disrupting breeding and therefore assisting in terrestrial pest control [47]. Using pheromone traps alongside feeding baited traps could be a future solution and although this method will not eradicate large populations it would reduce the volume of breeding males and also indicate presence of specific species [*47*]. With current invasions of the shore crab *Carcinus maenas* (known as green crab) in North and South America, South Africa and New Zealand, the replacement of shore crabs by the Asian shore crab *Hemigrapsus takanoi* [*48*], blue crabs, *Callinectes sapidus* in the Mediterranean and Chinese mitten crabs, *Eriocheir sinensis* throughout Europe and the US causing significant ecological and economical damage the use of natural chemical cues capable to trap or chase away is only going to increase, especially when changing ocean temperatures and climate bring about major species shifts globally.

References

1. T.R. Seastedt, and D.A. Crossley. (1984). *Bioscience* **34**: 157–161.
2. R., Kumar, O. Prakash, and R. Kumar, A.K. (2021). *In*: Nath, M., D. Bhatt, P. Bhargava, and D.K. Choudhary [eds.]. *Microbial Metatranscriptomics Belowground*. (Springer, Singapore, 2021).
3. C.A. Oi, J.S. van Zweden, R.C. Oliveira, A. Van Oystaeyen, F.S. Nascimento, and T. Wenseleers. (2015). *Bioessays* **37**: 808–821.
4. M. Thiel, and T. Breithaupt. (2011). *Chemical communication in crustaceans* (Springer Science 2011).
5. C. Navarro-Barranco, M. Ross, J.M. Tierno de Figueroa, and J.M. Guerra-Garcia. (2020). *Fisheries and Aquaculture vol. 9.* (Oxford University Press, 2020).
6. E. Ross, and D. Behringer. (2019). *Sci Rep.* **9**: 4375.
7. T. Wyatt. (2014). *Pheromones and Animal Behaviour*. (Cambridge Univ. Press 2014).
8. B.A. Hazlett. (2011). pp. 355–370. *In*: Thiel, M., and T. Breithaupt [eds.]. *Chemical communication of crustaceans*, [4] (Springer Science 2011).
9. B.D. Wisenden. (2003). pp. 236–251. *In*: Collin, S.P., and N.J. Marshall [eds.]. *Sensory Processing in Aquatic Environments*. (Springer, New York, 2003).
10. A.M. Draper, and M.J. Weissburg. (2019). *Front Ecol. Evol.* **7**: 72.
11. L.G. Weiss et al. (2018). *Nat. Chem. Biol.* **14**: 1133–1139.
12. R.X. Poulin, et al. (2018). *Proc. Natl. Acad. Sci. USA.* **115(4)**: 662–667.
13. M. Kamio, and C.D. Derby. (2017). *Nat. Prod. Rep.* **34**: 514–528.
14. C.D. Derby, and M.J. Weissburg. (2014). pp. 263–292. *In*: Derby, C.D. and M. Thiel [eds.]. *Nervous systems and control of behaviour*. (Oxford University Press. 2014).
15. F.C. Berry, and T. Breithaupt. (2010). *BMC Biol.* **8**: 25.
16. J.P. Lord et al. (2021). *Biol. Bull.* **240**: 2–15.
17. T. Breithaupt, F. Gherardi, L. Aquiloni, and E. Tricarico. (2010). pp. 132–170. *In*: Longshaw, M., and P. Stebbing [eds.]. *Biology and Ecology of Crayfish*. CRC Press, Taylor & Francis.
18. T. Breithaupt. (2011). *In*: Breithaupt, T., and M. Thiel [eds.]. Chemical Communication in Crustaceans. (Springer, New York, NY).
19. M. Briffa, and R. Williams. (2006). *Behaviour* **143**: 1281–1290.
20. J.D. Hardege, and J.A. Terschak. (2010). pp. 373–392. *In*: Breithaupt, T., and M. Thiel [eds.]. Chemical Communication in Crustaceans. (Springer, New York 2010),
21. N., Asai, N. Fusetani, and S. Matsunaga. (2001). *J. Nat. Prod.* **64(9)**: 1210–1215.
22. J.H., Ting, and T.W. Snell. (2003). Purification and sequencing of a mate-recognition protein from the copepod *Tigriopus japonicus*. *Mar. Biol.* **143**: 1–8.
23. M. Kamio, and C.D. Derby. (2010). pp. 393–412. *In*: Breithaupt, T., and Thiel, M. [eds.]. Chemical Communication in Crustaceans. (Springer, New York 2010).
24. J.S. Kittredge, M. Terry, and F.T. Takahashi. (1971). *Fish Bull.* **69**: 337–343.
25. D. Hayden et al. (2007). *Horm. Behav.* **52**: 162–168.
26. J.D. Hardege et al. (2011). *Mar. Ecol. Progr. Ser.* **436**: 177–189.
27. D. Zhang, J. Lin, M. Harley, J.A. Terschak, and J.D. Hardege. (2020). *Mar. Ecol. Progr. Ser.* **640**: 139–146.
28. D. Zhang, J.A. Terschak, M.A. Harley, J. Lin, and J.D. Hardege. (2011). *PLoS One* **6**(4): e17720.
29. J. Greenshields, P. Schirrmacher, and J.D. Hardege. (2021). *Mar. Poll. Bull.* **169**: 112533.

30. R. Bublitz et al. (2008). *Behaviour* **145:** 1465–1478.
31. P.J. Bushmann, and J. Atema. (2000). *J. Chem. Ecol.* **26:** 883–899.
32. C. Tropea, S.M.L. Lavarias, and L.S.L. Greco. (2018). *Zoology* **130:** 57–66.
33. F.W. Grasso, and J.A. Basil. (2002). *Curr. Opin. Neurobiol.* **12(6):** 721–727.
34. S.D. Bamber, and E. Naylor. (1996). *Mar. Biol.* **125:** 483–488.
35. J.A. Baeza, and M. Fernandez. (2002). *Funct. Ecol.* **16(2):** 241–251.
36. C.M. DeMille, S.E. Arnott, and G.G. Pyle. (2016). *Ecotoxicol. Environ. Saf.* **126:** 264–272.
37. B. Trotter, A.F.R.M. Ramsperger, P. Raab, J. Haberstroh, and C. Laforsch. (2019). *Sci. Rep.* **9:** 5889.
38. A.S. Krång, and M. Ekerholm. (2006). *Aquat. Toxicol.* **80(1):** 60–69.
39. M.S. Savoca, M.E. Wohlfeil, S.E. Ebeler, and G.A. Nevitt. (2016). *Sci. Adv.* **2(11):** e1600395.
40 M.A. Browne, A. Dissanayake, T.S. Galloway, and D.M. Lowes, R.C. Thompson. (2008). *Environ. Sci. Technol.* **42(13):** 5026–5031.
41 A.J.R. Watts et al. (2014). *Environ. Sci. Technol.* **48:** 8823–8830.
42. P. Schirrmacher, C.C. Roggatz, D.M. Benoit, J.D. Hardege. (2021). *J. Chem. Ecol.* **47(10-11):** 859–876.
43. FAO. 2020. *The State of World Fisheries and Aquaculture 2020. Sustainability in action.* Rome.
44. S. Barki, C. Jones, and I. Karplus. (2010). *In:* Breithaupt, T., and M. Thiel [eds.] Chemical Communication in Crustaceans. (Springer, New York. 2010).
45. G. Komala, R.R. Manda, and D. Seram. (2021). *J. Entomol. Res.* **6(2):** 247–253.
46. L. Aquiloni, and F. Gherardi. (2010). *Hydrobiologia* **649:** 249–254.
47. P.D. Stebbing et al. (2004). *En. Nat. Res. Rep.* **578**.
48. A. Cornelius, K. Wagner, and C. Buschbaum. (2021). *Mar. Biodiv.* **51**.

CHAPTER 5

Crustaceans as Good Marine Model Organisms to Study Stress Responses by –Omics Approaches

Maria Costantini,[1,*] *Roberta Esposito*[1,2#] and *Nadia Ruocco*[3,#]

5.1 Introduction on stress response in crustaceans

Marine organisms are continuously exposed to physical, environmental and physiological stressors, which force them to react against these disturbances by activating several molecular and biochemical modifications devoted to the restoration of internal homeostasis. Stress response may involve the immune system that, in crustaceans and in all invertebrates, is primarily innate [1]. The invertebrate immunological response is based on both humoral and cellular mechanisms and could be considered a suitable tool for understanding the ecological and evolutionary impact of stressors. In crustaceans, stress response to several environmental changes, is typically represented by a massive release of the hyperglycemic hormones (CHHs) from the X-organ/sinus gland complex localized in the eyestalk [2]. This event has been demonstrated when individuals with removed eyestalks did not display increasing CHH levels and hyperglycemia after the exposure to several physical and chemical stressors, such as heavy metals, lipopolysaccharides, bacteria, hypoxia, temperature, handling and commercial transport. This latter point is particularly important since crustaceans are often subjected to extreme conditions during capture, storage and transport for the use in human food chain. Some of these conditions,

---

[1] Stazione Zoologica Anton Dohrn, Department of Ecosustainable Marine Biotechnology, Villa Comunale, 80121 Napoli, Italy.
[2] Department of Biology, University of Naples Federico Íl, Complesso Universitario di Monte Sant'Angelo, Via Cinthia 21, 80126 Naples, Italy.
[3] Stazione Zoologica Anton Dohrn, Department of Marine Animal Conservation and Public Engagement, Amendolara Excellence Centre, C.da Torre Spaccata, 87071 Amendolara (CS) - Italy.
* Corresponding author: maria.costantini@szn.it
# These authors contributed equally to this work.

such as being caught in a trawl or held for long periods in close confinement with other individuals, induce a strong stress response. Several studies have evaluated the stress response in crustaceans by measuring the levels of CHH hormones in the haemolymph [*3*]. The CHH hormone has been described as the equivalent of cortisol and/or corticosterone neuroendocrine signals, promoting the conversion of glycogen to glucose in response to stress conditions. As upstream modulators triggering CHHs release from neuroendocrine tissue, several biogenic amines and enkephalins were described in crustacean stress response. For instance, serotonine (5-HT) has potent hyperglycemic effect in several crustaceans, whereas dopamine and enkephalins revealed controversial responses depending on crustacean species. Very few information is available on downstream CHH receptors, mostly belonging to a membrane (class II) guanylyl cyclase. Analysing the eyestalk and haemocyte transcriptome from *Procambarus clarkii*, two CHH variants (CHH-1 and CHH-2) were isolated. These isoforms exerted diverse functions, since CHH-1 was found six times more expressed in eyestalk [*4*]. In addition to CHH hormone, the concentrations in the haemolymph of other stress indicators have been found in crustaceans such as, inorganic ions, oxygen, metabolites, waste products and several proteins.

General stress biomarkers are heat shock proteins (HSPs), which have been widely used for assessing the stress response in several marine organisms, including crustaceans. HSPs belong to a wide family of proteins, involved in the innate immune response of several species of crustaceans, firstly associated to high temperature stress response in *Drosophila melanogaster*. These proteins are called molecular chaperones, since they ensure a correct protein folding and functioning. A well-known example of improper protein molecular aggregation is the development of Alzheimer's and Parkinson's disease. HSPs are classified according their molecular weight and are mostly located in cellular compartments, such as mitochondria, reticulum and cytosol, but can also be detected in the extracellular environment. The 70 kDa HSP proteins (HSP70) are the most studied for its high sensitivity and abundance, as well as, being ubiquitously expressed and associated with all subcellular compartments. These proteins actively participate in the homeostasis of all living organisms and exert a rapid response to any agent stressor representing a good indicator for assessing environmental pollution/contamination processes. Moreover, HSP70 is strongly conserved along the evolutionary processes, constituting an optimal molecular marker for diverse model organisms [*5,6*]. Additional factors are able to trigger the gene expression of this protein, such as the exposure to chemical substances and physical or biological stresses. For this reason, many works used this protein as a molecular biomarker, revealing possible environmental stresses by detecting its gene up- or down-regulation [*7*]. In crustaceans the synthesis of HSPs has been observed in the diapause process of the marine copepod *Calamus finmarchius* and when exposed to stress. Eight HSPs have been identified plus four isoforms of HSP70 (HSP70 A, B, C and D) characterized under stress, manipulation processes and in association with diapause.

In the case of anthropogenic pollutants and natural toxins, the first line of defence is the reduction of their intracellular inflow by the activation of a multixenobiotic resistance system (MXR), which consists of a group of plasma membrane proteins, belonging to the adenosine triphosphate (ATP)-binding cassette (ABC) family. The

MXR response was described in several marine organisms, such as, sponges, mussels, oysters, crabs, worms, sea stars, clams and fishes [8]. A second line of defence is the enzymatic transformation of toxic compounds by several reactions (oxidation, reduction, hydrolysis, for example), in order to promote the removal from tissues. To evaluate the possible hazard triggered by environmental pollutants, some biochemical biomarkers, including enzyme activities and other sub-cellular components, were considered the most promising tool for ecotoxicological applications. Regarding invertebrate response to environmental toxicants, a lot of attention was devoted to three detoxification enzymes: catalase (CAT), glutathione S-transferase (GST) and cholinesterase (ChE).

Another crustacean stress biomarker are hemocyanins, which are highly sensitive to external abiotic factors, such as pH, temperature and salinity. As a multifunctional protein, hemocyanin is involved in several physiological processes, including protein storage, immune defense and ecdysone transport. A recent study has suggested that hemocyanins are associated to shrimp nitrite stress [9].

The relationship between stress, depression of the immune system and disease was demonstrated in a great number of studies. Free radicals play an important role in crustaceans innate immune reaction. The accumulation of free radicals and some reactive or toxic intermediates, named reactive oxygen species (ROS), is an important source of stress, which is commonly referred to oxidative stress. ROS species are able to damage DNA, RNA, lipids and proteins when present in high quantities and its accumulation has been linked to several events such as, exposure to stressful environmental conditions or pathogen infections [10]. To convert these deleterious molecules into less active compounds, cells produce some detoxification enzymes, such as the already mentioned CAT, superoxide dismutase (SOD) and glutathione peroxidase (GPx), and some scavenger molecules (glutathione, carotenes, ascorbic acid and tocopherols), which act as cellular antioxidants preventing ROS damage. These enzymes and the products related to DNA damage, lipid peroxidation and protein carbonylation are common molecular markers useful to evaluate the presence of oxidative stress. For instance, protein oxidation, revealed by the accumulation of protein carbonyls, was widely used as marker of oxidative stress in several invertebrates, including crustaceans. In fact, contrarily to lipid peroxidation products, which are detoxified within minutes, cells degrade oxidized proteins from several hours to days. However, a frequent marker of oxidative stress due to lipid peroxidation consist in the quantification of the end products, malonic dialdehyde (MDA) [11]. In addition to CAT, SOD and GPx, another antioxidant enzyme, called peroxiredoxin (Prx), was a good marker of oxidative stress triggered by hyper-osmoregulation in the crab *Eurypanopeus depressus*. These class of ubiquitous enzymatic antioxidants are characterized by one or two conserved cysteines that reduce hydroperoxides in the presence of thiol. Several investigations revealed that Prxs are abundant in various tissues of crustaceans and their gene expression levels were often correlated to oxidative stress, caused by temperature stress and viral/bacterial infections.

In aquatic systems, hypoxia stress is often defined as dissolved oxygen levels below 2 mg/L. Variation in dissolved oxygen concentrations is able to affect the survival and fitness of aquatic organisms, including the reduction in growth and reproduction and can even be lethal in sensitive organisms [12].

Recently, some autophagy-related (ATG) protein-encoding genes were described as new potential molecular marker to evaluate the stress induced by hypoxia in the prawn *Macrobrachium nipponense*, thus suggesting that autophagy represents an adaptive response to hypoxia stress in prawns.

Alterations of water salinity and ionic concentrations, caused by climate changes and other stresses, may negatively influence the population genetic structure, physiology and overall biology of individual species. Osmoregulation is the principal process that maintains the ionic balance between the body fluid and the surrounding water column in case of aquatic salinity variations [*13*]. Marine organisms used various osmoregulatory strategies to prevent any stressful condition due to changes in external ionic content. Therefore, osmo-regulatory capacity (OC) was used as a nonspecific stress indicator in several aquatic animals exposed to various environmental stressors. The most important enzymes involved in ion uptake in crustaceans are the basolaterally located Na+/K+-ATPase (NKA) and the apically located V-H+-ATPase (VHA). The OC resulted a potent biomarker to evaluate the stress response to metal contamination, having heavy metals significant impact on the osmoregulation machinery in crustaceans. For instance, a significant over-expression of the salt stress-related genes, NKA and VHA, was found in recent studies evaluating the effects of cadmium and sulfide on the OC of the crustaceans *G. fossarum* and *Litopenaeus vannamei* [*14*].

Finally, as most of ecotoxicological studies, several works have studied the effects of environmental stressors, including xenobiotics, dietary supplementation, high temperature or salinity, by evaluating the effects on the growth, maturation, reproduction, survival and behaviors of the organisms [*15*].

Among invertebrates, a suitable environmental indicator in ecotoxicology is the freshwater crustacean *Daphnia magna* [*16*]. Very recently, some stress related genes mentioned before (CAT, GST and ABC), plus the cytochrome P450 (CYP450) enzyme family, have been found from the assembled *D. magna* genome, confirming this species a good model organism to deeper explore the molecular mechanisms induced by environmental stressors. Overall, acute and chronic tests to study the stress response in invertebrates was mostly performed in *D. magna* because this species is relatively easy to culture, possess a short lifecycle and the costs for maintenance in laboratory are very low. Nevertheless, *Daphnia* is not a suitable model for ecosystems with rising chloride levels, since a high sensitivity to high salt conditions was observed. A good alternative was represented by *Artemia franciscana*, able to live in environments with high salinity range and meeting all the criteria of a toxicity indicator for their abundance and detectable responses to environmental stress. Moreover, the *Artemia* transcriptome was recently published and used for high-throughput differential expression to find salt response stress genes in a wide salinity range. Another good model species is the crayfish *P. clarkii*, an invasive inveretbrate because of its high adaptability, tolerance and fecundity. This crustacean is used as a model organism in environmental stress and toxicity studies, particularly focused on evaluating the effects of metals and pesticides [*17*].

Regarding the methodologies applied for studying the stress response in crustaceans, several approaches were developed. For example, the enzyme-linked immunosorbent assay (ELISA) was used as a tool for the quantification of the

CHH neuropeptide in the hemolymph of crustaceans exposed to various acute environmental stresses [*18*]. On the other hand, to assess the genotoxicity due to oxidative stress, Comet assay is a suitable method used in marine invertebrates for the detection and quantification of DNA strand breaks, which are commonly caused by ROS increasing levels. Moreover, for the evaluation of ROS products (protein carbonyls and MDA), a simple and sensitive method is the spectrophotometric assay, which is particularly convenient since it requires no special or expensive equipments.

Overall, molecular approaches are mostly considered a suitable tool for understanding the physiological responses of marine organisms to environmental stresses, including *Real Time qPCR*, microarray, microRNA (miRNA) and transcriptomic approaches (Figure 1).

Changes in the transcriptome may be used to predict the whole response of the organisms, while gene expression detected by *Real Time qPCR* and microarrays can be applied to identify specific molecular biomarkers in response to stressful conditions [*19*].

The small non-coding RNAs (about 22 nucleotides), named miRNAs, negatively regulate gene expression by binding the 3'UTR region of the messenger RNAs (mRNAs). These RNA fragments regulate several biological processes including, embryonic development, cell differentiation, cell survival, cell cycle control, fat metabolism, immune responses and diseases. In crustaceans, some works explored the molecular mechanisms of miRNA in regulating the stress response to several disturbances.

In addition to transcriptomic analyses, genomics and proteomics, were also used as potent approaches in ecotoxicological studies [*20*]. In particular, they provide information concerning the toxic impact of contaminants on aquatic organisms by measuring the shifts in gene and protein expressions and their related functions, following the exposure to a chemical. Many environmental studies promoted this method to investigate the mechanism of action of known and unknown pollutants.

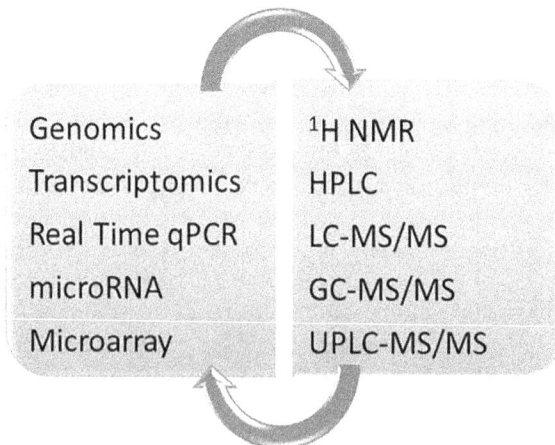

Variation in gene expression   **Change of endogenous metabolites**

Genomics	¹H NMR
Transcriptomics	HPLC
Real Time qPCR	LC-MS/MS
microRNA	GC-MS/MS
Microarray	UPLC-MS/MS

Figure 1. Techniques used to study variation in gene expression and changes of endogenous metabolites.

Finally, another "omic" approach, which is completely new in the investigation of the stress response of marine organisms exposed to various stressors, is metabolomics (Figure 1). Contrarily to older methods, metabolomics applies a shotgun approach for the identification of endogenous metabolites, within cells, tissues or organisms, in response to environmental disturbances [21]. Metabolomics focuses on the quantification of small molecules (less than 1000 Da), which participate in biochemical and metabolic reactions, offering precious information on the physiological state of organisms in any given time. In addition to its useful application for the establishment of the whole metabolome, this technique provides a powerful tool in discriminating the biochemical mode of action of organisms in response to a particular stressor. The influence of environmental stresses on physiology of some crustaceans was partially described through metabolomic methods. Since a stress status is firstly evident in the metabolome [22], metabolomics is extremely sensitive as indicator of stress response and it has gained a huge application in environmental sciences and in toxicity risk assessment. Furthermore, metabolomics gives an easier interpretation of biological data in comparison to standard genomics or proteomics approaches, because several metabolites possess similar chemical structures and functional roles across taxa. As mentioned before, some endpoints normally evaluated in ecotoxicological studies, such as mortality, reproductive dysfunction, aberrant growth and behaviors, are not sufficient to estimate the negative effects of stressors in trace levels. Conversely, metabolomic approaches display a high sensitivity, since they are able to evaluate the toxic effects of polycyclic aromatic hydrocarbons (PAH), polychlorinated biphenyls (PCB) and heavy metals at both low and sub-lethal concentrations.

The main platforms setting-up for the metabolite profiling are Nuclear Magnetic Resonance (^{1}H NMR) spectroscopy, high-performance liquid chromatography (HPLC), Gas Chromatography-Mass Spectrometry (GC-MS) and Liquid Chromatography-Mass Spectrometry (LC-MS). Among them, ^{1}H NMR is a high-throughput platform mostly used for the metabolome analysis for its highly reproducibility and easy preparation of samples. NMR data in combination with multivariate analyses, were considered efficient analytical tools to evaluate the metabolic profiles in both vegetal and animal samples, despite a poor sensitivity in case of metabolites with low abundance [23, 24]. This technique was also applied for the assessment of the stress response in crustaceans. For instance, ^{1}H NMR-based metabolomics was successfully applied on the freshwater crustacean *D. magna* in response to the stress of several environmental toxicants such as, PAH, heavy metals, flame-retardants and silver nitrate and coated silver nanoparticles. LC-MS approach, due to its high sensitivity, was applied to study the metabolome in several groups of organisms including yeast, plants, mice and aquatic invertebrates. In addition, the quadrupole time of flight mass spectrometry (Q-TOF) received increasing attention in metabolomic approaches for its high resolution and capability to provide an accurate mass and MS/MS information of the structure of unknown compounds very useful for biotechnological applications. When applied in combination to gas chromatography and liquid chromatography (GC-Q-TOF/MS or LC-Q-TOF/MS), the Q-TOF system allows a precise metabolic profiling and screening of biomarkers. Recently, a more sensitive platform, the ultrahigh-performance liquid chromatography-tandem mass spectrometry (UPLC-MS/MS) was applied to evaluate the metabolic variation in response to ammonia exposure in *Macrobrachium rosenbergii* [25].

Taking into consideration all the topics discussed above, in stress response studies, including those applied on crustaceans, the need to link the gene expression to the metabolite changes has become extremely evident for a better understanding of cellular biology [26]. Thus, multi-omics approaches that combine transcriptomics, proteomics and metabolomics data, give a global insight on the molecular changes triggered by a toxicant and a significant support for the screening of potential biomarkers in the environmental monitoring. In the following paragraphs, we report on the application of molecular biology and metabolomics approaches to study the stress response in the crustaceans.

5.2 Stress response in crustaceans by molecular biology approaches

Despite the lack of extensive genomic data, crustaceans were traditionally used in ecotoxicology studied by molecular approaches, reported as ecotoxicogenomic research. Crustaceans are exposed to several chemical and physical stressful agents in their natural environment.

5.2.1 Chemical stressors

Among chemical stressors studies on crustaceans concerned microplastics, sulfide and nitrite, heavy metals, nanoparticles endocrine disruptor, chemical products for therapeutic use, natural toxins and toxic diets, as reported below (see Table 1).

5.2.2 Heavy metals

Heavy metals represent the major cause of the aquatic environment, deriving from agricultural and industrial activities. Metals can enter the food chain and, as a result of bioaccumulation, cause serious health problems to humans. The sensitivity of crustaceans to heavy metals is well documented [27,28]; biotic factors, which seem to modify the sensitivity of crustaceans to heavy metals, include life stage, size, reproductive status, molting stage, and nutritional condition.

Copper, silver, zinc, lead manganese, arsenic and cadmium represent the most common heavy metals found in the marine environment [29]. It is important to consider that heavy metal-induced stress is closely associated with the induction of oxidative stress, detected in several crustaceans, including the freshwater crayfish *Procambarus clarkia* and prawns *Hyalella azteca*, *G. fossarum* and *D. magna* [30]. Many studies demonstrated that cadmium activated the genes involved in the metabolism of carbohydrates, lipids, proteins, transcription and translation processes, iron transport and genes coding for metallothioneins. A multi-generational exposure to sub-lethal cadmium concentrations induced tolerance in *Daphnia pulex* via lower mortality rates, which are in turn associated with decreased reproductive rates and size of organisms, suggesting that cadmium may affect calcium regulatory pathways reducing the size. No differences were observed between HO (hemolymph osmolality) mean values of controls and of exposed specimens in three populations.

Table 1. Crustacean species and effects induced by chemical stressors are reported.

Species	Heavy metals
Scylla serrata	Activation of genes involved in carbohydrates, lipids, proteins and transcription and translation processes
Palaemon elegans	Increased blood sugar
Palaemon elegans	Release of CHH
Ceriodaphnia dubia Daphnia magna Daphnia ambigua Daphnia pulex	Determine lethal effects concentrations
Daphnia pulex	Reduction of reproduction and size
Daphnia magna	Activation of genes involved in carbohydrates, lipids, proteins and transcription and translation processes
Calanus finmarchicus	Activation of antioxidant enzymes
Chironomus riparius Daphnia magna	Oxidative stress
Sinopotamon yangtsekiense	Activation of genes involved in carbohydrates, lipids, proteins and transcription and translation processes
Macrobrachium borellii	Activation of antioxidant enzymes
Cherax quadricarinatus	Activation of genes involved in carbohydrates, lipids, proteins and transcription and translation processes
Hyalella azteca	Oxidative stress
Procambarus clarkii	Oxidative stress
Sinopotamon henanense	Activation of genes involved in carbohydrates, lipids, proteins and transcription and translation processes
Daphnia pulex	Reduction of reproduction and size
Gammarus fossarum	Changes in hemolymphosmolality
Gammarus fossarum	Oxidative stress
Procambarus clarkii	Oxidative stress
Eriocheir sinensis	Weight reduction
Microplastics	
Hyas areneus	Increase of multixenibiotic proteins
Artemia franciscana	Changes in growth, mortality and swimming activity
Daphnia magna	Changes in growth, mortality and swimming activity
Daphnia magna	Changes in growth, mortality and swimming activity
Daphnia magna Gammarus fossarum Eucyclops serrulatus Niphargopsis casparyi Proasellus slavus	Changes in growth, mortality and swimming activity
Aeromonas hydrophila	Oxidative stress
Sulfide and Nitrite	
Litopenaeus vannamei	Oxidative stress
Litopenaeus vannamei	Oxidative stress
Litopenaeus vannamei	Changes in the synthesis of hemocyanin
Procambarus clarkii	Reduction of genera of intestinal microbite

Table 1 contd. ...

...Table 1 contd.

Species	Heavy metals
Endocrine Disruptors	
Homarus americanus	Ecdysis delay and increased expression levels of CYP45 and HSP70
Hyalella azteca *Daphnia magna*	All test organisms were dead within 24 h for the highest concentrations
Daphnia magna	Increased mortality
Procambarus clarkii	Increase in ROS and MDA
Daphnia magna	Damage to DNA
Pharmaceutical products	
Artemia parthenogenetica	Alterations on the cellular redox state
Daphnia magna	Antagonistic effect between fluoxetine and propranolol
Daphnia similis	Molting inhibition, delayed reproduction and reduced fertility
Nanoparticles	
Daphnia magna	Altered sensory development
Ceriodaphnia cornuta Moina micrura	Loss of antennae and armor
Natural toxins	
Mantide squilla *Crangon crangon Palaemon elegans* *Nephrops norvegicus Munida rugose* *Paguristes oculatu Pilumnus hirtellus* *Macropipus vernalis* *Parthenope massena* *Ilia nucleus*	Sugar increase
Litopenaeus vannamei	Reduction in growth performance, deterioration of health and hepatopancreas dysfunction
Eriocheir sinensis	Reduction in growth performance, deterioration of health and hepatopancreas dysfunction

However, high inter-individual variations of HO were observed in cadmium-exposed gammarids, which are translated by shifts in HO distribution between controls and exposed conditions. Surprisingly, *D. magna* responded differently from the other daphnid species when exposed to cadmium and zinc individually or in combination.

The mercury was the most toxic metal, followed by cadmium, copper and zinc. Its toxicity for *Palaemon elegans* caused a significant increase in blood sugar. In *P. elegans* serotonin (5-HT) induced a rapid and massive release of CHH (hyperglycemic hormone) from the eye into the hemolymph. In contrast, dopamine (DA) did not significantly affect CHH release. It has also been confirmed that the release of 5-HT, after copper exposure, preceded the release of CHH, underling its role as a neurotransmitter that acts on neuroendocrine cells [*31*].

5.2.3 Microplastics

Plastics become ubiquitous in our life, consistently increasing global plastic production over the years and currently stands at about 380 million tons being produced in 2015 (Plastics Europe https://www.plasticseurope.org). Due to increasing use and the durability of these artificial materials, plastics are becoming a global issue especially in marine environments. In fact, according to recent estimates, the magnitude of the plastic waste is at present 5800 million tons (Mt). The total accumulated mass of plastic waste in the world's oceans and land is expected to exceed 25 000 Mt by the year 2050. Plastic particles smaller than 5 millimetres are known as microplastics (MP), and, although practically invisible, are of particular concern [32]. The synthetic phenol 2,2 Bis (4-hydroxylphenyl)-propane (Bisphenol-A, BPA) is used as additive in the production of plastic materials (mostly polycarbonate), phenol and epoxy resins, polyesters and polyacrylates. It is also found in many products of daily life, such as polycarbonate bottles, thermal receipts, coatings of food tins and drinking water pipes, resin based dental sealing and epoxy glues. BPA is worldwide produced in huge amounts (e.g., 410.000 t/y in Germany) and occurs ubiquitous, even in dust and human urine samples. BPA is released into the environment through sewage treatment effluents, landfill leachates and natural degradation of polycarbonate plastics. Many studies reported on the negative influence of microplastics on different crustaceans, such as *D. magna*, *A. franciscana*, *G. fossarum*, *Eucyclops serrulatus*, *Niphargopsis casparyi* and *Proasellus slavus*, inducing mortality and altering growth and swimming capacity [33]. Studies on crab *Hyas areneus* demonstrated that 2,20,4,40-tetra bromo diphenyl ether (BPDE), BPA and diallyl phthalte (DPA) induced an increase in expression levels of multixenibiotic proteins (MXR). To study the effect on the immune system of crustaceans, *P. clarkii* samples were utilized to detect the immune related indicators after 1 week of exposure to 225 µg/L BPA. A significant increase of ROS level was observed together with the inhibition of antioxidant-related enzymes (SOD, POD, and CAT), thereby causing the oxidative stress. The enzyme activities of alkaline phosphatase (AKP), acid phosphatase (ACP) and lysozyme in hepatopancreas after BPA exposure were also depressed even after *Aeromonas hydrophila* infections. The relative expression profiles of immune-related genes after BPA exposure and bacterial infection showed suppressed trends of most selected genes with adverse effects on *P. clarkii* immune ability.

5.2.4 Sulfide and nitrite

Sulfide is generated from the anaerobic decomposition of the organic wastes in the bottom layer and sediments of shrimp pond [34]. Sulfide is usually included in water soluble hydrogen sulfide (H_2S), bisulfide anion (HS^-) and sulfide anion (S^{2-}). A decisive role in sulfide stress was played by H_2S, which is able to inhibit the electron transport chain of cytochrome oxidase, inactivate glutathione through combining with the thiol of glutathione, thus affecting the biological oxidation process, blocking the cell respiration and inducing hypoxia *in vivo*. Although sulfide concentration in natural water environment is typically low, low dose exposure can cause the immune depression and increase the pathogen susceptibility of the shrimp. Nitrite is widely present as a common toxic substance in aquatic systems and is not only a toxic intermediate produced during ammonia nitrification but also a product of denitrification of nitrate by bacteria during nitrogen cycling. Nitrite concentrations in coastal seawater are approximately 10–15 nM (0.14–0.21 NO_2–N µg/L); however, in middle- or late-stage cultures, the nitrite concentration can reach 1.43 mM (20 NO^2–N mg/L) and seriously affect the health of farmed animals. There is evidence that high concentrations of nitrite and sulfide have inhibitory effects on the immune system of aquatic animals. Nitrite- and sulfide-related stress leads to the excessive production and accumulation of ROS, which negatively influences the growth, survival and physiological activities of the organisms, such as the osmoregulation, respiratory metabolic and immune capacity and antioxidant capacity [35]. Studies on the microbial composition of the gut of *P. clarkii* and on the levels of hemocyanin and energy metabolism of the shrimp *L. vannamei* showed that sulphide and nitrite exposure led to a reduction of genera forming the intestinal microbiota. On the contrary, the exposure to nitrite alone can improve the accumulation of nitrite in hemolymphine and therefore reduce oxygenation and hemocyanin synthesis, leading to tissue hypoxia, accelerating anaerobic metabolism and inhibiting aerobic metabolism. The effects of nitrite stress on hemocyanin synthesis and energy metabolism may be one of the reasons for the mortality of *L. vannamei* in culture systems.

5.2.5 Endocrine disruptors

Endocrine disruptor chemicals (EDCs) are compounds that mimic natural hormones, inhibiting their activity or altering their normal regulatory function within the immune, nervous, and endocrine systems. These chemicals are of ecotoxicological significance due to their tendency to be absorbed onto humic material or into aquatic organisms, to accumulate, and to persist in water or food web for a long time. Thus, their effects can cause a long-term stress in aquatic organisms. To date, several EDCs, such as pesticides, bisphenol A, phathalates, dioxins, and phytoestrogens, can interact with the female reproductive system and lead to endocrine disruption [36]. Endosulfan and deltamethrin are commonly used pesticides in shrimp farms. Moreover, endosulfan is often used as a broad-spectrum insecticide mainly in agriculture. The most toxic effect was reported when these species were subjected to a combination of α+β-endosulfan+endosulfan sulfate products. Insecticide is considered to be highly toxic to aquatic organisms. EDC generate in the exposed crustaceans, *Homarus americanus*, *D. magna* and *H. azteca*, different responses including ecdysis delay, increased expression levels of CYP45 and HSP70, DNA damage, increased mortality, oxidative stress (MDA).

5.2.6 Pharmaceutical products

Thanks to the advances in environmental analysis technology in recent years, a new class of environmental pollutants, represented by pharmaceuticals and personal care products has begun to receive widespread attention. In the environment most of these chemical compounds have low concentrations, complex structures, and difficult degradation and accumulation characteristics [37]. Although the concentrations are low in the environment, long-term pollution of PPCPs may cause endocrine disruption or reproductive toxicity to aquatic organisms, induce changes in biochemical functions of aquatic habitats, and do great harm to the environment. Besides biological activity, these compounds are also resistant to metabolic degradation, and, usually, their lipophylicity is a basic requirement to be well absorbed by the organism. These factors contribute for the overall persistence in the environment. Other aspect to take into account when studying the environmental fate and effects of pharmaceuticals and personal care products is related to the variety of metabolites formed and to possible toxicological interactions (e.g., synergistic, additive, antagonistic effects) among pharmaceutical residues in the wild. Therefore, a considerable part of pharmaceuticals and personal care products may be considered as active, effective and persistent environmentally unfriendly compounds. Furthermore, high contamination values are reported for several classes of therapeutic agents, which can consequently lead to acute effects over organisms. Many studies evaluated acute and chronic toxicity in various cruastaceans, *Artemia parthenogenetica*, *D. magna* and *Daphnia similis*, exposed to therapeutic agents such as diazepam (anxiolytic), sodium dodecyl sulphate (detergent), fluoxetine (antidepressant), propranolol (antihypertensive) and carbamazepine (CBZ, anticonvulsant). Diazepam and dodecyl sulphate induced alterations on the cellular redox state in *A. parthenogenetica* [*38*]. In addition, diazepam had the ability to interfere with neurotransmission through inhibition of ChE (soluble cholinesterases). CBZ can cause molting inhibition, delayed reproduction and reduced fertility in crustaceans.

5.2.7 *Nanoparticles*

The development and commercialization of nanomaterials is proceeding rapidly, which revolutionize several industrial sectors, including building materials, health and medicine, electronics, and green technologies (Nanotechnology White Paper; EPA 100/B-07/001 February 2007, www.epa.gov/osa). Nanoparticles (NPs) are defined as supramolecular compounds having at least one dimension < 100 nm. Although they are composed of materials well studied in toxicology, their small size often alters their chemical and physical properties and may result in unexpected toxicity (Nanotechnology White Paper; EPA 100/B-07/001 February 2007, www.epa.gov/osa). NPs constitute a different range of compounds, which poses unique challenges to ecotoxicologists who must assess the risks posed by each of these particles and understand how chemical alterations affect this risk. NPs are expected to make their way into the aquatic environment where water chemistry influence their fate and transport. Metal oxide nanoparticles (NPs) (e.g., ZnO, TiO2, AgNPs) are the most widely commercialized nanomaterials. Because of their antimicrobial properties, they are found in medical devices, and a host of common household products, such as clothing, hygiene products, disinfectants, and washing machines [*39*]. The rapid growth of silver nanotechnology is particularly concerning because of the high potency of silver to aquatic organisms. Main cellular processes such as protein metabolism, signal transduction and in particular sensory development were altered by AgNPs. Zinc nanoparticles (NP ZnO) led to the loss of antennae and carapace in freshwater crustaceans *Ceriodaphnia cornuta* and *Moina micrura*. This study confirms the protective effect of chitosan against the dangerous effect of NP ZnO significantly reducing mortality and improving survival.

5.2.8 Natural toxins

Toxins are molecules produced by an animal, plant or microbial organism, which is harmful to some species. There are several types of toxins such as the lipopolysaccharide (LPS) endotoxin induced in mammals septic shock and the activation by LPS of hormone release through the hypothalamo–pituitary axis is well known. In crustaceans, an increase in circulating crustacean hyperglycemic hormone and hyperglycemia are reported after exposure to several environmental stressors, but the metabolic and hormonal effects of LPS *in vivo* are still unknown. T-2 toxin is a trichoticenic mycotoxin highly toxic for aquatic animals, but little is known about its toxicity effect in crustaceans. Aflatoxins are a family of toxins produced by the fungi *Aspergillus flavus* or *Aspergillus parasiticus*. They can cause serious health problems in shrimp and therefore reduced the yield and profitability of shrimp crops. It has been shown that the injection of a sub-lethal dose of LPS into different shellfish species (*Mantide squilla, Crangon crangon, P. elegans, N. norvegica, M. rugosa, P. oculatus, P. hirtellus, M. vernalis, P. massena* and *Ilia nucleus*) induced a 2- to 15-fold increase in blood sugar compared to controls (treated with saline). The injection of LPS introduced into the eyepiece of eyeless animals did not elicit a significant hyperglycemic response. Several studies reported the effects of T-2 toxin and aflatoxin 1 on crabs *Eriocheir sinensis* and shrimps *L. vannamei*, detecting a reduction in their growth performance, deterioration of health and hepatopancreas dysfunction [*40*]. Particularly, for shrimps an up-regulation of genes *HKII* (*hexokinase type 2*) and *FASN* (*fatty acid synthase*) was observed.

5.3 Physical stressors

Among physical stressors studies on crustaceans concerned temperature, salinity, anoxia, hypoxia, osmosis, manual declawing, as reported below (see Table 2).

5.3.1 Temperature and salinity

Stress is a state where organismal homeostasis is threatened or interrupted by intrinsic and/or extrinsic stressors. Habitat salinity is one of the most important abiotic factors influencing not only the distribution and abundance of crustaceans, but also their general physiology and wellbeing. Salinity changes cause salt stress because they can interfere with physiological homeostasis and routine biological processes. At an ecological level, superior salt stress tolerance is an evolutionary advantage in areas with fluctuating salinity, a condition predicted to increase due to climate change. At an economic level, global crustacean aquaculture is a growing multi-million-dollar industry [*41*]. Abiotic stress reduces crustacean immunity, increasing vulnerability to diseases. Most aquatic animals tolerate salt stress to a certain degree (low for stenohaline, high for euryhaline species). Most species are stenohaline and restricted to stable marine (30–40 g/l) or freshwater (< 0.5 g/l) habitats. Under

Table 2. Crustacean species and effects induced by physical stressors are reported.

Species	Physical stressors
Temperature/salinity	
Homarus americans	Changes in the expression levels of *HSP70* and *CHH*
Palaemonetes varians	Changes in the expression levels of *HSP70* and *CHH*
Artemia franciscana	Changes in the expression levels of *HSP70* and *CHH*
Callinectes sapidus	Increased glucose and lactate
Carcinus maenas	Increased glucose and lactate
Litopenaeus vannamei	Oxidative stress
Anoxia/Hypoxia	
Fenneropenaeus chinensis	Altered levels of expression of detoxification enzymes
Niphargus rhenorhodanensis	Increase in TBARS and SOD
Macrobrachium nipponense	Altered levels of expression of detoxification enzymes
Macrobrachium nipponense	Metabolic changes
Macrobrachium nipponense	Autophagy mechanism
Osmosis	
Eurypanopeus depressus	Up-regulation of *EdPrx-1*

salt stress, euryhaline aquatic animals use two main strategies: (i) osmoconformity, where the animal's body fluids are isosmotic to its external environment, and salt toxicity is avoided by internal conversion and concentration of organic osmolytes; (ii) osmoregulation, where the animal's body fluids are not isosmotic its environment, but regulated by active ion transport via specialized organs. Osmoregulation is divided in hyper-osmoregulation (keeping body fluids hyper-osmotical to the environment, in fresh water) and hypo-osmoregulation (keeping body fluids hypo-osmotical to the environment, in salt water). Most crustaceans adapt themselves to salt stress by osmoregulation using primarily the gills, but also the maxillary or antennal glands (mainly urine production), and the gut. Hydrostatic pressure and temperature are two key stressors, which may act as physiological limits on the ability of shallow-water organisms to tolerate deep-sea conditions. As thermodynamic variables, they affect biochemical equilibria and rates of biological processes in similar ways. Low temperature and high pressure both result in reduced fluidity of bio-membranes, principally caused by an increased or tightened packing of fatty acyl chains. This tightly packed, and highly ordered, configuration restricts molecular motion. Low temperatures and high pressures both decreased membrane fluidity in shallow-water adapted organisms. Additionally, observed reductions in oxygen consumption of organisms subjected to acute pressure exposures may reflect a drop in metabolism because of compromised membrane and enzyme functionality. Conversely, high temperature may act to ameliorate or reduce the effects of high pressure by the same thermodynamic principles. Consequently, physiological limits to pressure and temperature tolerance may be closely entwined and affected by shifts in either factor. An increase in ocean surface water temperature, and subsequent deep-water temperature increase [*42*], may actually facilitate bathymetric range shifts of shallow-water organisms by ameliorating any physiological limitations imposed by increasing pressure. Both hyposalinity/hypersalinity and different temperature conditions led to significant alterations in the expression levels of *HSP70* and *CHH* in crustaceans *Homarus americans*, *Palaemonetes varians* and *Artemia franciscana*. *Callinectes sapidus* crabs, exposed to hyperthermic conditions and hypoxia, showed an increase in glucose and lactate just like *Carcinus maenas* crabs subjected to electric shock. Low temperatures induced oxidative stress, DNA damage, lipid peroxidation and changes in osmolality in Pacific white shrimp *L. vannamei*.

5.3.2 Anoxia/hypoxia

Dissolved oxygen (DO) increases during the daytime due to the photosynthesis of phytoplankton and macrophytes, whereas decreases due to respiration of the organisms and decomposition of accumulated organic matter of unconsumed feed and feces during night time at the bottom layer of pond waters where shrimp spend most of their time. This hypoxic condition can certainly cause significant stress to shrimp. It was reported that hypoxia stunted normal growth of shrimp, including reduced molting frequency and retarded growth. The slow growth rate caused by hypoxic stress leads to a decrease of shrimp production directly. Hypoxic conditions cause stress, inhibiting the optimal development of crustaceans and resulting in lower frequency of molts, metabolic changes, avoidance behavior, slow growth, suppression of immune function and even death [43, 44]. Several studies reported on the molecular response to hypoxia and chronic hypoxia on chinese shrimp *Fenneropenaeus chinensis* and *M. nipponense* by altering the expression of many genes, such as *fatty acid binding protein 10*, *carbonyl reductase 1*, *aldehyde reductase 1*, *oncoprotein nm23*, *arginine kinase*, *phosphopyruvate hydratase*, *formylglutathione hydrolase* and *cytosolic manganese superoxide dismutase*, *cytochrome c oxidase subunit I*, *cytochrome oxidase I*, *NADH dehydrogenase subunit 1* and *carbonic anhydrase I*. An increase in thiobarbituric acid reactive substances during recovery from severe hypoxia and hyperactivation of superoxide dismutase after an anoxic stress of 24 h in the crustacean *Niphargus rhenorhodanensis* was detected. Autophagy is a crucial process for the maintenance of cell homeostasis, is under the control of numerous autophagyrates and is highly conserved in most animals and appears to play an important role in the adaptive response against hypoxia toxicity in crustaceans as *M. nipponense*. In particular, liver levels of ATG8 (autophagy-related proteins) can be directly indicative of acute hypoxia in shrimp and provide information on the time at which exposure to hypoxia occurs.

5.3.3 Osmosis

Osmotic stress is the most common environmental stress factor for aquatic organisms. Osmoregulation plays an important role being one of the most important regulatory functions in aquatic organisms to maintain osmotic homeostasis. Several studies also reported the influence of stressors as temperature or salinity on organisms by altering the osmoregulation capability via elevated levels of Na+-K+ ATPase activity or heat shock protein [45] to sustain relative osmotic haemolymph homeostasis. Prxs are a family of ubiquitous proteins that help in minimizing the harmful effects of oxidative stress by catalyzing the reduction of hydrogen peroxide and organic hydroperoxides to less harmful forms. EdPrx-1 gene is expressed at low level in the gill, hypodermis and hepatopancreas tissues of mud crab *E. depressus* in non-stressful conditions. However, its expression is approximately three times higher in the gills under hypo-osmotic stress. This suggests a possible role in the protection from oxidative stress by the metabolic activities associated with hyperosmoregulation.

5.4 Stress response in crustaceans by metabolomic approach

As reported above, metabolomics is a systematic study of the endogenous, small-molecule metabolites involved in specific biological processes, providing an assessment of the physiological status of an organism. Metabolomic analysis is currently applied in medicine, toxicology and environmental sciences. The development of metabolomics as an environmental research tool holds great promise for contributions to environmental risk analysis, discovery of new biological insights, and for developing environmental system models (see Table 3).

One of the first metabolomics studies applied to crustaceans dates back to 2010, in which NMR-based approach was applied to assess the response of the Atlantic blue crab *Callinectes sapidus* to different oxidative stressor agents. This crab is of a great interest from economical and ecological point of views, considering that its population is currently decreasing because of fishing, environmental pollutants and warming waters [*46*]. In fact, in the natural environment, crabs are constantly exposed to anthropogenic stressors and natural stressors, such as variations in salinity, temperature, oxygen and viral/bacterial infections. In addition, experimental approach devoted to evaluate the health of these crustaceans are important because reflect in turn the health of coastal ecosystem and may be considered as indicators of environmental pollution. NMR spectroscopy was used to compare the response of *C. sapidus* metabolome to decrease of aerobic metabolism through the injection of the bacterium *Vibrio campbellii*, versus the increase of aerobic metabolism after treatment with 2,4-dinitrophenol (DNP), a known uncoupler of oxidative phosphorylation. Changes in metabolic profile were detected in crab hemolymph after 30 min of injection with *V. campbellii* and DNP. The largest variations in the metabolomes concerned glucose, considered a good indicator in crustaceans for biological stress, and lactate, a metabolite involved in anaerobic respiration. This analysis provided insight into the biochemical pathways involved in crustacean hemolymph, which represents an unexplored biochemical resource for toxicological studies. Metabolic and molecular approaches were combiend to study responses to cadmium of *Daphnia magna*, important model organisms in both ecology and toxicology due to their wide geographic distribution, able to adapt to different habitats and high sensitivity to anthropogenic chemicals. Chronic toxicity in *D. magna* was detected after 24-hours sublethal cadmium exposure (18 μg/L, corresponding to 1/10 LC50). Metabolites were detected and identified in small volumes by resonance mass spectrometry and NMR spectroscopy. Moreover, by mass spectrometry based metabolomics of hemolymph disruption to two major classes of metabolites, amino acids and fatty acids, were detected. This data helped in understanding of how cadmium can disrupt nutrient uptake and metabolism, resulting in decreased energy reserves and chronic

Table 3. Crustacean species and effects induced by different stressors are reported.

Species	Stress response
Litopenaeus vannamei	Changes in glucose, lactate, norepinephrine and dopamine levels
Callinectes sapidus	Metabolic changes in hemolymph
Daphnia magna	Metabolic changes in hemolymph
Diporeia spp.	Down-regulation of polyunsaturated fatty acids, phospholipids, and amino acids and their derivatives
Procambarus clarkii	Oxidative stress, metabolic dysfunction, and dyslipidemia
Daphnia magna	Identification of 18 different metabolites between control and treated animals
Daphnia magna	Changes in glucose, lactate and amino acids levels
Callinectes sapidus, Eriphia verrucosa and *Cancer pagurus*	Changes in fatty acids, lactate amino acids levels
Daphnia magna	Neonates and adults respond uniquely to sub-lethal contaminant exposure
Eriocheir sinensis	Changes in glucose, lipids and amino acids levels
Portunus pelagicus and *Scylla tranquebarica*	Metabolites with anti-inflammatory, antioxidant and antibacteria activity
Litopenaeus vannamei	Changes in gluconeogenesis, protein synthesis and energy metabolism
Daphnia similis	Changes in sugars, fatty acids and amino acids levels
Procambarus clarkii	Changes in amino acids levels
Litopenaeus vannamei	Synergistic effect on enhancing immunity and disease resistance
Euphausia superba	Changes in amino acids and sugars levels
Litopenaeus vannamei	Changes in glucose and fatty acids levels
Macrobrachium rosenbergii	Changes in purine metabolism, amino sugar and glutathione metabolism, and phosphonate and phosphate metabolism, and on the terpenoid biosynthesis
Macrobrachium nipponense	Changes in glycerophospholipid metabolism and sphingolipid metabolism

toxicity also giving insights on the use of hemolymph, which as other biofluids are intimately linked to the cellular function of the organs.

Another study reported on the application of NMR-based metabolomics on energetic impairments of *D. magna* exposed for 48 hours to increasing sub-lethal concentrations of two organophosphates, diazinon or malathion, or bisphenol-A (BPA). Principal component analysis (PCA) showed aberrant metabolomic profiles after diazinon exposure at all exposure concentrations tested (0.009–0.135 µg/L), otherwise malathion at the two highest exposure concentrations tested (0.32 µg/L and 0.47 µg/L) caused significant shifts from the control. These metabolic changes induced by both organophosphates indicated that the response was not exposure-dependent but exposure severity-dependent. For example, at intermediate concentrations of diazinon (0.045 µg g/L and 0.09 µg g/L) and malathion (0.08 µg g/L) a decrease in amino acids, such as leucine, valine, arginine, glycine, lysine, glutamate, glutamine, pheny-lalanine and tyrosine, was observed together with the increase in glucose and lactate, so suggesting a mobilization of energy resources to counteract the stress. At the highest exposure concentrations for both organophosphates metabolic activity was stopped, with the increasing of the same amino acids and decreasing of glucose and lactate, suggesting a slowdown in protein synthesis and depletion of energy stocks. These data suggested a similar response in the metabolome between two organophosphates with changes in the metabolome at intermediate and severe stress levels. In the case of BPA exposure, a significant change in metabolome was detected at 0.1 mg/L, 1.4 mg/L and 2.1 mg/L. Starting from 0.7 to 2.1 mg/L of BPA exposure aminoacids such as alanine, valine, isoleucine, leucine, arginine, phenylalanine and tyrosine increased. These metabolite changes were correlated with decreases in glucose and lactate. This response was similar at the highest organophosphate exposure, leading to the hypothesis that in *D. magna* stress response could be linked to altered energy dynamics.

Until the 1990's the holarctic amphipod *Diporeia* spp. represented the most abundant benthic macroinvertebrate in the offshore region of the Laurentian Great Lakes basin. In subsequent years the presence of *Diporeia* precipitously declined, due to the competition for food with mussels. At this point it became necessary to have more information on how *Diporeia* responded to starvation. Liquid chromatography (LC) coupled with time-of-flight mass spectrometry (TOFMS) was applied to study its metabolite profiles during starvation. For this purpose, *Diporeia* samples were collected from Lake Michigan and starved for up to 60 days in laboratory, determining metabolite levels at 12-day intervals and comparing them with data at the day 0. A down-regulation of some metabolites was observed, including polyunsaturated fatty acids, phospholipids, and amino acids and their derivatives. Overall, starved organisms predominantly activated glycerophospolipid metabolism and protein-based catabolism for energy production. These findings clearly demonstrated that LC-MS based metabolomics was useful to assess physiological status, showing that these freshwater amphipods can use unique metabolite profiles in starvation condition and underlying the cause of Diporeia's decline.

The crab *P. clarkii* was proposed to monitor the contamination in Doñana National Park (southwest Spain) using conventional biomarkers [47]. A metabolomic approach based on direct infusion mass spectrometry, allowing an easy and very fast study of metabolites in a single run, associated to metal accumulation in tissues were used for pollution assessment of this area and the response of this crab to contamination. Several metabolites changed in response to pollution: carnosine, alanine, niacinamide, acetoacetate, pantothenic acid, ascorbate, glucose-6-phosphate, arginine, glucose, lactate, phospholipids and tryglicerides decreased their levels; acetyl carnitine, phosphocholine, choline and uric acid increased. These metabolites were considered as potential biomarkers of pollution.

Comparative metabolic studies were performed on the blue crab *C. sapidus*, captured in the Acquatina lagoon (Italy), an autochthonous *Eriphia verrucosa* and to a commercial crab species *Cancer pagurus*, aiming to support the commercial exploitation and the integration of the blue crab in human diet of European countries as an healthy and valuable seafood. Both lipid and aqueous extracts of raw claw muscle were analyzed by ^{1}H NMR spectroscopy and MVA (multivariate data analysis). In details, higher levels of glutamate, alanine and glycine were found in the aqueous extract of *C. sapidus*, while homarine, lactate, betaine and taurine in those from *E. verrucosa* and *C. pagurus*. Moreover, signals of only monounsaturated fatty acids signals were distinguished in the lipid profiles of the three crab species.

Chinese mitten-handed crab, *E. sinensis*, is one of the most important economic crustacean species cultivated in China [*48*]. Recently, the requirements of key nutrients including protein and lipid have been evaluated on this species. The research consensus is that Chinese mitten-handed crab requires high dietary protein to maintain rapid growth. However, excess dietary protein level not only increases dietary cost but also causes high nitrogen excretion to the environment. A number of studies have demonstrated that appropriate use of dietary lipid would efficiently provide physiological energy to spare dietary protein in animal nutrition.

Nutritional values of animal diets is strictly dependent on the type of oil in the feed ingredient, but the underlying metabolic mechanisms of dietary oil in animal feed were not clarified in aquatic animals, mainly in crustaceans. The first metabolomics study to identify the key pathways and crucial metabolites as biomarkers to differentiate the metabolic mechanisms of crustaceans fed contrasting dietary oils was GC–MS-based and very recently published. In this study metabolomics and nutritional parameters was recently reported in order to investigate the possible metabolic mechanisms between juvenile Chinese mitten crabs *E. sinensis* fed olive oil containing 69% oleic acid (OA) and perilla oil containing 56% linolenic acid (LNA). Lower concentrations of hepatic glycogen, triglycerides and peroxidation products were found in crab fed OA respect those fed LNA, displaying the first ones faster growing rate. In particular, six metabolites related to glycolysis and TCA (tricarboxylic acid) cycle (pyruvate, succinic acid, lactose, malic acid, D-glyceric acid and threitol), methionine, 2-keto-isovaleric acid (intermediate for valine and leucine synthesis) and 2-hydroxybutanoic acid (intermediate for glutathione synthesis) were higher in the OA group than in the LNA group. Only glutaconic acid, which is the intermediate of ketogenic amino acids, was higher in the LNA group. It was possible to conclude that crab in the OA group increased degradation of glucose and lipids to provide energy for growth respect to LNA group.

Astacus leptodactylus is a freshwater decapod crustacean, a widespread species distributed throughout Europe, eastern Russia and the Middle East and actively commercialized in Europe, where crayfish are considered as luxury food.

^{1}H NMR analysis of the hepatopancreas, muscle and haemolymph of this decapod after feeding with polyphenol-enriched diet reported on changes in metabolome, demonstrating that it is useful tool to study the metabolomics in relation to diets. Lipophilic extracts spectra showed the presence of cholesterol, fatty acid residues, phospholipids and triglycerides, whereas that of aqueous extracts identified 35 metabolites in the hepatopancreas, 31 in the muscle and 22 in the haemolymph. Furthermore, (i) a total of 20 metabolites including amino acids and their derivatives were present in the hepatopancreas, the muscle and the haemolymph; (ii) a total of 10 metabolites were present in both the hepatopancreas and the muscle, including five amino acids, 2-hydroxybutyrate, choline, myo-inositol, glycogen and uracil. 2-Hydroxyisobutyrate and creatine were present in both the hepatopancreas and the haemolymph. Phosphorylethanolamine, phosphocholine and fumarate were present only in the hepatopancreas and isoleucine only in the muscle.

Sensitivity of metabolites to changes in food availability and dissolved oxygen concentrations were measured in the red swamp crayfish *P. clarkii* by magnetic resonance-based metabolomics [49]. Samples were exposed to one of three food availability or high, normal, low levels of dissolved oxygen treatment. Starved crayfish showed lower amounts of amino acids than fed animals, suggesting catabolic effects of starvation on tail muscle tissue for energy requirements. In contrast, crayfish exposed to hypoxic conditions experienced changes in abundance of metabolites primarily associated with energy metabolism.

Stress response in crustaceans is mediated by neuroendocrine mechanisms, which are very different and less explored than those of vertebrates, even if indirect evidence suggests catecholamines (CA) involvement but their levels are not measured. The hemolymph and tissue levels of catecholamines were detected in response to handling stress in white leg shrimp *Litopenaeus vannamei*, representing the most important shrimp for high-density farming in ponds worldwide. Shrimps were stressed by transfer, chasing, and confinement and their responses were analyzed at different intervals of time (10–240 minutes). In hemolymph lactate levels increased from 10 to 30 minutes, glucose increased at 60 minutes, both returning to basal levels at 240 minutes. These data suggested a possible activation of gluconeogenesis due to the increasing lactate levels. In hepatopancreas was observed an increased level of epinephrine from 10 to 20 minutes, indicating its involvement in lactate uptake, glucogenogenesis and lipid catabolism in this organ. On the other hand, norepinephrine levels decreased by 92% between 20 and 30 minutes in the eyestalks, suggesting that its release in the X-organ sinus gland complex mediate lactate effects in this organ. Furthermore, *L. vannamei* is inevitably affected by exposure to sulfide released from pond sediment, considering that starts its benthic life post-larvae. Despite its importance in aquaculture farming the knowledge on the toxic mechanism of sulfide in *L. vannamei* is poor [50]. The toxicity and mechanisms of sulfide in *L. vannamei* was studied by energy metabolism and metabolomics approach at 24, 48, 72 and 96 hours at the concentrations of 425.5 and 851 µg/L. This study indicated that chronic exposure of shrimp to sulfide can decrease health and lower survival through functional changes in gluconeogenesis, protein synthesis and energy metabolism. In fact, shrimp survival decreased at the higher sulfide concentration of 851 µg/L, not affecting weight gain or the hepatopancreas index. Metabolomics assays showed that shrimp exposed to sulfide at the concentration of 425.5 µg/L had lower amounts of serum pyruvic acid, succinic acid, glycine, alanine, and proline respect the control; whereas when exposed to 851 µg/L lower amount were detected for phosphate, succinic acid, beta-alanine, serine and l-histidine. Moreover, chronic sulfide exposure affected protein synthesis in shrimp but enhancing gluconeogenesis and substrate absorption for ATP synthesis and tricarboxylic acid cycles in order to supply extra energy to counteract sulfide stress. High levels of serum n-ethylmaleamic acid, pyroglutamicacid, aspartic acid and phenylalanine were also detected, indicating that chronic sulfide exposure adversely affected the health status of *L. vannamei*.

The white shrimps *L. vannamei* were also used to study how a synbiotic improved their immunity. Shrimps were fed four experimental diets for 60 days, including a basal diet with no galactooligosaccharide or probiotic (control), 0.4% galactooligosaccharide, probiotic, and 0.4% galactooligosaccharide in combination with the probiotic. Results showed that the galactooligosaccharide in combination with the probiotic diet significantly increased survival of *L. vannamei* after 24 hours after *Vibrio alginolyticus* injection. ^{1}H NMR analyses revealed that twenty-two hepatopancreas metabolites were matched and identified between the galactooligosaccharide in combination with the probiotic diet and control groups, among which three metabolites, inosine monophosphate, valine, and betaine, significantly increased. These data were confirmed using RP-HPLC and spectrophotometric methods showed that inosine monophosphate presented high amounts in the hepatopancreas, but not in the plasma of shrimp. In contrast, valine and betaine metabolites were in high concentrations in both the hepatopancreas and plasma. Al, together these results suggested that this diet had a synergistic effect on enhancing immunity and disease resistance of *L. vannamei* against *V. alginolyticus* infection, thanks to the syntheses of a nucleotide (inosine monophosphate), a branched amino acid (valine), and a methyl group donor (betaine) in the hepatopancreas. These metabolites were in turn released into the plasma and directly taken up by hemocytes, resulting in a triggering of melanization and phagocytosis processes in cells.

The effects of dietary fatty acids on shrimp *L. vannamei* was tested in low-salinity culture [51]. Individuals were fed with three diets (coconut oil, fish oil, or an equal mixture of both) of differing fatty acid profiles for eight weeks, being maintained at two different salinities (3 or 30 practical salinity unit, psu). Then, a GC-TOF/MS-based metabolomics analysis was performed to reveal the regulatory roles of fatty acids in shrimp growth, survival and osmoregulation at two salinities. The shrimp fed with the mixed oil diet had higher weight and survival, whereas those fed on the diet with only fish oil added showed higher unsaturated fatty acid levels in the gills and hepatopancreas independent from the salinity level. When coconut oil was the only lipid source, shrimp tissues contained higher tyrosine, lysine and serine levels at 3 psu than at 30 psu. In contrast, the shrimp fed fish oil as the sole lipid source had higher glucose at lower salinity. This study indicates that appropriate supplementation of dietary unsaturated and saturated fatty acids can improve shrimp osmoregulation capacity, promoting their growth and survival in low-salinity water.

Crabs are also considered good candidates as sources of bioactive compounds that revolutionized treatment of serious diseases, including antibacterial, antifungal and antiviral metabolites isolated from various tissues and organs. The first attempt to investigate and compare the natural antibacterial properties from whole extract of marine blue swimmer crab was represented by the studies done on *Portunus pelagicus* and mud crab *Scylla tranquebarica* against fish pathogenic bacteria [52]. Liquid chromatography/mass spectrometry was used to characterize the variation in secondary metabolite production in both these crabs. Different metabolites are evaluated in both crab species using LC/MS-QTOF, containing anti-inflammatory and antibacterial properties down-regulated in *S. tranquebarica* and up-regulated in *P. pelagicus* samples. Furthermore, methanolic *P. pelagicus* extracts showed the best antimicrobial response against Gram positive bacteria *Streptococcus agalactiae* and Gram-negative bacteria *Vibrio alginolyticus*, *Klebsiella pneumoniae* and *Escherichia coli*, respect to *S. tranquebarica* extract. These findings assumed a great significance in pharmaceutical field, suggesting crab metabolites as potential antibacterial drugs. On this line, it is important to consider that these animals are usually exposed in their natural environment to various pathogens, thus forced to develop strategies to defend themselves from pathogenic.

Poor diet quality and nutritional state represent a strong constrains for animal growth and reproduction, able to alter their population dynamics. In fact, the identification and measurement of nutritional state represent a key step in studying the connections between diet and animal ecology. The nutritional state of *D. magna* was determined by analyzing its endogenous metabolites by metabolomic approach [*53*]. In particular, the metabolite composition of these crustaceans was measured under control conditions (indicating normal food without environmental stress), under different diets (at low quantity, nitrogen and phosphorus limited), and in the presence of two common environmental stressors (bacterial infection and salt stress). Eighteen metabolites were found significantly different between the control condition and with one limiting food type or environmental stressor, indicating that metabolite composition can be promising indicators of diet-induced nutritional stress.

The pharmaceuticals represent contaminants for the sea water, altering the aquatic ecosystem health. ¹H NMR approach was also applied to *D. magna* to study the metabolic responses of these sub-lethal contaminant exposure. Daphnids samples aged day 0 and 18 were exposed to 28% of the lethal concentration of 50% of organisms tested of atrazine, propranolol and perfluorooctanesulfonic acid (PFOS) for 48 hours. Significant separation of contaminants from the control was found in both neonates and adults exposed to propranolol and PFOS. In the case of atrazine exposure, a separation from the controls was found only in *D. magna* adults.

In the last decade myriad engineered nanoparticles (ENPs) were developed thanks to the advent of nanotechnology. ENPs were applied in very different fields, including environmental biotechnology, nutraceuticals, cosmetics, textiles and biomedicine. Among them silver nanoparticles (AgNPs) received great attention, being incorporated in many consumer products, such as bactericides, food storage bins and bandages. Moreover, their so wide applications increased the release of AgNPs in the aquatic environment, representing potential risks for environment and aquatic organism health. Non-targeted mass spectrometry-based metabolomics approach was for the first time used to study chronic toxicity tests and subsequently the metabolic changes of the growing crustacean *Daphna similis* after exposure to 0, 0.02, and 1 ppb AgNPs. A dynamic kinetic pattern of the metabolome was observed from day 0 to day 21. In more detail, metabolite levels gradually increased from day 0 to day 13, before to reach the baseline level of day 0 on day 21. Even if no morphological or structural changes were observed, numerous metabolites changed, such as the amino acids serine, threonine and tyrosine, sugars D-allose and fatty acids arachidonic acid.

Metabolomic approach was also applied to study the response to pollution in Antarctic krill *Euphasia superba*, one of the most successful species on Earth also being the largest potential protein bank for both Antarctic animals and humans [*54*]. GC-MS method was used to identify 293 compounds divided into several classes, including amino acids, sugars and polyols, metabolic intermediates, small molecules and other metabolites. A serious problem was represented by both 2-hydroxybiphenyl and dioctyl phthalate detected in these krill.

Tissues, as well as some intermediates of pesticides and medicinal or chemical compound products. These findings represented a very important starting point for the development of international contracts or policies to protect the pristine ocean.

M. rosenbergii is a crustacean introduced in China in 1976, initially farmed in South China and then extended to the Yangtze River Basin. The overwhelming majority of *M. rosenbergii* is farmed in pond monoculture, which is strongly dependent on high feed input. The uneaten food and animal waste can cause water eutrophication and increase the NH_3 concentration, which is one of the commonest problems in aquaculture process because it accumulates in the animal. Previous studies showed that chronic ammonia exposure reduced crustacean development and growth, causing immunological responses, and histological damage. It can lead to reduced oxyhemocyanin in *L. vannamei*.

Naqvi et al. [55] reported that the *M. rosenbergii* growth of at various NH_3 levels was significantly lower than that of the control. Little is known about the ammonia metabolism and detoxification strategy of prawn, an only very recently the effects of ammonia-N (0, 0.108, 0.216, 0.324, or 0.54 mg L^{-1}) was evalueted on growth and metabolizing enzymes in hepatopancreas of *M. rosenbergii*, such as glutamine synthetase, alanine aminotransferase, aspartate aminotransferase, and glutamate dehydrogenase. The metabolome of its muscle was also analyzed after exposure to ammonia-N (0, 0.108, 0.324, or 0.54 mg L^{-1}) for 20 days. The results showed very clearly that the survival rate of *M. rosenbergii* decreased significantly after treatment with ammonia-N 0.54 mg L^{-1} compared with the other concentrations. Ammonia-N had significant effects on the lipid, carbohydrate and protein metabolism of these prawns, including purine metabolism, amino sugar and nucleotide sugar metabolism, α-linolenic acid metabolism, arginine and proline metabolism, glutathione metabolism, and phosphonate and phosphate metabolism, and on the terpenoid biosynthesis, lysine degradation, and lysine biosynthesis pathways. High concentrations of ammonia-N stress also increased the content of glutamate and arginine, involved in the urea cycle to synthesizes glutamine or urea to eliminate ammonia toxicity. An integrated approache was used to better understanding of the mechanisms underlying the male sexual differentiation of *M. nipponense*. In particular, strong candidate sex-related metabolic pathways and genes in *M. nipponense* were identified by integrated metabolomics and transcriptome analyses of the testis in response to different temperatures and illumination times, as confirmed by PCR analysis and in situ hybridization.

5.5 Concluding remarks

Marine crustaceans represent widely used food and feed supplement, also providing an important contribution to capture fisheries and aquaculture worldwide. In addition, crustaceans are becoming the new target of world fisheries to cope the increasing demand due to the growth of human populations at a global scale and substitute high trophic level species depleted by overfishing.

In the understanding of stress response, molecular biology-based assays have already provided insight into the mode of toxicity of different stressors. Ecotoxicogenomics is a rapidly growing field, linking the molecular, cellular, whole organism and population responses to stressors, remaining one of the great challenges in ecology and ecotoxicology. Linking molecular changes with relevant ecological responses will greatly improve the predictive powers of tests based on molecular responses. In fact, variation of gene expression in response to environmental stress can be useful to predict chronic effects at individual and population level. Thus, gene pathways underlying physiological processes can be identified and investigated, giving further elucidation on the mechanism of action of stressors. On this line, stressor-specific signatures in gene expression profiles offer a diagnostic approach to identify the consequences of an aquatic pollution event. A great step forward has been made in the last few years with the advent of metabolomics techniques. Metabolomics have been proven to provide in-depth insights into the biochemistry of diets, toxicity, medicine, physiology and pathology. Thanks to high sensitivity, metabolomics-based studies also help to understand possible mechanisms by identifying metabolites and their toxicological pathways. Different from genomics, proteomics and transcriptomics, which test endogenous substances, metabolomics can also detect exogenous metabolites. The combination of metabolic and transcriptional changes allows for a more complete understanding of how marine organisms, and in particular crustaceans, can counteract to stress.

In conclusion, omic technologies offer unprecedented opportunities to better understand toxicity and downstream its secondary effects by providing a holistic view of the molecular changes underlying physiological disruption. Additionally, omic tools can potentially help screen for biomarkers of sublethal toxicity in environmental monitoring. The combination of transcriptional and metabolic measurements represents a challenging potential to reveal both regulatory processes as well as those more closely linked to organism fitness.

References

1. R.P. Ellis, H. Parry, J.I. Spicer, T.H. Hutchinson, R.K. Pipe et al. (2011). *Fish Shellfish Immunol.* **30:** 1209–1222.
2. M.L. Fanjul-Moles. (2006). *Comp. Biochem. Physiol. Part C Toxicol. Pharmacol.* **142:** 390–400.
3. C. Manfrin, A. Pallavicini, S. Battistella, S. Lorenzon, and P.G. Giulianini. (2016). pp. 107–116. *In:* Lessons in immunity - From single-cell organisms to mammals (Academic Press, Elsevier, 2016).
4. C. Manfrin, L. Peruzza, L. Bonzi, A. Pallavicini, and P. Giulianini. (2015). *Invertebr. Surviv. J.* **12:** 29–37.
5. H. Jonsson, D. Schiedek, and A. Goksøyr. (2006). *Aquat Toxicol.* **78S:** S42–S48.
6. T. Liu, L. Pan, Y. Cai, and J. Miao. (2015). *Gene* **555:** 108–118.
7. M.P.M. Coelho, C. Moreira-de-Sousa, R.B. Souza, Y. Ansoar-Rodríguez, E.C.M. Silva-Zacarin et al. (2017). *Environ. Sci. Pollut. Res.* **24:** 22007–22017.
8. C. Minier, J. Forget-Leray, and A. Bjørnstad, L. Camus. 2008. *Mar. Pollut. Bull.* **56:** 1410–1415.
9. Z.S. Li, S. Ma, H.W. Shan, T. Wang, W. Xiao. (2019). *Ecotox. Environ. Safe.* **186:** 109753.
10. T. Zhou, J. Wei, J. Su, Z. Hu, Y Li, et al. (2019). *Environ Toxicol.* **34:** 223–232.
11. V.I. Lushchak. (2011). *Aquat. Toxicol.* **101:** 13–30.
12. R. Sussarellu, T. Dudongnon, C. Fabioux, P. Soudant, D. Moraga et al. (2013). *J. Exp. Biol.* **216:** 1561–1569.
13. J.C. McNamara, and S.C. Faria. (2012). *J. Comp. Physiol.* **182B:** 997–1014.
14. Y. Duan, Y. Wang, H. Dong, H. Li, Q. Liu et al. (2018). *Fish. Shellfish Immun.* **81:** 161–167.
15. T. Bosker, G. Olthof, M.G. Vijver, J. Baas, and S.H. Barmentlo. (2019). *Environ. Pollut.* **250:** 669–675.
16. J.R. Shaw, M.E. Pfrender, B.D. Eads, R. Klaper, and A. Callaghan. (2008). *Adv. Exp. Biol.* **2:** 327–328.

17. C. Pueyo, J.L. Gómez-Ariza, M.A. Bello-López, R. Fernández-Torrez, N. Abril et al. (2011). pp. 167–196. In Pesticides in the modern world - trends in pesticides analysis (InTech, RijeKa, Croatia. 2011).
18. E.S. Chang. (2005). *Integr. Comp. Biol.* **45**: 43–50.
19. S. Sun, H. Fu, J. Zhu, X. Ge, X. Wu et al. (2018). *Int. J. Mol. Sci.* **19**, 1990.
20. L. Calzolai, W. Ansorge, E. Calabrese, N. Denslow, P. Part, and T. Lettieri. (2007). *Comp. Biochem. Physiol. D: Genomics Proteomics* **2**: 245–249.
21. O.K. Adeyemo, K.J. Kroll, and N.D. Denslow. (2015). *Aquat. Toxicol.* **168**: 60–71.
22. Y.S. Keum, J. Kim, and Q.X. Li. (2010). pp. 627–643. In Handbook of Pesticide Toxicology (Academic Press, New York, 2010).
23. S. Moco, J. Vervoort, S. Moco, R.J. Bino, R.C.H. De Vos et al. (2007). *Trend Anal. Chem.* **26**: 855–866.
24. E.S. Ong, C.F. Chor, L. Zou, and C.N. Ong. (2009). *Mol. Biosyst.* **5**: 288–298.
25. X. Dong, Q. Liu, D. Kan, W. Zhao, H. Guo et al. (2020). *Ecotox. Environ. Saf.* **189**: 110046.
26. K. Saito, and F. Matsuda. (2010). *Annu. Rev. Plant Biol.* **61**: 463–489.
27. M. Ahsanullah, D.S. Negilski, and M.C. Mobley. (1981). *Mar. Biol.* **64**: 299–304.
28. L. Migliore, and M. de Nicola Giudici. (1990). *Hydrobiologia* **203**: 155–164.
29. A. Yamuna, V. Kabila, and P. Geraldine. (2000). *Indian J. Exp. Biol.* 38: 921–925.
30. P.T. Gauthier, W.P. Norwood, and E.E. Prepas. (2015). *Environ. Sci. Technol.* **49(19)**: 11780–11788.
31. S. Lorenzon, P. Edomi, P.G. Giulianini, R. Mettulio, and E.A. Ferrero. (2005). *J. Exp. Biol.* **208**: 3341–3347.
32. R. Geyer, J.R. Jambeck, and K.L. Law. (2017). *Sci. Adv.* **3(7)**: e1700782.
33. A. Gerhardt. (2019). *J. Toxicol. Risk Assess.* **5**: 2017.
34. S.W. Hsu, and J.C. Chen. (2007). *Aquaculture* **271**: 61–69.
35. L. Jiang, J. Feng, R. Ying, F. Yin, S. Pei et al. (2019). *Fish Shellfish Immun.* **92**: 230–240.
36. E.M. Costa, P.M. Spritzer, A. Hohl, and T.A. Bachega. (2014). *Arq. Bras. Endocrinol.* **58**: 153–161.
37. J.L. Liu, and M.H. Wong. (2013). *Environ. Int.* **59**: 208–224.
38. B. Nunes, F. Carvalho, and L. Guilhermino. (2006). Chemosphere **62**: 581–594.
39. H.C. Poynton, J.M. Lazorchak, C.A. Impellitteri, B.J. Blalock, and K. Rogers. (2012). *Environ. Sci. Technol.* **46**: 6288–6296.
40. W. Zhao, M. Wang, L. Wang, M. Liu, K. Jiang et al. (2018). *Aquaculture* **485**: 191–196.
41. S. De Vos, G. Van Stappen, P. Sorgeloos, M. Vuylsteke, S. Rombatus et al. (2019). *Aquaculture* **500**: 305–314.
42. J.P. Morris, S. Thatje, J. Ravaux, B. Shillito, D. Fernando et al. (2015). *Comp. Biochem. Physiol. Part A* **181**: 9–17.
43. J. Qiu, W.N. Wang, L.J. Wang, Y.F. Liu, and A.L. Wang. (2011). *Comp. Biochem. Phys. C* **154**: 36–41.
44. M. Sun, Y.T. Li, Y. Liu, S.C. Lee, and L. Wang. (2016). *Sci. Rep.* **6**: 19405.
45. M. Sun, K. Jiang, F. Zhang, D. Zhang, A. Shen et al. (2012). *Genet. Mol. Res.* **11(2)**: 978–986.
46. Y. Zohar, A.H. Hines, O. Zmora, E.G. Johnson, and R.N. Lipcius. (2008). *Rev. Fish Sci.* **16**: 24–34.
47. A. Gago-Tinoco, R. González-Domínguez, T. García-Barrera, J. Blasco-Moreno, M.J. Bebianno, et al. (2014). *Environ. Sci. Pollut. Res.* **21**: 13315–13323.
48. X.P. Ying, W.X. Yang, and Y.P. Zhang. (2006). *Aquaculture* **256**: 617–623.
49. N.M. Izral, R.B. Brua, J.M. Culp, and A.G. Yates. (2018). *Environ. Sci. Pollut. R.* **25**: 36184–36193.
50. G. Ritvo, J.B. Dixon, W.H. Neill, T.M. Samocha, and A.L. Lawrence. (2007). *J. Word Aquacult. Soc.* **31**: 381–389.
51. K. Chen, E. Li, C. Xu, X. Wang, H. Li et al. (2019). *Aquaculture* **499**: 329–340.
52. A.A. Laith, M. Ambak, A.B. Abol-Munafi, W.W.I. Nurhafizah, and M. Najiah. (2017). *Aquacult. Rep.* **7**: 7–15.
53. N.D. Wagner, B.P. Lankadurai, M.J. Simpson, A.J. Simpson, and P.C. Frost. (2015). *Physiol. Biochem. Zool.* **88**: 43–52.
54. F. Zhang, M. Zhao, C. Ma, L. Wang, C. Feng et al. (2018). *Crustaceana* **91**: 961–999.
55. A.A. Naqvi, S. Adhikari, B.R. Pillai, and N. Sarangi. (2007). *Aquacult. Res.* **38**: 847–851.

PART 2
Ecology and Taxonomy

CHAPTER 6

Crustacean Decapods are Models to Describe the General Trends of Biodiversity According to Ocean Acidification

*Valerio Zupo** and *Emanuele Somma*

6.1 Introduction

Food webs are considered, according to Winemiller and Polis [*1*], "*networks of interactions among species, groups of organisms, populations, or aggregate trophic units*". Scientists followed several approaches, depending both on the characters of individual ecosystems, and the available techniques for their investigation [*2*]. Food webs are a well-established topic of ecology and they were initially considered in terms of "food chains". However, food chains have been deprived of importance, although some of these frameworks are still considered and discussed, as a demonstration of the importance of theories joining food webs and ecological interactions among organisms. In fact, the struggle to reveal a balance and a simple theory [*3*] to synthesize the ecological complexity of natural systems, is still alive. In this perspective, crustaceans represent key organisms, due to their central roles in planktonic and benthic food webs and remembering the importance of copepods, for example, in planktonic food webs and that of decapod crustaceans for fisheries and export of energy from most benthic communities.

When we describe food webs, the initial idea of Elton [*4*] taking into account key interactions between prey and predators, progresses rapidly into the Lindeman [*5*] concept of prisms of energy and matter, creating a common currency to study

Stazione Zoologica Anton Dohrn, Department of Ecosustainable Biotechnology, Villa Comunale. 80121 Naples, Italy.
* Corresponding author: vzupo@szn.it

and compare aquatic ecosystems. Synthetically, their visions produced the bases to investigate the concept of "ecosystem", as the connection among abiotic and biotic components of animal and plant communities. Further developments of these concepts indicated that biodiversity is a key factor assuring the stability of natural communities [6]: food webs and biodiversity are basic concepts indispensable to evaluate the stability and diversity [7] of ecosystems. Further studies demonstrated various emergent properties of ecosystems realized according to the structure of their food webs. Among these, we should consider a remarkable inverse relationship between "*connectance*" and diversity. Connectance is a measure of the food web complexity, based on the proportion of *realized* ecological interactions among *potential* ones [8] and these patterns were further investigated and confirmed by various scientists [9].

Interestingly, structural properties of networks, as statistically investigated, affect their stability. In particular, an increased number of nodes, as well as increased complexity and interaction strengths, represent destabilization factors for natural networks as ecosystems. Consequently, we may forecast an inverse relationship between connectance and species diversity. Similar assumptions were investigated and obtained partial confirmation on some ecosystems [10]. In this chapter, we will apply these relationships to test a general model of biodiversity trends based on the responses of crustacean decapods to the abundance of feeding sources, in a range of environments variably impacted by O.A. The conclusions reached within this chapter will demonstrate consistent properties characterizing the assemblages of aquatic creatures, and extensible to various structural levels, from single cells to the largest ecosystems [11].

In particular, we aim at identifying universal patterns of biodiversity to be applicable to a range of plant and animal associations, as some previous authors attempted, from Riede et al. [12] to Layman et al. [13]. Thus, in the next paragraphs, we will introduce a model for rapid prediction of the available trophic resources (RAFI) and we will conclude, using a mathematical representation based on empirical data on the distribution of decapod crustaceans, that biodiversity is inversely related to the abundance of resources, and that such law is valid for any natural system, at any level of complexity.

6.2 Trophic interactions and biodiversity

A remarkable lack of punctual and comparable data on the availability of trophic resources characterizes most studies relating biodiversity and food webs. Consequently, studies relating the effect of trophic resource availability on the biodiversity of consumers are almost absent, because the abundance of trophic resources was investigated over a range of methods, measures and volumes or area units [14]. However, the availability of a statistical method to evaluate the abundance of available resources [15] and the application of decapod crustacean data over various types of environments, permitted to reach intriguing results in this never-ending issue.

It is evident that animals interfere each other through chemical, physical or other interactions, and that community dynamics and structures are largely organised by trophic interactions [16]. Experimental investigations employed total community biomass (Btot) as a measure for functioning, while species richness (S) may represent a direct measure of biodiversity. Such models demonstrated a close relationship between biodiversity and biomass/productivity of ecosystems. The same relationships are well correlated with the stability of ecosystems [17].

Several useful models of complex ecosystems exist, although we could repeat, as suggested by Box [18]: "*models are all wrong but some of them are useful*". Simple models, as the Lotka-Volterra simulation for competition, may be effective exemplifying scenarios of community assemblage using elegant simplifications as:

$$\frac{dB_j}{dt} = (s_j - \sum\nolimits_k C_{jk} B_k)B_j$$

where sj and Cjk $\geq$ 0 for all j, k.

The model is based on the logistic growth model and indicates that several competing species hamper each other's growth. Periodic invasion-fitness distributions may complicate the model and permit evaluations of the trends of biodiversity in further times. Previous investigations [19] demonstrated that a benthic community is chemically defended by the attack of invader species (competition avoidance), because its members share knowledge—ignored by the possible invaders—about the toxicity of some algal prey. However, even when saturated and chemically protected, communities are not closed to invasions and an extirpation may eventually be due to the direct knock-out of a resident species by an invader or it may result from complex interactions within the community, facilitating the process. All this explains that the natural turnover of communities is a result of continuous processes of invasions and defences, leading to progressive changes of the species composition, including possible changes in the total biodiversity. Due to the influence of climate changes and increased shipping activities, further and faster invasions of alien species complicate the relationships within the ecological communities. Environmental modifications due to climate changes may influence the chemical relationships among organisms and increase the vulnerability of chemical defences [19], so impacting the pristine associations [20] and facilitating an easier admission of invaders. Fixed mathematical limits constrain the number of species naturally assembled in a community, even if species composition was progressively modified by climate changes: the biodiversity has space constraints. Consequently, since there is less space at higher latitudes than at lower ones, less species may be predicted to globally co-exist, as the planet warms up and the oceans acidify [20].

6.3 Crustacean decapods as a mean to detect trends of biodiversity

Most theories provide measures of biomass to detect the effects of biodiversity on the productivity of ecosystems and evaluate their services. For example, various mathematical simulations aim at evaluating the relationships between the total biomass of a community (B_{tot}) and its species diversity (S). In this view, according to Wilson et al. [21] and Wilson and Lundberg [22], we can apply a technique named the *mean-field approximation,* and forecast that:

$$B_{tot} = \frac{SK}{1 + (S-1)IC}$$

where I indicates the invasion rates and C the competition rates. This indicates a continuous increase of species richness (up to a plateau, normally expected in any natural community) until the rate of competitive exclusions of species is balanced to the rate of invasions. This prediction represents an indirect demonstration that a higher species richness (S) corresponds to a higher total biomass of the community (Btot). Consequently, the increase of species richness is naturally correlated to an increase of biomass of the community, up to a plateau [23] imposed by space constraints and niche resources [24].

However, this does not explain if the relative abundance of trophic resources influences S (both positively or negatively). For example, will an increase of available plant biomass trigger an explosion of species diversity for vegetarians? Will a higher abundance of organic detritus sustain a larger diversity of detritus feeders? From a simple thermodynamic point of view, a higher availability of "prey" could both induce an increase of species diversity or an increase of individuals deriving from a few species. Both in the case of an increase of individuals of a few species, or an increase of the total number of species "S", a larger availability of feeding sources should lead to a higher biomass of consumers, but the effects on the general levels of diversity might be contrasting [25]. A third possible output, taking into account mainly the chemical communications [26], could result in a null effect of the abundance of feeding sources on the biomass of consumers, because it could be only partially consumed (if chemically defended) and the theory of feeding interactions might be applied. The question is not trivial, since it will determine, for example, if an inorganic fertilization process (increased resources for plants) will lead to a simple increased production of the species already cultivated (increased exploitation of shared resources), or conduct to an increase/decrease of biodiversity of the plant community [27], according to an increase of resource-mediated competition process.

From a mathematical point of view, food webs are networks, because each consumer cannot feed on all the sympatric species. Different models may take into account the chemical composition of bodies, the spatial structure of food-webs or specific features of networks, as the *connectance,* or the seasonal variability, the trophic transfer efficiency between levels and the number of trophic levels. The "species richness" S (i.e., the number of species living in a given ecosystem) is only partially related to the number of trophic links L. In parallel, the maintenance of biodiversity implies the maintenance of individual populations within a community. A mathematical analysis of niche models implies that it generates distinctive distributions of in- and out-degrees, according to the distribution of scaled numbers of resources k/Z [28,29]. Following to this perspective, the probability that a species has m consumers is the following:

$$P[m\ consumers] = \frac{1}{2Z} \int_0^{2Z} \frac{t^m e^{-t}}{m!} dt$$

and the probability for it to reach k resources is:

$$P[k\ Resources] = \frac{1}{2Z} E_1 \left(\frac{k}{2Z}\right)$$

From these assumptions an integral exponential function may be derived, for the evaluation of resources:

$$E_1(x) = \int_x^\infty \frac{e^{-1}}{t} dt$$

These general mathematical laws may be applied to the scaled number of consumers m/Z. For similar reasons, a relationship has been proposed [28,30] to predict the functions explaining the probability to increase the number of trophic links (cumulative distributions) and the number of resources, as well as the diversity of consumers and the abundance of resources:

$$Cumulative\ distributions = exp^{(-x)} - xE_1(x)$$

where, $x = k/(2Z)$.

According to these considerations, the number of trophic links (closely related to the diversity S of consumers), depends on the abundance of resources according to the $exp^{(-x)}$. Natural associations of decapod crustaceans were employed to test this model, through the computation of feeding indices (RAFI) [15]. In addition, since ecological communities are dynamical systems, just like assemblages of cells within organized tissues, we expect that the diversity of forms observed along a range of resources may be reproduced at any organizational level with similar patterns of distributions, derived by the mathematical forecasts for such networks.

To test this hypothesis, we collected crustacean decapods at five stations around the Island of Ischia (Bay of Naples, Italy), consistently represented by seagrass ecosystems (*Posidonia oceanica*) meadows but characterised by different conditions: two stations in a meadow off Castello d'Ischia hosting a *P. oceanica* meadow in normal ecological conditions (pH 8.1), two stations in a meadow located a few hundred of meters apart, characterized by the presence of CO_2 vents, inducing an acidification of the seawater up to a pH of 7.6, and a control site located off Lacco Ameno d'Ischia, at normal (pH 8.1) conditions (Table 1). We analysed the taxonomical composition of decapod crustaceans at the three sites and compared both their numerical abundances and the species compositions. Finally, we analysed their distributions in trophic groups, to determine their abundances as compared to their diets and food sources.

In parallel, we evaluated the RAFI indices [15] for each of the considered site, in each meadow, and produced a table of numerical abundance of species as ascribed

Table 1. Features of the sites considered for this study (categorized according to Zupo et al., 2017). The correspondence of each character is indicated by "X".

Sites	High canopy seagrass	Exposed	Sheltered	Eutrophic	Anthropogenic impacted	Natural impacts (e.g., estuaries)	Shallow
Lacco Ameno (8.1)	X	X			X		X
Castello N1 (8.1)	X	X					X
Castello S1 (8.1)	X		X	X			X
Castello S2 (7.8)	X		X	X		X	X
Castello N3 (7.8)	X	X				X	X

to each trophic group *vs.* the abundance of resources in each group. Tables of percent abundance were obtained for each of five sites. The points characterized by y (species diversity, S) and x (abundance of resources available for each trophic group, RAFI; Table 2) were then plotted in a trophic space to detect the shape of the best fitting functions. In particular, we checked if the above hypothesized exponential decay function might fit their relationships.

Table 2. Evaluation of the relative abundance of trophic resources (RAFI%) in each site as a percentage of available biomass obtained by means of the RAFI index [15] as compared to the abundance of decapod crustacean species abundance (alpha diversity).

	Castello N3 (7.8)		Castello S2		Castello N1		Castello S1		Lacco Ameno (8.1)	
Trophic groups	Rafi%	Alpha div. %	Rafi%	Alpha div. %	Rafi%	Alpha div. %	Rafi%	Alpha div. %	Rafi%	Alpha div. %
mCa	5,76	20,00	5,33	16,67	4,66	17,65	4,33	23,53	5,63	17,65
Ca	1,75	10,00	1,08	5,56	1,87	17,65	1,16	29,41	4,22	23,53
mHe	3,93	20,00	2,18	16,67	8,75	11,76	4,87	5,88	5,28	11,76
He	17,28	0,00	11,98	5,56	16,79	0,00	11,70	0,00	10,14	0,00
mOm	1,28	30,00	0,79	11,11	2,07	29,41	1,28	23,53	4,69	29,41
Om	13,09	5,00	4,84	22,22	11,66	5,88	4,33	11,76	14,08	0,00
mDeF	5,59	10,00	5,16	11,11	3,32	11,76	3,08	0,00	3,75	11,76
DeF	24,74	0,00	34,86	0,00	22,04	0,00	31,19	0,00	22,17	0,00
DeFS	4,71	5,00	8,71	11,11	7,00	5,88	13,00	5,88	8,45	5,88
DeFHe	20,95	0,00	24,21	0,00	20,52	0,00	23,83	0,00	18,58	0,00
FF	0,93	0,00	0,86	0,00	1,33	0,00	1,23	0,00	3,00	0,00

6.4 The shape of biodiversity according to decapod crustaceans

The distribution of food items available for each trophic group (Figure 1) is similar in the different sites, considering they are all characterized by *Posidonia oceanica* environments.

However, some remarkable differences are due to the exposition, to the ambient pH and to the location of meadows. In particular, the meadows off Castello d'Ischia are characterized by a higher abundance of organic detritus that produce higher abundance of trophic resources available for DeF and DefS. In addition, the meadows at normal pH (8.1) exhibit a more uniform availability of resources with respect to the acidified (7.6) ones. Besides these differences, we must take into account that decapod crustaceans are characterized by various trophic specializations. Most species share the ability to adapt to a range of diets according to the availability (opportunistic carnivores, herbivores-detritivores and omnivores) while some trophic categories (e.g., filter feeders and deposit feeders) are almost absent in this specific group of organisms.

Notwithstanding these constraints, the decapods associated to *Posidonia* meadows represent good descriptors of the general trends of biodiversity and they are distributed in an inverse relationship with respect to the available resources, as forecasted according to the introductive concepts above expressed. In fact, in correspondence to the least abundant food categories (e.g., food available for micro-carnivores, carnivores and micro-omnivores) the highest biodiversity is observed. In contrast, the largest availability of food for herbivores and detritus feeders corresponds to the lowest levels of species diversity. The distribution of biodiversity according to the trophic groups (Figure 2) definitely permits to define the shape of the biodiversity and its relationship with feeding resources.

This concept should not be confused with the biomass of consumers, which may follow different trends according to the abundance of individual species. For this reason, a higher numerical abundance actually characterizes the trophic groups feeding on abundant items and this permits a reasonable and efficient exploitation of the available resources. However, species diversity is inversely related to the abundance of resources according to an exponential decay relationship and decapod crustaceans may exemplify this remarkable correlation.

6.5 Conclusions

Biodiversity is heterogeneously distributed across the Earth [*31*] and current ecological theory [*32*] suggests that food webs are structured by the interactions of resources and their consumers [*33*]. In this chapter, we hypothesized that patterns of biodiversity are primarily driven by the heterogeneity in the amount of energy available, or the primary productivity, occurring in each ecosystem, according to Mittelbach et al. [*34*] and tested this hypothesis on communities of decapod crustaceans. In fact, the relationships between productivity and species diversity (taking into account various animal taxa) may have variable shapes [*35*]; however, both the abundance and the availability of food may influence the patterns of biodiversity in any ecosystem [*36*].

Here, evidences from decapod crustacean communities associated to *Posidonia oceanica* indicated that species diversity is correlated to a hump-shaped food

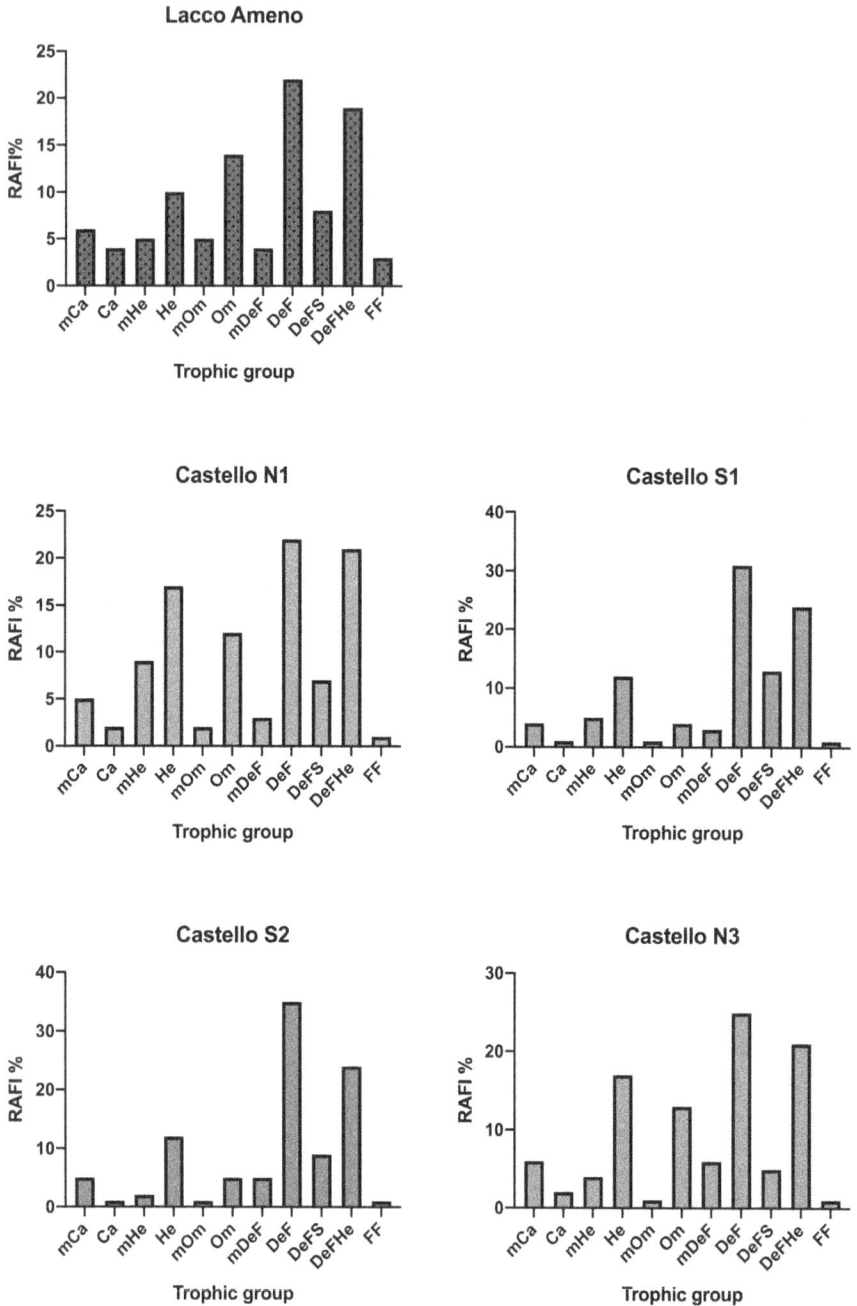

Figure 1. Distribution of food item availability in each of the considered sites. The structures of food webs are similar because the same environment is considered (*Posidonia oceanica* meadows) but they vary according to local features of the environments and pressures, as indicated in Table 1.

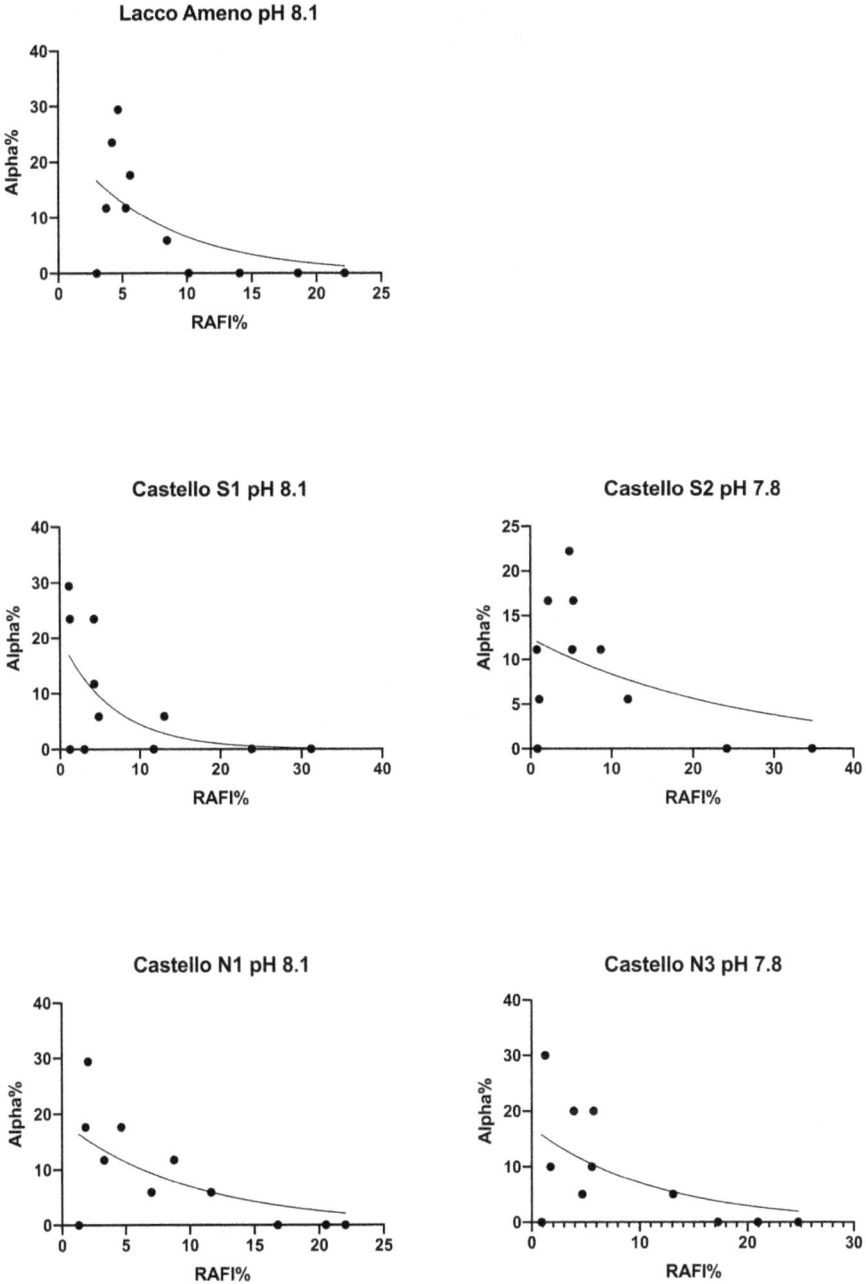

Figure 2. The shape of the biodiversity as expressed by the community of decapod crustaceans associated to a range of *Posidonia oceanica* meadows. An inverse exponential relationship was imposed obtaining good fit of RAFI% with the alpha diversity (S) of feeding groups, percentually represented in each environment.

availability-diversity relationship (as also demonstrated in terrestrial plants [*37*]) and that diversity is a unimodal function of food supply rates. Productivity, trophic resource availability and body size are interrelated in natural systems [*38,39*]; however previous research, through a top-down approach [*33*], were focused mainly on the influences of predators on populations of their prey [*40*]. In this chapter we explored the relationships between species diversity and the abundance of trophic resources at various sites located around the island of Ischia, taking into account the decapod communities associated to *Posidonia oceanica* meadows, along a range of anthropic impacts and acidification pressures. We hypothesized that a higher availability of trophic resources (RAFI%) must correspond to lower levels of alpha biodiversity (S). As a matter of fact, applying our hypothesis to the communities of crustacean decapods identified during a field study, we demonstrated that this key relationship is roughly represented by a function of exponential decay:

$$Diversity\ of\ consumers = \text{trophic resources} * exp^{(-k*x)}$$

Substantially, this chapter answers three basic ecology questions: (a) The level of functional redundancy in a given system are predictable according to the relative abundance of the main prey available for each trophic group? (b) The abundance of trophic resources in different marine systems is influenced by the diversity of species assemblages? (c) Does compartmentalization and diversification of trophic resources allow for a spread of species diversity within a given ecosystem?

Here we demonstrated that species diversity exhibits an inverse exponential relationship with food abundance and this is consistently demonstrated in all the considered sites: the highest diversity levels are obtained in groups of species feeding on less abundant food items, and *vice-versa*. This evidence confirms the hypotheses based on mathematical models and decapod crustaceans, given their wide distribution into trophic groups, are definitely good models for such food web theoretical investigations. These results, on the whole, may be viewed also in the light of body size allometric constraints and may be explained in the light of the arguments raised by Bagchi and Ritchie [*42*]: biodiversity is inversely correlated to the abundance of food items, according to an inverse exponential relationship.

These conclusions open the possibility for testing further hypotheses in selected ecosystems, adopting different crustacean groups, both in benthic and in planktonic environments. Even more interestingly, considering that this model of biodiversity was based on mathematical properties of networks, the observed trends of biodiversity might be transferable to other biological systems in which any network of cells or organisms is ruled by trophic relationships. For example, we could forecast that the addition of fertilizers in agriculture fields will trigger larger production of a few crops, while a scarcity of resources will facilitate an increase of biodiversity, with the introduction of several (generally undesirable) species. Similarly, we may forecast that, in a given organism, an abundance of trophic resources available for individual cells (e.g., plenty of sugars, and fatty acids) will facilitate the production of a single type of cells (eventually pathological?), while a scarcity of trophic resources (less food in their diet!) might promote a "biodiversity" in the tissue omeostasis, i.e., an organic development of the individual.

All these hypotheses and several others might be tested in further investigations, in order to establish the actual limits of the model herein developed, linking the biodiversity levels of crustacean decapods to the availability of trophic resources and based on the above stated function of exponential decay. This will facilitate a deeper comprehension of the biodiversity patterns observed in any ecosystem, by comparing the trophic structures of various groups of crustaceans.

References

1. K.O. Winemiller, and G.A. Polis. (1996). *Food Webs: Integration of Patterns and Processes.* Chapman & Hall, New York, 1–22.
2. J.A. Dunne. (2009). *Encyclopedia of Complexity and Systems Science.* Springer, New York, 3661–3682.
3. E.P. Odum. (1953). *Fundamentals of Ecology.* W.B. Saunders Company, Philadelphia, 384 pp.
4. C.S. Elton. (1927). *Animal Ecology.* Sidgewick & Jackson, London, 209 pp.
5. R.L. Lindeman. (1942). *Ecology* 23: 399–417.
6. KS. McCann. (2005). *Ecological Paradigms Lost: Routes of Theory Change.* Elsevier Academic Press, Burlington, MA, U.S.A., 183–200.
7. K.S. McCann. (2000). *Nature* 405: 228–233.
8. T. Poisot, and D. Gravel. (2014). *Peer J.* 2: e251.
9. J.E. Cohen, and C.M. Newman. (1985). *Proc. R. Soc. Lond. Ser. B* 224: 421–448.
10. S.L. Pimm. (1982–2002). *Food Webs.* Chapman and Hall, pp. 432.
11. T.W. Schoener. (1989). *Ecology* 70: 1559–1589.
12. J.O. Riede, U. Brose, B. Ebenman, U. Jacob et al. (2011). *Ecosystems* 14: 169–178.
13. C.A. Layman, S.T. Giery, S. Buhler, R. Rossi et al. (2015). *Food Webs* 4: 14–24.
14. R.M. May. (1973). *Stability and Complexity in Model Ecosystems.* Princeton University Press, 304 pp.
15. V. Zupo, T. Alexander, and G. Edgar. (2017). *R. Soc. Open Sci.* 4: 160515.
16. A.G. Rossberg, H. Matsuda, T. Amemiya, and K. Itoh. (2005). *Ecol. Complex* 2: 312–321.
17. R. Hassan, R. Scholes, and N. Ash. (2005). *The Millennium Ecosystem Assessment Series,* vol. 1. Island Press, Washington, DC.
18. G.E.P. Box. (1979). *Robustness in Statistics: Proceedings of a Workshop.* Academic Press, New York, 201–236.
19. C. Maibam, G. Romano, M.C. Buia, M.C. Gambi et al. (2014). *Mar. Biol* 161: 1639–1652.
20. V. Zupo, C. Maibam, M.C. Buia et al. (2015). *J. Chem. Ecol.* 41(8): 766–79.
21. W.G. Wilson, P. Lundberg, D.P. Vazquez et al. (2003). *Ecol. Lett.* 6: 944–952.
22. W.G. Wilson, and P. Lundberg. (2004). *Proc. Royal Soc. B* 271: 1977–1984.
23. R.M. May. (1972). *Nature* 238: 413–414.
24. M.R. Gardner, and W.R. Ashby. (1970). *Nature* 228: 784–785.
25. R.H. MacArthur. (1955). *Ecology* 36: 533–536.
26. C. Maibam, P. Fink, G. Romano, M.C. Buia et al. (2015). *Mar. Ecol. Evol. Persp.* 36(3): 819–835.
27. A.M. Makarieva, V.G. Gorshkov, B.L. Li et al. (2008). *Proc. Natl. Acad. Sci. USA* 105: 16994.
28. J. Camacho, R. Guimera, and L.A.N. Amaral. (2002). *Phys. Rev.* 65: 030901R.
29. J. Camacho, R. Guimera, and L.A.N. Amaral. (2002). *Phys. Rev.* 88: 228102.
30. D.B. Stouffer, J. Camacho, W. Jiang, and L.A.N. Amaral. (2007). *Proc. Royal Soc. B* 1621: 1931–1940.
31. K.J. Gaston. (2000). *Nature* 405: 220–227.
32. L. Oksanen, and T. Oksanen. (2000). *Am. Nat.* 155(6): 703–723.
33. GA. Polis. (1999). *Oikos* 86(1): 3–15.
34. G.G. Mittelbach, C.F. Steiner, S.M. Scheiner et al. (2001). *Ecology* 82(9): 2381–2396.
35. K.L. Gross, M.R. Willig, L. Gough, R. Inouye, and S.B. Cox. (2000). *Oikos* 89(3): 417–427.

36. A. Sokolowski, M. Wolowicz, H. Asmus et al. (2012). *Estuar. Coast Shelf Sci.* **108:** 76–86.
37. D. Tilman, and S. Pacala. (1993). *Species diversity in ecological communities*. University of Chicago Press. Chicago, 13–25.
38. T.W. Schoener. (1976). *Theor. Pop. Biol.* **10:** 309–333.
39. A. Schroeder, K.A. Nilsson, L. Persson et al. (2009). *J. Anim. Ecol.* **78:** 1152–1162.
40. R.R. Braga, H. Bornatowski, and J.R.S. Vitule. (2012). *Rev. Fish Biol.* **22(4):** 915–929.
41. S. Bagchi, and M.E. Ritchie. (2012). *Theor. Ecol.* **5:** 141–151.

CHAPTER 7

Isopod Crustaceans as Seagrass Consumers
A Mediterranean Perspective

Maurizio Lorenti

7.1 Introduction

The Mediterranean Sea hosts at present six species of true marine seagrasses of which the most common are the large endemic *Posidonia oceanica* (L.) DELILE and the Atlanto-Mediterranean *Cymodocea nodosa* (UCRIA) ASCHERSON. They share, although at a different degree, the ability of forming complex habitats and to dominate ecosystems characterized by a high level of plant-animal relationships [1,2]. In terms of trophic resources for herbivores, the plant material available to consumers is plentiful in seagrass systems ranging from micro- and macroepiphytes on leaves and rhizomes to the live and dead tissues of the structuring seagrass itself. Among other ecosystem functions supported, the abundance and qualitative variety of food items may have the dual ecological role of reducing the competition within the plant feeder guild and of somehow protecting the seagrass from overgrazing. The extent of seagrass tissue consumption in this network of interactions has been unclear for long and generally deemed to be minor except perhaps for the detrital material [3]. However, the generalization of the low-feeding paradigm is being re-evaluated on a global scale [4] and specific attention is given to the recognition of single species or of groups of species capable of consuming seagrass tissues directly. In a Mediterranean context, increasing evidence of the substantial role of macroconsumers of seagrass biomass such as fishes and sea urchins has accumulated. On the other hand, a clearer picture of the impact of smaller-sized invertebrate grazers (mesograzers or mesoherbivores *sensu* [5]), mostly including crustaceans, molluscs and polychaetes, is still needed despite a number of species are recognized as seagrass consumers [1].

Department of Integrative Marine Ecology, Ischia Marine Centre, Stazione Zoologica Anton Dohrn, Punta San Pietro, 80077, Ischia (Naples), Italy.
Email: mlorenti@szn.it

Present knowledge indicates that isopod crustaceans are significantly involved in the consumption of both living and detrital seagrass mass worldwide [6,7]. This particular aspect of organism interactions will be approached in this review in relation to Mediterranean seagrass systems.

7.2 Isopods in Mediterranean seagrass communities: a tale of diversity

Isopod crustaceans constitute a common and characteristic component of the motile epifauna associated with Mediterranean seagrass systems [8,9,10]. Although generally collected in lower total abundance than dominant taxonomic groups such as amphipods and molluscs, except for their substantial occurrence in detrital leaf litter [11] and nighttime-collected samples [12], isopod assemblages encompass an outstanding range of morphological and functional forms.

From a trophic point of view, they feature a variety of feeding types ranging from carnivore scavengers (Cirolanidae) and parasites (*Gnathia* juveniles) to deposit (most Asellota) and deposit/suspension (*Astacilla*) feeders. However, the bulk of isopod populations is constituted by plant feeders in whose diet live and detrital material is most often mixed [9]. Major representatives of this group are idoteid and holognathid valviferans (genera *Idotea*, *Stenosoma* and *Cleantis*), sphaeromatids of the genera *Cymodoce* and *Dynamene* and seagrass tissue burrowers of the genus *Limnoria*, with each species occupying preferentially a specific subhabitat within seagrass formations (Figure 1). They can all be assigned to the category of mesograzers, which comprise forms with body lengths of up to a few centimeters characterized by a suite of traits including a cryptic lifestyle and a diet typically determined by the trade-off between the need for shelter from predators and food availability/quality.

Isopod grazers are known to play a crucial role in the control of plant epiphytes in seagrass systems according to a "mutualistic mesograzer model" implying reciprocal benefit between the seagrass and the consumer [13]. From this point of view, the resources provided to isopods by the structuring plant would be fundamentally in terms of shelter and substrate for other food items. However, in the case of a number of species, interactions may extend to the direct consumption of the seagrass tissues.

7.3 Isopod consumers of live seagrass tissue: what and why?

Living tissue of Mediterranean seagrasses (especially of the best studied of them, the large Neptune grass *Posidonia oceanica*) has been stated to be scarcely consumed by invertebrates of mesograzer size [1]. Even lesser is, reportedly, its contribution to the nutrition of this faunal fraction in terms of assimilated matter [14]. A variety of seagrass traits (toughness, poor nutritional quality, high content in lignocellulose and feeding-deterrent compounds) have been implicated in the resistance to grazing. However, on a global scale, a number of isopod species, belonging in particular to the family Idoteidae, have been since long recognized as live seagrass consumers, sometimes exerting a sensible impact on standing biomass [7]. In principle, live seagrass tissue is an abundant and temporally stable resource for animals which have

Figure 1. Major plant-feeding isopods and their preferential subhabitat in a *Posidonia oceanica* bed. Species (specimens not in scale): a – *Idotea hectica*, b – *Stenosoma appendiculatum*, c – *Idotea balthica*, d – *Cleantis prismatica*, e – *Cymodoce* spp., f – *Limnoria mazzellae*. Subhabitats: 1 – leaves, 2 – litter, 3 – rhizomes, 4 – scales. Photos: M. Lorenti.

the morphofunctional tools to deal with relatively large and tough food masses, as idoteids do. In fact, they are characterized by heavily chitinized mouthparts able to bite, abrade and triturate the food material [*15*] and can reach as adults a body size that falls in the upper range for mesograzers (Table 1).

Three species belonging to Idoteidae are commonly found in Mediterranean seagrass communities, namely *Idotea balthica* (PALLAS), *Idotea hectica* (PALLAS) and *Stenosoma appendiculatum* (RISSO), plus *Cleantis prismatica* (RISSO) which belongs in the very close family Holognathidae [*3,10*]. However, this thus far established nomenclature may change as the taxonomic and phylogenetic position of Mediterranean idoteids is under scrutiny. *I. hectica* has been recently moved to the genus *Synischia* although this action is debated [*16*]. More complex is the case with *I. balthica*, which has since long been considered a polytypic species with distinct geographic subspecies occurring in the Northern Atlantic and Baltic Seas as well as in the Mediterranean and Black Seas [*17*]. Actually, the Mediterranean form (*I. balthica basteri*) has been proposed as a possible good species based on a phylogenetic reconstruction [*16*]. Hereafter, I shall retain the "traditional" classification of *Idotea* species for clarity purposes assuming that even *I. balthica*, be it a subspecies mosaic or an unresolved species complex, constitutes a functional unit with shared basic properties relative to feeding ecology and physiology all over its distribution range.

Table 1. Characteristics of the major isopods plant feeders in Mediterranean seagrass systems (main sources: Naylor, 1955; Dumay, 1971; Lorenti and Scipione, 1990; Lorenti, 2006).

Species	Family	Max. length (mm)	Preferential subhabitat	Distinctive behaviour	Feeding mechanism (simplified)
Idotea hectica	Idoteidae	35	Canopy	Clinging	Biting
Idotea balthica	Idoteidae	30	Litter	Swimming/ clinging	Biting
Stenosoma appendiculatum	Idoteidae	25	Canopy	Clinging	Biting
Cleantis prismatica	Holognathidae	14	Rhizome layer	Crawling	Biting
Cymodoce spp.	Sphaeromatidae	11	Rhizomes-canopy	Migratory	Scraping
Limnoria mazzellae	Limnoriidae	2.5	*P. oceanica* scales	Burrowing	Grinding
Limnoria sp.	Limnoriidae	2.5	*C. nodosa* leaves	Burrowing	Grinding

A widespread inhabitant of shallow vegetated habitats, *I. balthica* displays an opportunistic feeding behaviour which may undergo, among other ones, ontogenetic and geographical variation [7,18]. In the Mediterranean, this species is commonly associated to *P. oceanica* litter accumulations both inside and outside meadows [3,11,19] whereas is rare or absent from the canopy proper [10,20]. *I. hectica*, while occurring also in litter [11,21], is a typical inhabitant of seagrass canopies, in particular of those formed by *P. oceanica* and *C. nodosa*, in which its body form is perfectly adapted to the leaf blade habitat by vertue of its coloration, shape and attachment capabilities. *Stenosoma appendiculatum* is similar to *I. hectica* in terms of homochromy and association with seagrass leaves whereas *C. prismatica* generally occurs in the rhizome layer and in the litter although showing nocturnal migrations toward the leaf canopy [12].

Both *I. balthica* and *I. hectica* have been shown to be able to consume living tissues of *P. oceanica* and *C. nodosa*, based on direct observations in laboratory experiments [22,23 and 24 as *Synischia hectica*]. Adult individuals of both species reared on living *P. oceanica* shoots were able to ingest large portions of both epiphytized and non-epiphytized healthy leaves (Figure 2). The feeding activity of idoteids in captivity was discontinuous but rates of tissue consumption by a single large individual might reach values as high as 10 milligrams of dry weight per day (author's data).

When force-fed on green leaf sections, also adults of *S. appendiculatum* left conspicuous bite marks on them (Figure 3) which would show the potential of the species for the removal of fair amounts of live seagrass tissue.

On the other side, few observations have been made of the feeding behaviour of Mediterranean idoteids in their natural habitat. *I. hectica* has been stated to be a typical consumer of leaves in *P. oceanica* stands based on an indirect method, i.e., the recognition of bite marks of peculiar shape which are attributed to the action of the species [25]. However, when investigated by means of stable isotope techniques, the

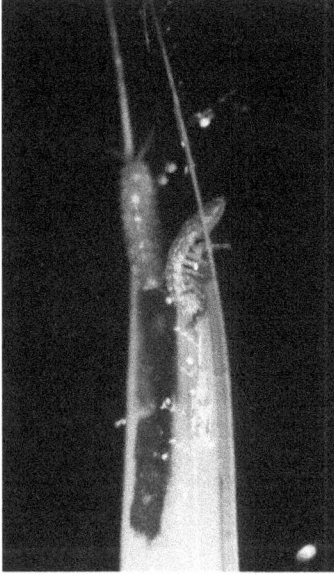

Figure 2. *Idotea balthica* individuals feeding on a *Posidonia oceanica* leaf. Photo: K. Wittmann.

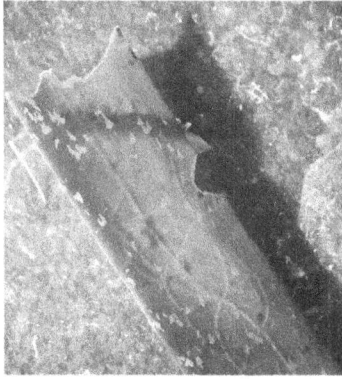

Figure 3. Bite marks made by an adult individual of *Stenosoma appendiculatum* on a living leaf of *Posidonia oceanica* (the animal can be discerned underneath the blade) (leaf width: 10 mm). Photo: M. Lorenti.

realized diet (or, at least, the assimilated and incorporated part of it) of individuals of *I. balthica* and *I. hectica* collected from leaf litter [21] or from meadows [20] of *P. oceanica* seems to consist mainly of macroalgae and epiphytes, with a higher tendency toward omnivority in *I. balthica*. One question, then, is to what extent a diet based on live seagrass tissues may support survival and growth of these animals. In no-choice trials, juveniles of *I. balthica* fed on living *P. oceanica* leaves showed on average 70% lower survival after 15 days as compared to individuals fed on macroalgae (*Ulva*) or on seagrass-associated fungi (*Corollospora*) (author's data). Similarly, a lower survival of *I. balthica* juveniles reared on *Zostera marina* was observed by [*18*] with respect to those fed on *Ulva*. In a simple comparison conducted with the ash method of Conover, assimilation of *P. oceanica* by adults of the same species was found to be 30% lower than for *Ulva*, which is the best preferred food in multiple-choice trials [*22*]. Moreover, a variable proportion of seagrass tissue egested by both *I. balthica* and *I. hectica* maintained its structure and, in part, resulted to be still viable in feces as demonstrated by residual chlorophyll fluorescence (personal observation). Breakdown of seagrass cells may be limited also in relation to a low efficiency of enzymes degrading structural carbohydrates in species of *Idotea* although there is some evidence of the occurrence of gut symbiotic bacteria in Northern *I. balthica* [*26*] feeding on seagrasses. On the whole, while a low assimilation efficiency is the rule for seagrass consumers, there exist conditions at which seagrass tissue may be more palatable and possibly more nutritive to idoteids. As a matter of fact, the quality of seagrasses as food do vary according to the nutrient content and to the age of leaves, as also correlating to their mechanical resistance in terms of toughness. For example, in feeding trials, *I. balthica* showed preference for meristematic tissues of *P. oceanica* [*22*] and *I. hectica* consumed preferentially young *C. nodosa* leaves [*23*]. Seagrass quality may also vary depending on environmental changes [*24*]. For example, seawater acidification may improve the quality of *P. oceanica* as food in terms of nitrogen content and carbohydrate allocation, increasing the consumption by fish and urchins [*27*]. The response of idoteids to such changes may however be different with respect to macrograzers. In mesocosm experiments, *I. hectica* was found by Tomas et al. [*24*] to prefer CO_2-enriched over non-enriched leaves of *C. nodosa* while being deterred by nutrient-enriched ones, in a reversed pattern compared to the urchin *Paracentrotus lividus*. Reasons for this were unclear and related by the authors to some change in the mechanical properties of the plant.

At any rate, assumed that normally living (green) seagrass tissue is not a nutritive food for idoteids, the question araises as to why it is consumed at all and even, on occasion, in large amounts. Actually, live seagrass tissue is eaten also in the presence of more palatable items as demonstrated in multiple choice experiments [*22*] as well as in gut content analyses of wild individuals. An appreciable proportion of *I. balthica* and *I. hectica* specimens examined by [*20*] and [*21*] contained living *P. oceanica* tissues but no relationship was found between the isotopic signature and the presence of seagrass material in the gut, further pointing to an uncoupling between consumption and assimilation in these animals.

A number of feeding strategies may be implicated in the consumption of living seagrass tissue by idoteids. How they may be realized in individual species, under what circumstances and with what synergy are all questions which need investigating. Such mechanisms may include:

- incidental ingestion of live seagrass in the process of feeding on associated epiphytes. This form of consumption would be regarded as particularly wasteful although involving considerable biomass removal,
- compensatory feeding in the shortage of more nutritive food [*28*]. This is more likely to occur in limited-choice or no-choice experiments than in nature as idoteids are motile, may move between different sources and undergo even diel

migrations, possibly to access adequate quality food [*12*]. However, at certain conditions, high consumption rates of poorly digestible food may provide sufficient nutrition to animals such as members of *Idotea* which show high food turnover in terms of gut residence time [*29*] and may rely on detoxification mechanisms against the accumulation of secondary metabolites [*30*].

- complementary feeding aiming at diet integration or incorporation of chemical defense and pigments. Mixed diets including nutritionally poor items may improve the performance of generalist herbivores by providing a better nutrient balance and/or a dilution of plant secondary compounds [*7*]. These mechanisms may apply to seagrass-associated idoteids as well. One possible use of plant-derived compounds may be related to the visual camouflage requirements of idoteids, which are dominant preys for carnivores and have to rely on effective behavioural and morphological traits for reducing predation susceptibility [*31*]. *I. balthica* constitutes a classical example of chromatic adaptation to the variety of vegetated habitats it may occupy by background matching of colour patterns [*32*]. A different case is that of *I. hectica*, which, seemingly due to the narrower habitat range, features two main colour morphs, a predominantly green one, when associated to the leafage of large seagrasses, and a brown one, when occurring in leaf litter. Some Northern Pacific species of *Idotea* (*Pentidotea*) show a similar double chromatism in connection with habitat shifts between seagrass beds and macroalgal stands. They seem to modify pigments (carotenoids) derived from the host plant for assuming a protective coloration [*33*]. Although debated as for its occurrence among idoteid species [*34*], a similar mechanism may explain in part the consumption of living seagrass tissue by *I. hectica* and probably by *S. appendiculatum* as well. [*21*] mention a green individual of *Idotea* sp. (seemingly *I. hectica*) in which a δ^{13}C value close to that of living *P. oceanica* leaves was found. Similarly, one explanation given by the same authors for the large consumption of detrital seagrass material by idoteids is the assimilation of brown pigments from dead leaves.

Isopods belonging to the family Sphaeromatidae can also have a role in the consumption of live seagrass tissue as observed in tropical and subtropical systems [*35,36*]. In the two genera (*Cymodoce* and *Dynamene*) commonly associated to Mediterranean seagrasses [*9,10,37*], active feeding is fundamentally an attribute of juveniles as adults are mostly confined to cryptic reproductive microhabitats. Members of the genus *Cymodoce*, which may be abundant in *P. oceanica* beds [*9,12*], act by scraping seagrass leaf surface tissues with the robust mouth pieces [*37*], similar to other sphaeromatid genera [*38*]. In laboratory experiments (personal observation), both juveniles and adult males of *C. hanseni* DUMAY were observed to progressively scrape epiphytized leaves of *P. oceanica* until producing a hole throughout the blade. On the whole, consumption of living tissue is likely to be mostly incidental in the process of feeding on epiphytes and associated particulate which would represent the main food sources for these isopods [*37*]. However, its occurrence in nature remains unexplored and may assume particular aspects. For example, a species of *Cymodoce* was found to attack seeds of a variety of Australian seagrasses [*35*]; it would be worth investigating this possible feeding behaviour in Mediterranean sphaeromatids.

A further category of live seagrass consumers is constituted by leaf burrowers belonging to the genera *Limnoria* and *Lynseia* (Limnoriidae) [6]. In beds of *Amphibolis griffithii* and *Posidonia* spp. in Australia, and of *Thalassia testudinum* in the Caribbean, *Limnoria* species can impact living plants in a proportion of infested leaves as high as 50% [7]. An unidentified species of *Limnoria* was found to excavate into leaves of *C. nodosa* at two locations around the island of Ischia (Italy) where burrows attributable to its action impacted 18% of the shoots sampled on one occasion [39]. As observed in the field and in the laboratory (personal observation), animals fed on mesophyll in a fashion typical of "leaf miners" *sensu* [40]. It is unlikely that the association between *Limnoria* and *C. nodosa* constitutes an isolated phenomenon. If confirmed widespread, such substantial consumption figures would have implications for the very structure of the impacted beds. Another species of *Limnoria* (*L. mazzellae* COOKSON AND LORENTI, misidentified in earlier works as *L. tuberculata*), a burrower into *P. oceanica* dead sheaths, may in some instances attack living tissues, although its impact seems to be negligible [41].

7.4 Isopod consumers of detrital seagrass material: working for teardown

Seagrass beds are major producers of detritus which is either retained in the system or exported to other systems [42] with all the implications this has for trophic webs in the coastal sea. In relation to the detrital status of seagrasses, a distinction can be made whether decaying tissues have been separated from the living plant (mostly as shed leaves) or they are still attached to it (senesced leaf portions and dead sheaths: debris *sensu* [3]). Isopods play a pivotal role in the processing of both categories, behaving as functional detritivores in the litter and as pseudograzers [3] towards debris.

Litter derived from seagrass stands mainly constituted by detached leaf accumulations is a variable substrate in terms of age, persistence, patch size, drift distance from the stand and association with other plant material (rhizome and root remains, algal clumps). The presence of *I. balthica* in such accumulations correlates with the known ability of the species to occupy mobile habitats such as drift macroalgae and detritus or floating seaweed rafts [29]. *I. balthica* contributes in reducing significantly the size of *P. oceanica* detritus fragments [19] seemingly as a result of both sloppy feeding and deposition of undigested pellets. In fact, a low fraction of the ingested material is assimilated [21]. This would confirm the notion that *I. balthica* is an inefficient utilizer of the food mass, however large, it consumes. Similarly, *I. hectica* and *C. prismatica* occurring in the litter habitat do not seem to assimilate seagrass material in a significant manner [21] despite it is largely ingested. Reasons commonly invoked for this are the low quality of seagrass detritus in terms of high C:N ratios and residual toughness, so that micro- and macro epigrowth and algae associated to the litter are stated to be the main nutritional sources for the isopods [19,21]. Sphaeromatids of the genera *Sphaeroma* and *Lekanesphaera* are involved in the fragmentation of coarse seagrass detritus in brackish environments along with *Idotea* spp. [38]. Members of this family belonging to the genera *Cymodoce* and

Dynamene do occur in litter of *P. oceanica* (pers. obs.) but the extent of their role in detritus processing is unclear.

The second category of isopod feeders on decaying seagrass tissue is constituted by consumers of debris as defined above. They are mainly represented by the species *Limnoria mazzellae*, misidentified on discovery as *L. tuberculata* [43], which burrows into *P. oceanica* scales, i.e., the dead sheaths that persist on the stem after the blade has been shed. Scales build up a substantial mass refractory to decomposition and may have an important role in protecting meristems from grazing and environmental disturbances. Contrary to polychaete borers, *L. mazzellae* attack the more recent and less degraded scales into which they excavate narrow galleries of varying length and branching patterns, often hosting more than one individual [44]. Overcoming the toughness of sheaths is permitted by the "rasp-and-file" structure of the mandibles, similar to that of limnoriid wood borers. Burrowing results in both microhabitat expansion and supply of food resources. In both processes, isopods may benefit from the conditioning of scale tissue by microflora and fungal growth [41] which in turn is enhanced by the deposition of fecal pellets inside the burrows [44]. In essence, the association between the isopod and the microbiota functions as a holobiont, a consortium of symbiotic organisms working as a whole for plant tissue degradation. This microhabitat-wide association would clearly differ from the interaction between one individual and its endosymbionts. Data available so far indicate a host specificity of *L. mazzellae* to *P. oceanica* beds along the whole longitudinal extension of the Mediterranean Sea [39], although discontinuously in regard to the local occurrence and the proportion of shoots infested in a single bed [41,45].

7.5 Ecosystem effects of seagrass consumption by isopods: players and influencers

Isopods can contribute to top-down control of seagrass biomass, sometimes substantially according to laboratory or field estimates [7]. Extant data do not clarify to what extent a similar effect may be exerted on Mediterranean seagrass stands as far as living shoots are concerned. Estimates based on indirect methods indicated that grazing by *Idotea* spp. on blade tips affected 15 to 100% (depending on season and depth) of total bitten *P. oceanica* leaves in a Ligurian meadow [46]. These estimates were however based on the shape of bite marks as depicted by [25] whose specificity to idoteids needs, in my opinion, to be confirmed (see, e.g., [38] for bite marks on seagrass leaves by different peracarids). There is a lack of data on the grazing pressure exerted by isopods on smaller, faster growing seagrasses, which may constitute a more palatable substrate than *P. oceanica*. In feeding trials, *I. balthica* showed preference for *C. nodosa* over *P. oceanica* adult leaves [22], but this was a qualitative observation. The single observation of burrowing into leaves of *C. nodosa* by *Limnoria* (ca. 18% of shoots affected) needs confirmation on a larger scale.

Consumption of necromass (*sensu* [42]) by isopods may be substantial. Laboratory measurements of the impact of peracarid shredders, including *I. balthica* and *C. prismatica*, revealed an up to 5.4-fold reduction in the size of detrital particles in *P. oceanica* dead leaf accumulations [19]. *I. balthica* may contribute massively in this process as it can locally reach in litter abundances as high as 520 individuals per square meter [*11*], ranking second only to amphipods. As regards debris consumption, estimates of the average proportion of a single scale biomass consumed by *Limnoria mazzellae* yielded a value around 11% [*6*]. Generalizing estimates of detritus consumption across different beds is difficult due to the ephemeral nature of the litter on the one hand, and to the high location-dependence of borer occurrence and prevalence on the other. The role of isopods in the processing of detritus from seagrasses other than *P. oceanica*, which can constitute a substantial amount, has not been explored in the true marine domain although it is possibly significant similar to what has been observed in brackish environments [*38*].

Summarizing, on the basis of present knowledge, isopod plant feeders enter the trophic web of seagrass systems in various ways. They mediate the transfer of primary production to secondary consumers via the assimilation of plant epiphytes and other micro- and macroalgal material with a purportedly minor contribution of live seagrass tissue. In the cycling of detritus, they are involved in the initial processing of litter and debris by disrupting the detritus structure and so favouring the action of smaller-sized detritivores and microorganisms. In addition, they produce a considerable amount of feces containing unassimilated seagrass tissue. Conditioning microorganisms may have a higher activity in fecal pellets than in other categories of detritus, thus accelerating decomposition processes. In this respect, the possible role of feces in the *Limnoria*-microbiota holobiont has been mentioned above. Interestingly, feces of *I. balthica* have been shown to contain active phosphatases of endogenous origin which may accelerate degradation and remineralisation processes in the surrounding environment [*47*]. Conditioned fecal pellets may also be exploited by coprophagous metazoan consumers, particularly when retaining their size for a certain time period. In the case of remains of living seagrass, egestion with feces may be a sort of shortcut for the input of metabolically active tissue into the detrital food chain. In the end, trophic relationships within the system will culminate in predation on isopods, which appear to be a valuable prey for intermediate carnivores, possibly triggering a cascade modulating their impact on living and dead seagrasses.

Seagrass tissue removal by isopods may have appreciable indirect effects on other consumers. An effect on the palatability may be through the induction of deterrent compounds in grazed tissues as shown for *I. hectica* feeding on *C. nodosa* [*23*]. Such a modification in seagrass tissue properties may limit its availability to further grazing. Another indirect effect may be associated with the release of odour cues by wounded (grazed) seagrass tissue. Chemical signals represented by VOCs (Volatile Organic Compounds) from potential food sources may have different effects, either by determining the attractiveness towards the consumer or by warning against the presence of predators. Extracts of VOCs from wounded leaves of *P. oceanica* were demonstrated to elicit different responses in invertebrates depending on compound concentration, invertebrate species and water acidification conditions [*48*]. Isopods may promote the release of VOCs by removing seagrass tissue and, in turn, be influenced by them. The sphaeromatid *Dynamene bifida*, sometimes found associated with seagrass meadows, was attracted to Posidonia extracts but only at low doses, possibly because high concentrations might be indicative of the presence of large omnivore grazers (potential predators) causing an extensive damage to leaves [*48*].

Consumption by isopods may also lead to the creation of novel habitats inside the very structure of the seagrass plant. Burrows excavated by *Limnoria* spp. both in *P. oceanica* scales and in *C. nodosa* leaves often host a variety of small invertebrates including tanaids, harpacticoid copepods, polychaetes and other isopod species [*39,44*]. In that, limnoriids act as microhabitat engineers, occupying a further tile in the mosaic of interactions between isopods and seagrasses.

7.6 Conclusions and outlook

While diverse among species, major plant-feeding isopods associated with Mediterranean seagrass systems show a number of features apt to enable seagrass consumption in a significant manner. Idoteids associated with both litter and living leaves have the behavioural, morphofunctional and physiological means for dealing with the structural and chemical complexity of seagrass bio/necromass, and the potential for high per-capita feeding rates. Limnoriid burrowers match an efficient feeding apparatus with the ability to promote the growth of a symbiotic microbiota in their galleries. Sphaeromatids play a more obscure role as seagrass consumers, yet their action may be of relevance, at least in terms of sloppy feeding.

However, a better knowledge is needed of the various ways herbivory has taken and can take in the plant-feeder guild of seagrass isopods. The feeding behaviour has to be investigated in single species, populations or even individuals. One example is the role in seagrass consumption of less studied forms such as *S. appendiculatum* and *Cymodoce* spp., which are often dominant in isopod taxocenes [*9*]. One fundamental issue is the actual extent of generalism and trophic plasticity in idoteids and shaeromatids which may afford mixing resources including seagrass tissue or switching between them according to their availability. Further research should also be directed to the evolution of trophic relationships in a changing environment. For example, acid-tolerant herbivore isopod species [*49*] may have access to seagrass tissues of a better quality than under current pCO_2 conditions. Moreover, introduced seagrass species may constitute a novel food source for native grazers.

At an ecosystem scale, how all the traits shown by isopod consumers translate into an appreciable grazing pressure on Mediterranean seagrass formations is a matter that is not adequately resolved. What is needed is *in primis* targeted field research following a natural history approach. An accurate estimate of isopod occurrence and density across systems built up by different seagrass species should be obtained by implementing efficient sampling techniques, also accounting for the subhabitat partitioning among species and for variation among locations, seasons and bed morphologies. A quali-quantitative assessment of the action of the isopod feeders should be conducted using observational and manipulative field experiments [*4*]. The field approach would hopefully provide a sound frame for contextualizing and supplementing what is known about the behaviour of individual species as assessed in the laboratory, with the prospect of merging it in the general network of interactions within the seagrass communities.

References

1. L. Mazzella et al. (1992). pp. 165–187. *In*: John, D.M., S.J. Hawkins, and J.H. Price [eds.]. *Plant-Animal Interactions in the Marine Benthos*. Systematics Associations Special Volume n° 46, Clarendon Press, Oxford.

2. L. Mazzella et al. (1993). pp. 103–116. *In*: Özhan, E. [ed.]. *Proceedings of The First Conference on the Mediterranean Coastal Environment*, MEDCOAST 93, November 2–5, 1993, Antalya (Turkey).

3. J.A. Ott. (1981). *P.S.Z.N. Mar. Ecol.* **2(2):** 113–158.

4. R.J. Nowicki, J.W. Fourqurean, and M.R. Heithaus. (2018). pp. 491–540. *In*: Larkum, A., G. Kendrick, and P. Ralph [eds.]. *Seagrasses of Australia*. Springer, The Netherlands.

5. S.H. Brawley. (1992). pp. 235–263. *In*: John, D.M., S.J. Hawkins, and J.H. Price [eds.]. *Plant-Animal Interactions in the Marine Benthos*. Systematics Associations Special Volume n° 46, Clarendon Press, Oxford.

6. M.C. Gambi, B.I. van Tussenbroek, and A. Brearley. (2003). *Aquat. Bot.* **76:** 65–77.

7. V. Jormalainen. (2015). pp. 502–534. *In*: Thiel, M., and L. Watling [eds.]. *The Life Styles and Feeding Biology of the Crustacea*, The Natural History of the Crustacea Series. Oxford Univ Press, Oxford), Vol. 2.

8. M. Ledoyer. (1966). *Recl trav Stn ma. Endoume* **41(57):** 135–164.

9. M.C. Gambi, M. Lorenti, G.F. Russo, M.B. Scipione, and V. Zupo. (1992). *P.S.Z.N. I: Mar. Ecol.* **13(1):** 17–39.

10. M.B. Scipione, M.C. Gambi, M. Lorenti, G.F. Russo, and V. Zupo. (1996). pp. 249–260. *In*: Kuo, J., R.C. Phillips, D.I. Walker, and H. Kirkman [eds.]. *Seagrass Biology: Proceedings of an International Workshop*, Rottnest Island, Western Australia, 25–29 January 1996.

11. M. Dimech, J.A. Borg, and P.J. Schembri. (2006). *Biol. Mar. Medit.* **13(4):** 130–133.

12. M. Lorenti, and M.B. Scipione. (1990). *Rapp. Comm. Int. Mer. Médit.* **32(1):** 17.

13. P.L. Reynolds, J.P. Richardson, and J.E. Duffy. (2014). *Limnol. Oceanogr.* **59:** 1053–1064.

14. S. Vizzini. (2009). *Bot. Mar.* **52:** 383–393.

15. E. Naylor. (1955). *J. Mar. Biol. Assoc. UK* **34:** 347–355.

16. A.I.L. Natal. (2015). Phylogeny of the marine isopod genus *Idotea* in the Northeast Atlantic Ocean and Mediterranean Sea *[PhD Thesis]*. University of Porto.

17. E. Tinturier-Hamelin. (1963). *Cah. Biol. Mar.* **4:** 473–591.

18. T.M. Bell, and E.E. Sotka. (2012). *Oecologia* **170(2):** 383–93.

19. K. Wittmann, M.B. Scipione, and E. Fresi. (1981). *Rapp. Comm. Int. Mer. Médit.* **27(2):** 205–206.

20. N. Sturaro. (2005). Diversité trophique des Idoteidae associés à l'herbier et à la litière de *Posidonia oceanica* (L.) Delile. *Thesis Université de Liège*, 55 pp.

21. N. Sturaro, S. Caut, S. Gobert, J.M. Bouquegneau, and G. Lepoint. (2010). *Mar. Biol.* **157:** 237–247.

22. M. Lorenti, and E. Fresi. (1983). *Rapp. Comm. Int. Mer. Médit.* **28(3):** 147–148.

23. B. Martínez-Crego et al. (2015). *PLoS One* **10(10):** e0141219.

24. F. Tomas, B. Martínez-Crego, G. Hernán, and R. Santos. (2015). *Glob. Chang. Biol.* **21:** 4021–4030.

25. C.F. Boudouresque, and A. Meinesz. (1982). *Découverte de l'herbier de Posidonie*. Cahier n.4. Parc National de Port-Cros, Hyères, 79 pp.

26. J.M. Mattila, M. Zimmer, O. Vesakoski, and V. Jormalainen. (2014). *J. Exp. Mar. Biol. Ecol.* **455:** 22–28.

27. A. Scartazza et al. (2017). *Sci. Tot. Environm.* **607, 608:** 954–964.

28. E. Cruz Rivera, and M.E. Hay. (2000). *Ecology* **81:** 201–219.

29. L. Gutow, J. Strahl, C. Wiencke, H.D. Franke, and R. Saborowski. (2006). *Mar. Biol.* **149:** 821–828.

30. P. De Wit, K. Yamada, M. Panova, C. André, and K. Johannesson. (2018). *Sci. Rep.* **8:** 1–8.

31. R.J. Best, and J.J. Stachowitz. (2012). *Mar. Ecol. Prog. Ser.* **456:** 29–42.

32. S.M. Guarino, C. Gambardella, M. Ianniruberto, and M. de Nicola. (1993). *J. Mar. Biol. Assoc. UK* **64:** 21–33.

33. W.L. Lee, and B.M. Gilchrist. (1972). *J. Exp. Mar. Biol. Ecol.* **10:** 1–27.

34. K.M. Hultgren, and H. Mittelstaedt. (2015). *Curr. Zool.* **61:** 739–748.

35. R.J. Orth, G.A. Kendrick, and S.R. Marion. (2006). *Mar. Ecol. Prog. Ser.* **313:** 105–114.

36. C.L. Lee, Y.H. Huang, C.Y. Chung, S.C. Hsiao, and H.J. Lin. (2015). *Mar. Ecol. Prog. Ser.* **525:** 65–80.

37. D. Dumay. (1971). *Téthys* **2(4):** 763–907.
38. G. Mancinelli. (2012). *Estuar. Coast. Shelf Sci.* **110:** 125–133.
39. M. Lorenti. (2006). *Biol. Mar. Medit.* **13(4):** 154–157.
40. A. Brearley, and D.I. Walker. (1995). *Aquat. Bot.* **52:** 163–181.
41. P. Guidetti. (2000). *J. Mar. Biol. Assoc. UK* **80:** 725–730.
42. C. Boudouresque et al. (2016). *Hydrobiologia* **781:** 25–42.
43. L.J. Cookson, and M. Lorenti. (2001). *Crustaceana* **74(4):** 339–346.
44. P. Guidetti, S. Bussotti, M.C. Gambi, and M. Lorenti. (1997). *Aquat. Bot.* **58:** 151–164.
45. M.C. Gambi, M. Lorenti, S. Bussotti, and P. Guidetti. (1997). *Biol. Mar. Medit.* **4(1):** 384–387.
46. A. Peirano, I. Niccolai, R. Mauro, and CN. Bianchi. (2001). *Sci. Mar.* **65:** 367–374.
47. I.M. Böök, and R. Saborowski. (2020). *Mar. Biol.* **167:** 1–14.
48. V. Zupo et al. (2015). *J. Chem. Ecol.* **41(8):** 766–779.
49. L.M. Turner et al. (2016). *Mar. Biol.* **163(10):** 211–220.

CHAPTER 8
Ecology and Ethology of Littoral Amphipods

Felicita Scapini

8.1 Beach environments

Beaches are both marine and terrestrial environments, being subjected to the marine influence in the "littoral active zone", from the shoreline up the dunes, where seawater, splash and aerosol have effects on habitats, and the influence of terrestrial inputs through river mouths, freshwater runoff, sediments and plant debris in the dunes. These effects concur in maintaining rich ecosystems, where abundant and diverse macrofauna communities live. Nowadays, the economic importance of coastal environments is globally increasing, and several human activities are carried out on beaches: fishery, maritime and leisure uses, which, added to an increasing urbanisation, may negatively affect the littoral environments and linked ecosystems [1].

8.1.1 Littoral habitats

Beaches are characterised by habitat zonation, from absence of vegetation in the backshore above waterline, to colonising plants and established vegetation landward. Geomorphological features (height over the sea level, beach extension), sediments (mineralogy, granulometry) and hydrology (swash, tides, currents) may influence plant and animal life. Littoral communities are adapted to such peculiar conditions; they are characterised by few abundant species, which exploit inputs of both marine and terrestrial origin, but also are subjected to severe and changing environmental dynamics. In temperate regions, talitrid amphipods (Amphipoda, Talitridae) are key dwellers of littoral habitats, consuming kelp and wrack of marine origin stranded in the intertidal zone and plant debris in the dunes. The abundant populations that live in restricted littoral habitats are predated by several species, both marine and terrestrial, particularly birds, thus substantially contributing to the food web, by linking sea and land inputs.

Università degli Studi di Firenze. Italy.
Email: felicita.scapini@unifi.it

8.1.2 Habitat variation in space and time

Beach habitat zonation is variable in time, depending on tides, seasons and meteorological events, up to extreme events, which are increasing in frequency under climate change. Beach zones can be enlarged, restricted or displaced across the beach, up to disappearance. Under the risky conditions of displacement seaward or landward away from the suitable zone, burrowing and mobility are prerequisite for animal life in beaches, coupled with resistance to dehydration, moisture and salinity changes, which are common in coastal environments for the influence of runoff and river discharge into the sea. On the beach, talitrids can move from one zone to another, from the shoreline to the upper beach and dunes, to fulfil life needs and avoid environmental potential stressors. Through moving about and migrating, talitrids may exploit different littoral habitats, depending on the time of day, tides and seasons, and their life stage. Moreover, the habitats may be modified by talitrid activities (e.g., digging, burrowing, fragmentation and decomposition of wrack and vegetation debris), which is of high importance for ecosystem services and resilience. In the context of climate change, negative drivers may threaten beach ecosystems on large spatial and temporal scales (e.g., sea level rise, increasing temperature and storminess), while direct human actions may affect coastal environments locally on smaller spatial and temporal scales and, consequently, macrofauna communities [2].

8.1.3 Predictability/unpredictability of changes

Beaches have been always subjected to changes, both periodical (tides, seasons) and a-periodical (waves, winds, storms), the effects of which depend on geomorphological and hydrological beach characteristics, such as exposure, extension and types of sediments. Beach species show peculiar adaptations to physical variation of their environment, including a flexibility of adaptation, which is typically expressed in behaviour. Environmental changeability is predictable, both in the short (tides, seasons) and long term (changed marine currents, climate change), and talitrids are adapted to face such environmental changes, as far as habitats are resilient. When habitats are displaced or restricted, populations may recover, both in the short and long term; however, populations and ecosystems cannot face habitat disappearance or their strong reduction [1,2].

8.2 Talitrids as key species in littoral environments

Crustaceans have evolved in marine environments, and several major groups have reached and successfully exploited terrestrial habitats. The ecological subdivisions of decapods into crawling (benthic, *reptantes*) and swimmers (pelagic, *natantes*) may apply also to amphipods, even if most amphipods share a shrimp-like shape, which apparently is more suitable to swim than crawl or run. This is the case of the large Talitridae family (Amphipoda, Gammaridea), which has successfully colonised all types of temperate beaches, terrestrial humid areas and also tropical forests. Talitrid way of life is a paradigmatic example of terrestrial adaptation of original marine species, which colonised terrestrial habitats from the benthic sublittoral to sandy beaches up the dunes, periodically inundated coastal plains or estuary banks.

Following an "ecological" classification based on movement, as mentioned above, sand-hoppers (e.g., *Talitrus, Britorchestia, Megalorchestia, Atlantorchestoidea, Notorchestia*) colonise sandy beaches and dig themselves into the sand, while beach-hoppers (e.g., *Orchestia, Deshayesorchestia, Platorchestia*) crawl between pebbles and wrack and hide beneath [3]. However, this group has remained linked to humid habitats, at difference with some truly terrestrial decapods and isopods. In talitrids, the morphological and physiological constraints due to their marine origin, have been overcome through physiological and behavioural adaptations, confirming that behaviour may be a motor of evolution.

8.2.1 *Morphological and physiological adaptation and constraints to life in littoral habitats*

Swimming, clinging, crawling, hopping and digging are movement adaptations shared by talitrids, facilitated by a protective cuticle, the absence of carapace, relatively short antennae, legs of different lengths, lack of any relief on the head corresponding to the eye cuticle (a smooth surface offers less friction against sand when the amphipod is burrowing). These features are a clear adaptation to crawling, clinging on algae and burrowing into soft sediments, as the absence of a carapace and leg articulations are an adaptation to hopping (Figure 1). Gills are protected on the ventral side of the animals, are relatively compact in shape and allow oxygen exchanges in the air, as far as sufficient moisture is maintained on their surface.

Day-night changes in air humidity is a major driver of talitrid adaptation on beaches, particularly in the zone devoid of vegetation. On the other side, they face the risk of being swept away into the open sea by tides, waves and currents. Under submersion, the possibility of oxygen exchanges in seawater is still maintained, but other activities may be impaired. Individuals displaced in the sea tend to cling to floating materials, drifting with them and eventually reach a beach again [4]. Talitrid osmoregulation appears rather powerful and they can colonise beaches at the edges of low salinity waterbodies. Regarding spectral sensitivity, in *Talitrus saltator* vision appears adapted to terrestrial life, showing similarities with insects, with a higher sensitivity to short blue wavelengths with respect to marine amphipods that are more sensitive to yellow lights, which indicates a terrestrial evolution of this species [5].

8.2.2 *Advantages of living in littoral habitats*

The ecological role of talitrids on beaches can be defined as opportunistic, as they consume marine wrack, terrestrial plant and animal debris. So abundant populations may develop, strongly contributing to both the marine and terrestrial trophic networks [6]. Littoral habitats are interconnected at the sea side, less so at the terrestrial one, where beaches are interrupted by river mouths or headlands, or by urbanization. When accidentally (more and more often in the present time) littoral habitats disappear, populations may colonise new habitats exploiting marine transportation by floating algae or other material. In this way, exchange of individuals and genes between populations may occur at large spatial scale, allowing the recovery of populations, when the environmental conditions are maintained, but not on highly urbanised small beaches.

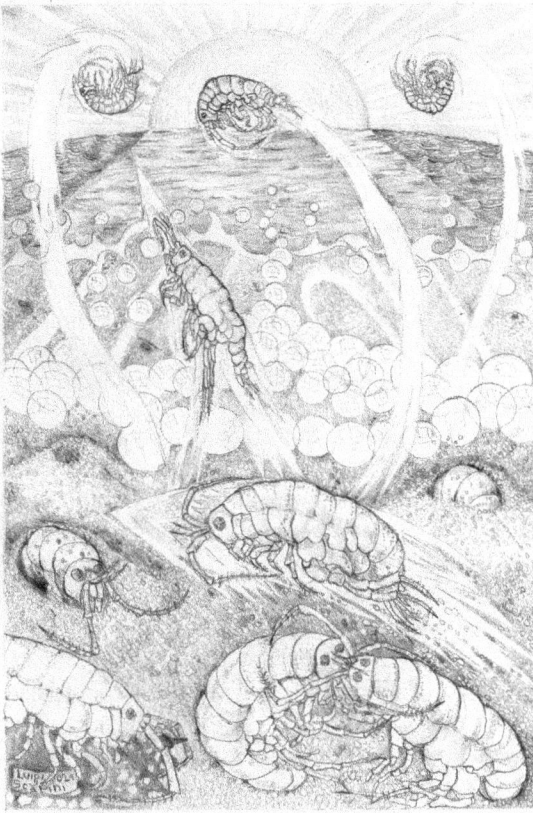

Figure 1. "What does it look like to be a sand-hopper or beach-hopper?" Crawling, hopping, digging, emerging, encountering a partner... Movement as key adaptation to "their world" at the land-sea interface. (vision of the artist Luigi Scapini).

8.2.3 *Trade-off: avoiding predation*

Abundant populations of relatively large, fragile and harmless talitrids, dwelling in specific zones of beaches, constitute a rich prey for many marine and terrestrial animals, particularly fishes, insects and birds. The strategies adopted by the prey to escape from predators take advantage of high mobility. Talitrids can hop high exploiting the wind, dig themselves rapidly into the sand and, when the substrate is hard, zigzag up and down, confounding potential predators. Talitrids may also adopt a freezing strategy, i.e., they suddenly stop moving when disturbed and are not-visible for their colour similar to the sand of the beach they live in. Under crowded conditions, which are common on narrow ecotones, small (young) or fragile (moulting) individuals may be cannibalistically eaten by con-specifics [7]. Avoidance strategies against cannibalism may be: not-overlapping rhythms of activity and different zonation of age classes [8].

8.3 Reproduction adaptations

The lack of larval stages and development within mother brood pouches are typical of all peracarids and are an important pre-requisite to life in extreme or border environments, such as beaches, which require specific adaptations. A K strategy of brood protection and release of young under optimal conditions, in the same habitat as the parents, may favour the establishment of populations in harsh and localised environments. This would favour the evolution of local adaptations, under relatively stable environmental conditions. On the other hand, as a consequence of habitat fragmentation, small populations may incur genetic bottle necks caused by inbreeding and risk disappearance in case of environmental changes, which for beaches is highly probable, under increasing urbanisation and climate change pressures [2].

8.3.1 Life cycle

Life cycle of talitrid species have been studied both in nature and under laboratory conditions for the abundance of natural populations and their relatively easy rearing. Specific points were addressed, like population dynamics and physical drivers, the duration of individual life, the possibility of a diapause, either in winter or summer, sex ratio and the determination of sex. Under natural conditions, population dynamics strongly depends on geographic and climatic conditions, as was analysed in populations living at different latitudes, on oceanic or Mediterranean or Baltic coasts [9]. Common features are the overlapping of cohorts, over-wintering adults and seasonal peaks of reproduction. In fact, new cohorts that appear earlier in spring reproduce in the same year, but do not survive the winter, while those that appear later in the summer, either reproduce in autumn, under favourable conditions, or the next spring following overwintering in the backshore. Single females breed monthly in correspondence of moulting, and, accordingly, population sampling should be made fortnightly. Moreover, individuals may overwinter, living up to two years, so that the populations should be followed for two years, to have complete data sets on the life cycle. These observations were confirmed under laboratory conditions, with controlled temperatures and light-dark cycle. The question is still open, whether the absence of reproduction in winter is a diapause, determined by the photoperiod. At high latitudes, short days coincide with lower temperatures, which reduce or stop (under 10°C) brood development [10]. In Mediterranean populations of *Talitrus saltator*, controlled rearing experiments under natural light cycle showed breeding also during the winter, thus excluding a diapause (personal observations, 1979–1985). Rearing single females and selected pairs under controlled conditions allowed to verify that females are inseminated during the moulting and do not store spermatozoa, which may have implications in genetic variability in relatively small populations. Males cling to moulting females, but, at difference from aquatic gammarids, do not guard inseminated females through a riding position, so that in nature females can be inseminated by a number of males (personal observations).

8.3.2 Constraints of the direct development and sex ratio

Females carrying embryos in their brood pouches, which become violet coloured, may appear more visible to day predators than males, but are also less mobile, remaining buried higher in the beach during the day. It is suggested that the behavioural adaptation of breeding females would be more efficient in avoiding risky conditions with respect to males. For example, breeding females not only occupy a safer zone in the backshore and dig themselves dipper, but also more rapidly recover the moist sand when displaced under dry conditions with respect to males; females also tend to display more precise activity rhythms and orientation. The evidences of such "better" behavioural adaptations (zonation, orientation and activity rhythms) are sparse, probably due to the overlapping of cohorts under natural conditions and the difficulty in distinguishing sexes when animals are sub-adults. However, when data were analysed separately for age classes, males and females, the evidence was univocal. This can be considered a sort of parental care; the question is still open whether females choose a specific zone of the beach and a more suitable time of the day to release the young. Under experimental conditions, females carrying broods do not release young when immersed in seawater [4]. There is also evidence that females live longer than males and are more abundant in the populations during non-breeding seasons, as in winter [9].

8.3.3 Geographical variation in life cycle

A major driver of the observed variation in life cycles is latitude, which causes seasonal and climatic differences, included gradients of temperature that may slow down or speed up development. One single peak of reproduction was shown in summer on British coasts, while two or three peaks were observed at smaller latitudes, on oceanic coasts in Portugal and on Mediterranean ones in Italy and Tunisia [9]. Univoltine to multivoltine populations have been observed across latitudinal gradients. Also, rainfall that reduces water and sediment salinity in coastal environments and favours pioneer plant growth may accelerate or block talitrid reproduction. In the Island of Gozo (Malta), a population was observed composed of a single cohort of non-active sub-adults deeply buried in a dry wadi (Scapini, personal observations, July 2001).

8.3.4 Population traits as bioindicators of ecosystem health

The monitoring of supratidal and intertidal talitrid populations has revealed their importance as bioindicators to assess beach ecological conditions. Life cycle characteristics, population structure, sex ratio, as well as development traits, such as fluctuating asymmetry in individual metrics, may be used as robust bioindicators of beach ecosystems [11,12]. The rationale is that healthy ecosystems allow the establishment of abundant populations that efficiently exploit the available resources, increase in size and reproduce. Intra- and inter-population variation will in turn favour a long-term maintenance of the ecosystem, in case of gradual changes of the beach system. The geographic diffusion of talitrids allows comparisons across spatial scales, which may replace time series, which are difficult to obtain on a long-term. Under severe impacts, the evidence of stressed populations may be achieved too late, when the recovery is no more possible: a comparative evidence, based on early warning bioindicators, may help to prevent ecosystem loss. Behavioural traits, observed in individuals, have been shown to be reliable bioindicators of ongoing environmental change, based on the idea that individuals "know" the environment where they live and "notice" changes, ready to contrast stressful factors [*13*].

8.4 Behavioural adaptation

Littoral amphipods are morphologically fragile, tiny and devoid of defensive arms, but express robust behavioural adaptation, particularly to face the physical risks of their habitat (Table 1).

Through behavioural adaptation (e.g., zonation, activity rhythms and migrations), they succeed in exploiting the whole beach-dune system and its habitats, from the intertidal zone up the dunes. Talitrids have become a model for orientation on beaches, genetics, development and plasticity of orientation [*14*].

8.4.1 Behavioural rhythms as adaptation to predictable changes

Tides show a high geographic variation that impacts on beaches, more on eastern Atlantic and Pacific oceanic coasts, but also on Mediterranean ones as in eastern Tunisia and Israel, in correspondence of flat plains. The impressive tidal range observed on the western coast of Europe may form extended humid habitats in the flat coasts, periodically inundated, hosting rich macrofauna communities that perform rhythmic vertical (up and down in the sediments) and horizontal (from one zone to another one) migrations. But also a few centimetres of tidal range may have influence on animals that live in and exploit the intertidal zone of the beach, as they risk displacement into the sea during ebbing tide and a severe compression of their habitat during rising tide. It is not surprising that intertidal animals evolved typical activity rhythms as adaptation to tides [*15*]. Rhythms allow to express behaviour at the most suitable time, avoiding risks and adapting to predictable changes. Night activity appears a major adaptation in talitrids, with strong endogeneity and persistence, both demonstrated under constant laboratory conditions. On sandy beaches, endogenous circadian rhythms may allow the recovery of the suitable activity after a period of inactivity of individuals buried in the sand. In the beach zone where vegetation is absent and superficial layers of sediments are subjected to dehydration during the sunny days, surface activity is safer during the night, when air humidity is high. Night-day rhythms have been observed in talitrids throughout the year, with variations depending on meteorological conditions and age [*16*]. Interestingly, emergence from the burrows within the sand was delayed in dry nights and anticipated in the young, which are more subjected to dehydration risk than adults; in rainy days, surface activity was observed also during daytime. Circadian activity rhythms, shown under constant laboratory conditions, tended to have periods slightly longer than 24 hours with a significant variation depending on the season:

Table 1. Life on sandy beaches: risks and adaptations of talitrids.

Risk	Causes	Adaptations	Evidences
Surface sediment dehydration	The heat from sun radiation, also favoured by sediment large grain size	Burrowing: the cuticle protects the body and eyes from sand friction, legs and antennae are used to dig. Seawards orientation, oriented by: sun, beach slope, wind, landscape vision	Direct observations; Sampling with corers/quadrats; Registrations of individual tracks; Orientation experiments using controlled arenas
Submersion and dislocation	Waves, tides, currents	Emergence from burrows; Landwards orientation on wet sand and in seawater; Zonation of surface activity; Endogenous activity rhythms	Direct observations; Registration of individual tracks in the intertidal zone; Orientation experiments with controlled arenas; Capture of surface-active individuals through pitfall traps across the beach; Registration of activity under controlled laboratory conditions
Osmotic stress	Salinity changes caused by runoff, tides, waves or evaporation	Choice of the suitable salinity; Daily and seasonal changes of zonation; Change of direction in orientation: seawards versus landwards	Choice experiments under controlled conditions; Capture across the beach; Orientation in water of decreasing salinity
Sediment compaction	Increased moisture and decreased grain size; Human trampling; Mechanical beach cleaning	Choice of the suitable moisture; Zonation of burrowing	Choice experiments under controlled conditions; Sampling across the beach with corers
Diminished food supply	Beach cleaning; beach erosion; changed currents;	Night migration to the dunes; Opportunistic change of diet; Cannibalism	Comparison of surface activity between species; Comparison of diet among species on the same beach; Comparison of different beaches; Observations under natural and laboratory conditions
Changes in temperature	Night-day; Season; Latitude; Long-term climate change	Changes in zonation; Life cycle adaptation: seasonal peaks of reproduction; Inactivity under non-suitable conditions (winter and/or summer)	Sampling across the beach; Population dynamics throughout the yearly cycle: sex-ratio, reproduction, identification of new cohorts, growth, mortality; Comparison of populations

in spring-summer 24 hour rhythms were more precise (in period, signal-to-noise ratio and day-to-day variation) than in autumn-winter; a variation in precision was observed within individuals, between individuals and among populations, with lower precision in the populations from more stable beaches, with less vegetal detritus, where amphipods could find a refuge [17]. Under natural conditions, an influence of tides on activity was also shown: these supralittoral animals avoid submersion by rising tide migrating towards the dunes, whereas feed in the intertidal zone during low tide. Under controlled laboratory conditions, a tidal component has been observed in the circadian activity rhythm, with higher evidence in populations from beaches subjected to large tidal ranges.

8.4.2 Zonation and choice of microhabitat

Behavioural adaptation permits to live in the most suitable microhabitat, by choosing specific physical-chemical conditions, which may be a challenge on continuously changing environments, such as sandy beaches. On beaches temperature and humidity, granulometric characteristics and salinity may show gradients from the waterline landwards, and periodic and a-periodic variation, depending on day time, season, tides, waves and meteorological events. Macrofauna community is zonated across gradients from the waterline up the dunes, and talitrids occupy a relatively narrow zone in the supralittoral fringe. Following displacement, they find the suitable place where to burrow into, choosing a particular grain size, moisture and salinity. When the substrate is hard and dry, displaced individuals circle around, or orient to reliable cues to recover the optimal zone and burrow into moist and soft sediments. The individual tracks become more and more straight with a growing experience of the environmental conditions [18]. Laboratory born individuals released on a beach made searching tracks with many directional changes, eventually orienting to the shoreline, while experienced individuals collected on the beach oriented promptly and showed straight tracks seawards, based on various environmental cues, both local (wind, landscape vision, beach slope) and universal (sun, polarised sky light, earth magnetic field).

8.4.3 Zonal orientation and zonal recovery: function and mechanisms

Zonal orientation in littoral amphipods, both sand-hoppers (*Talitrus saltator*) and beach-hoppers (*Orchestia gammarellus* and *O. montagui*), is observed in displaced individuals, which rapidly and directly crawl towards the waterline, where they would find moist sand, even if the vision of the sea is hidden by a screen or the individuals are released behind the dune. It was experimentally demonstrated that the animals use the sun or the polarized sky light to orient when displaced [19]. A sun compass mechanism had been firstly observed in bees and birds and it was surprising that the vision of amphipods was adapted to fulfil a similar task, which implies a recognition of the sun in the sky and an endogenous clock to compensate for the apparent sun movement throughout the day, varying with the season and latitude. Even more surprising was the orientation performance of sandhoppers at night, based on moon movement. These pioneer observations on astronomic orientation were confirmed in different populations and species, at different latitudes, both on Mediterranean and oceanic shores [*14*]. The link of sun orientation with the endogenous clock was analysed at individual (physiological) level, confirming the survival importance of the orientation mechanism [*20*]. Evidence was later obtained of a separate lunar clock [*21*]. Under natural conditions, various orientation mechanisms are individually calibrated to local cues, such as landscape characteristics, beach slope, dominant winds, swash and tides, which ensures survival in face of risky conditions. This is a learning process, favoured by the relatively long individual life on the same beach and the frequent (tidal and daily) migrations, whereas a genetic adaptation may be established across generations, if the shoreline maintains the same characteristics for a long time. The question is still open on the time required for the behavioural adaptation observed in natural populations. Long term and large (geographic) scale comparisons across talitrid species, populations and beaches have shown that sun orientation behaviour is widespread and persistent also under varying meteorological and beach geomorphological conditions [*22*]. These results confirm the importance of the behavioural adaptation in talitrids, which will possibly allow population survival and further evolution under climatic change pressure, if sandy beach habitats will persist.

8.4.4 Development of efficiency in orientation

Under natural conditions, the most risky stage in the development of littoral amphipods is when young are released from the mother brood-pouch to start an independent life on the beach; at this point, the parental care has fulfilled its function. Primarily, an inborn orientation mechanism would preserve the young from dehydration risk; the inborn mechanism would later be adjusted to the actual environmental context, by calibrating the universal and efficient sun compass with other simpler (however less reliable because locally variable) mechanisms, such as orientation to the most bright zone (the sea), shadow (kelp piles above watermark; dune silhouette), or beach slope [*14*].

8.5 Population genetics, speciation and evolution

In talitrids, the direct development, site fidelity and behavioural flexibility, to face environmental changes, large geographic diffusion and variation of habitats, have favoured genetic differentiation across populations, within and between species. At population level, site fidelity is a consequence of direct development, but also a result of the linear (one-directional) extension of beaches: mating occurs preferentially with neighbour con-specific along the shoreline, not in all directions. Moreover, individuals actively migrate across the beach, from shoreline to the dunes and vice versa, maintaining a precise location, partially contrasting the passive drifting by shoreline currents, which, however, occurs within a beach delimited by headlands or other natural/anthropogenic barriers.

Talitrus saltator (Montagu 1808) was morphologically re-analysed across eastern Atlantic and Mediterranean coasts and it was proposed that it is in fact a complex of species, with a geographic variation particularly between Atlantic and Mediterranean shores: on Atlantic and Baltic coasts the species *Talitrus saltator* is present, whereas *T. pachycheles* and *T. cloqueti* are distributed on Mediterranean shores [*23*]. Similar conclusions were reached using isoenzyme analyses at population level [*24*]. The highest genetic differentiation of *Talitrus* was observed across Mediterranean coasts, with distinct groups within the same species in Aegean, North-African, Adriatic and Tyrrhenian coasts, while the Atlantic populations were more similar to the Tyrrhenian than to the Adriatic populations. It is suggested that this group differentiated across the coasts of the Mediterranean, a sea delimited by three continents, and from there colonised the Atlantic coasts through the Gibraltar Strait. The differentiation of *T. pachycheles* (formerly *saltator*) has likely not occurred beyond the species level, as was shown by the broods obtained in the laboratory from crosses between Adriatic and Tyrrhenian populations, which, under natural circumstances, do never come into contact (personal observations, 1990). It is suggested that the group is still evolving, favoured by beach fragmentation. The meta-population structure of the group was confirmed at mitochondrial DNA level and a low gene flow between beaches of the same coast was observed [*25*].

8.5.1 Genetics of orientation and evolution

Daily migration of sandhoppers on beaches occurs on a relatively small spatial scale, from the backshore to the intertidal zone and vice versa, still it is of survival importance and has been subjected to evolution, likely through a severe selection. In talitrids, it was demonstrated for the first time the genetic basis of orientation behaviour, which, during individual life, would adjust to the real environment through experience [*19*]. Cross-breeding of differently oriented parents gave broods that oriented to an intermediate direction. Genetic variation was observed within natural populations, which would favour the establishment of new adaptations in case of shoreline changes and/or colonisation of new coasts: selected pairs gave differently oriented broods, confirming a high inheritance of this behaviour and intra-population variation [*19*]. When these evidences were first published, the debate was on the time scale necessary for the evolution of changes in behaviour, being evolutionary times in biology much longer than shoreline changes. Clear evidence was provided of a relationship between population heterozygosity and genetic determination of orientation, with populations with higher heterozygosity levels showing also a more precise orientation; these populations also lived on more stable beaches, thus suggesting a "genetic homeostasis" *sensu* Lerner [*26*]. Genetic homeostasis would favour the maintenance of genetic diversity in well-established abundant populations, whereas small populations, subjected to genetic bottle necks and high inbreeding would rely on orientation to local cues and/or learning by trials-and-errors [*19,27*].

Genetic diversity in natural populations is a pre-requisite for evolutionary changes, following environmental modifications. Behavioural adaptations may favour survival under changing environmental conditions, and contribute to maintain abundant populations, possibly structured in sub-populations, which has been observed in talitrids along coastlines on a large geographic scale. Future research on epigenetics and gene regulation might clarify the evolutionary mechanisms of adaptation in talitrids, which is critical when considering environmental impacts under climate change and ecosystem resilience through behavioural adaptations and plasticity [*2*].

References

1. A. McLachlan, and O. Defeo. (2018). *The Ecology of Sandy Shores*. (Elsevier, Academic Press, London).
2. F. Scapini, E. Innocenti Degli, and O. Defeo. (2019). *Estuar. Coast. Shelf. Sci.* **225:** 106236.
3. E.L. Bousfield. (1982). *Publ. Biol. Oceanogr.* **11:** 1–73.
4. L. Fanini, and J. Lowry. (2014). *J. Exp. Mar. Biol. Ecol.* **457:** 120–127.
5. A. Ugolini, G. Borgioli, G. Galanti, L. Mercatelli, and T. Hariyama. (2010). *Biol. Bull.* **219:** 72–79.
6. I. Colombini, M.A. Mateo, O. Serrano, M. Fallaci, E. Gagnarli, L. Serrano, and L. Chelazzi. (2009). *Acta Oecol.* **35:** 32–44.
7. C. Duarte, E. Jaramillo, H. Contreras, and K. Acuña. (2010). *J. Sea Res.* **74:** 417–421.
8. F. Kennedy, E. Naylor, and E. Jaramillo. (2000). *Mar. Biol.* **137:** 511–517.
9. J. Marques, S. Gonçalves, M. Pardal, L. Chelazzi, I. Colombini, M. Fallaci, M. Bouslama, M. ElGtari, F. Charfi-Cheikhrouha, and F. Scapini. (2003). *Estuar. Coast. Shelf Sci.* **58** (supplement): 127–148.
10. A. Ingólfsson, Ó.P. Ólafsson, and D. Morritt. (2007). *Mar. Biol.* **150:** 1333–1343.
11. T.A. Schlacher, D.S. Schoeman, A.R. Jones, J.E. Dugan, D.M. Hubbard, O. Defeo, C.H. Peterson, M.A. Weston, B. Maslo, A.D. Olds, F. Scapini, R. Nel, L.R. Harris, S. Lucrezi, M. Lastra, C.M. Huijbers, and R.M. Connolly. (2014). *J. Environ. Manage.* **144:** 322–335.
12. G. Pereira Frota, T.M.B. Cabrini, and R.S. Cardoso. (2019). *Estuar. Coast. Shelf Sci.* **223:** 138–146.
13. F. Scapini, L. Fanini, S. Gambineri, D. Nourisson, and C. Rossano. (2013). *Crustaceana* **86:** 932–954.
14. F. Scapini. (2014). *Estuar. Coast. Shelf Sci.* **150:** 36–44.
15. E. Naylor. (2010). *Chronobiology of marine organisms* (Cambridge University Press, Cambridge UK).
16. I. Colombini, M. Fallaci, E. Gagnarli, C. Rossano, F. Scapini, and L. Chelazzi. (2013). *Estuar. Coast. Shelf Sci.* **117:** 37–47.
17. C. Rossano, E. Morgan, and F. Scapini. (2008). *Chronobiol. Int.* **25:** 511–532.
18. S. Gambineri, F. Scapini. (2015). *Behav. Process.* **113:** 13–23.
19. F. Scapini. (2006). *Mar. Freshw. Behav. Physiol.* **39:** 73–85.
20. F. Scapini, C. Rossano, G. Marchetti, and E. Morgan. (2005). Animal. *Behav.* **69:** 835–843.
21. A. Ugolini, L. Hoelters, A. Ciofini, V. Pasquali, and D.C. Wilcockson. (2016). *Sci. Rep.* **6:** 35575.
22. F. Scapini, F. Bessa, S. Gambineri, and F. Bozzeda. (2019). *Estuar. Coast. Shelf Sci.* **220:** 25–37.
23. J.K. Lowry, and A.A. Myers. (2019). *Zootaxa* **4664:** 451–480.
24. V. Ketmaier, F. Scapini, and E. De Matthaeis. (2003). *Estuar. Coast. Shelf Sci.* **58** (supplement), 159–167.
25. V. Ketmaier, E. De Matthaeis, L. Fanini, C. Rossano, and F. Scapini. (2010). *Ethol. Ecol. Evol.* **22:** 17–35.
26. F. Scapini, M. Buiatti, E. De Matthaeis, and M. Mattoccia. (1995). *J. Evol. Biol.* **8:** 43–52.
27. F. Scapini, M. Buiatti, and O. Ottaviano. (1988). *J. Comp. Physiol. A* **163:** 739–747.

Chapter 9

Ethology of Crustaceans Influencing their Ecology

Giuseppe Mazza[1] and *Elena Tricarico*[2],*

9.1 Introduction

Crustaceans, particularly decapods, are considered good model organisms to study behaviour for their species richness and diversity, relatively large size, widespread availability and their suitability for rearing in laboratory conditions [1–3]. They have been extensively used to investigate relevant ethological theories, such as sexual selection, animal contests, resource assessment or social recognition [2,3], providing valuable inputs on these themes. Their behaviour has been studied also to improve production in aquaculture and gain more information on their ecology for conservation purposes [4]. Moreover, behaviour is an important factor in invasion ecology, being many marine and freshwater crustaceans currently notable invaders [5,6]. Given the importance of crustaceans outlined above, it is crucial to know and assess their behaviour to better understand their ecology and to promote more efficacious management practices. This chapter will review three important aspects of crustacean behaviour (agonistic, mating and antipredatory). It will demonstrate its diversity in relation to species, morphology and environment, and show how recent research has found new intriguing behavioural aspects in this taxon (e.g., personality, individual recognition). The majority of the published studies are on decapods, which are the most active and widely farmed order within the crustaceans, thus providing more opportunities for ethological research.

---

[1] CREA –Research Centre for Plant Protection and Certification (CREA-DC), via di Lanciola 12/a, Cascine del Riccio, 50125 Florence, Italy.
[2] Department of Biology, University of Florence, Via Madonna del Piano 6, 50019 Sesto Fiorentino (FI), Italy.
* Corresponding author: elena.tricarico@unifi.it

9.2 Agonistic behaviour

Agonistic behaviour is largely studied in crustaceans for several reasons: they have relevant weapons to fight (i.e., the claws), they use a variety of communication signals (chemical, visual, tactile) in a multimodal way, and they are aggressive [1,2,7]. The agonistic interactions are characterised by ritualised patterns that can be similar or different across species and orders [1,2,7,8], and with differences in levels of aggression depending on the species, the morphology and the ecology. For example, stomatopods are highly aggressive crustaceans, defending burrows with their raptorial appendages that can inflict lethal wounds. Within this order, mantis shrimps that have raptorial smashing appendages show more complex and aggressive agonistic patterns than those without [8]. Another example of the above-mentioned diversity is provided by the different used sensory media according to crustacean ecology and morphology: species living in turbid waters, caves or burrows or that are nocturnal rely more on chemical than visual cues during fights and for other important behaviours, such as social recognition and mating [1,7].

Intrinsic (e.g., neurochemical state, individual physical characteristics, fighting ability) and extrinsic factors (e.g., signals, value and type of resource, isolation, temperature, previous experience) can influence the agonistic behaviour and the encounter outcomes [1,2,9]. Hermit crab fighting behaviour exemplifies the combination of these two factors: individual survival, growth, and reproduction strictly depend on the occupancy of gastropod shells of appropriate size and shape, and shell value strongly influences the agonistic encounters [10]. In European populations of common hermit crab *Pagurus bernhardus* (Linnaeus, 1758), escalated shell fights occur when the shell at stake is of a higher quality than the attacker's shell (see in [2]). In contrast, motivations to acquire a new shell, in North American populations of long-wristed hermit crab *Pagurus longicarpus* Say, 1817, are exclusively influenced by the value of the shell it inhabits rather than by the quality of the shell offered in laboratory conditions [10]. Recent studies looking at outcomes of contests showed that spatial skills play a role in determining success of hermit crab fights: Lane and Briffa [9] found "that spatial components of skill (accuracy and precision) contribute to the chance of victory in contests. In fighting hermit crabs, the ability to hit a specific area of the defender's shell repeatedly is clearly an important determinant of victory" (Figure 1).

Studies on neuroendocrine control of aggression in crustaceans revealed that biogenic amines (e.g., serotonin, octopamine) and neurohormones (e.g., crustacean Hyperglycemic Hormone) are involved in the agonistic behaviour. Manipulation of these chemical levels can induce a temporary inversion of the hierarchical rank (and thus a change in the agonistic behaviour and aggressiveness) demonstrating again the importance of several intrinsic and extrinsic factors on the outcome of encounters (see in [1]).

Finally, the presence of parasites can also affect the agonistic behaviour in crustaceans. Males of the lesser swimming crab *Charybdis longicollis* Leene, 1938 infected by rhizocephalan barnacle *Heterosaccus dollfusi* Boschma, 1960 performed less and fewer aggressive patterns of agonistic behaviour when compared with uninfected males, suggesting that the parasite reduced aggression, possibly in order

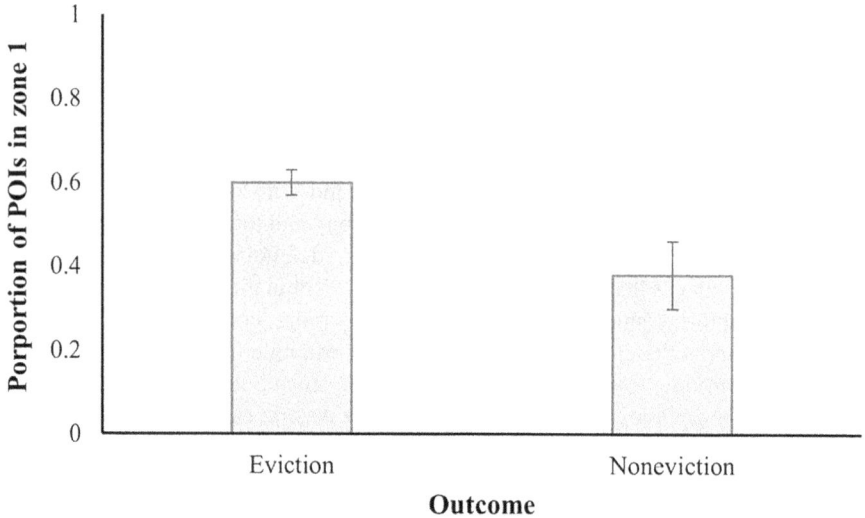

Figure 1. Proportion of points of impact (POIs) in zone 1 (= the body whorl adjacent to the upward-facing aperture) on the likelihood of eliciting an eviction in contests of the common hermit crab *Pagurus bernhardus*. Figure provides mean values ± standard errors. Modified after Lane and Briffa [9].

to increase the life expectancy of both host and parasite [*11*]. Similarly, the infection caused by the microsporidian *Cucumispora dikerogammari* Ovcharenko, Bacela, Wilkinson, Ironside, Rigaud & Wattier, 2009 reduced the intra- and interspecific competitivity of two Ponto-Caspian amphipods, the killer shrimp *Dikerogammarus villosus* (Sowinsky, 1894) and the demon shrimp *D. haemobaphes* (Eichwald, 1841), meaning they were less capable of defending their shelters [*12*].

Even in crustaceans, fights usually progressively escalate until one of the opponent retreats, with individuals establishing dominance hierarchies quite quickly (e.g., ten minutes in stomatopods: [*8*]). The following, but not mutually exclusive, mechanisms are responsible for the establishment and maintenance of dominance hierarchies also in crustaceans: the "winner & loser effect" (an animal behaves in accordance with its own experience independently of the rival); status recognition (an animal can recognise the status of the opponent by a pheromone, a posture or a behaviour, without any previous direct experience with it); and individual recognition (an animal can recognise the previously encountered opponents from chemical or visual cues exclusive to them—"true individual recognition"—or proper of one of two categories, such as mate vs. a non-mate, a higher-ranking vs. a lower-ranking opponent, revised in [*13*]). Among crustaceans, a few species seem able of individually recognising the opponent in agonistic contexts: the Australian crayfish *Cherax destructor* (Clark, 1936) and *C. dispar* (Riek, 1951), the American lobster *Homarus americanus* (Milne Edwards, 1837), the dwelling stomatopod *Neogonodactylus* (=*Gonodactylus*) *festae* (Nobili, 1901), the long-wristed hermit crab *P. longicarpus*, the fiddler crab *Uca capricornis* (Crane, 1975) and the eusocial shrimp *Synalpheus regalis* Duffy, 1996 (revised in [*13*]). This ability is particularly linked to the dear enemy phenomenon, that is the difference in the level of aggression displayed by territorial animals toward neighbours and non-neighbours (see in [*13*]). The presence of individual recognition, considered for long time a prerequisite of vertebrates, not only highlights the cognitive ability of crustaceans but also explains why in certain species individuals can live in burrows in close proximity, without continually fighting.

Agonistic interactions can occur within the same species or between different species of crustaceans. This latter aspect is particularly relevant for alien species (e.g., species accidentally or intentionally introduced by humans outside their native range) occupying similar niches or competing for the same resources (e.g., food or shelter) with native species. Many laboratory and field studies show that alien crustaceans are usually more competitive and aggressive than their native counterparts, leading to a population decrease or even displacement and extinction of the native species, both in marine and freshwater environments. This effect is seen across different crustacean orders (e.g., crabs and crayfish: [5,6]; amphipods: [14]). Moreover, by studying the interspecific interactions among alien crustaceans native to different biogeographic areas, it is possible to assess which species are more invasive and can thus exert highly negative impacts on invaded ecosystems [6,12].

Another factor playing a key role in agonistic encounters is personality (i.e., the consistent variation in behaviour among individuals) that has been found in several crustaceans (marine crabs, crayfish, freshwater amphipods and hermit crabs, reviewed in [15]). Bolder individuals (i.e., individuals usually reacting rapidly to a situation perceived as dangerous) tend to be more aggressive and prone to acquire a resource more quickly and/or explore an area. Indeed, bold males of the swimming crab *Portunus trituberculatus* (Miers, 1876) had higher fighting willingness and ability than shy individuals, this difference being closely correlated to energy metabolism [16]. Boldness is a highly variable trait in crustaceans and is related on species, population, sex and life stage, and differences in environmental conditions. For example, individuals of the Atlantic mud crab *Panopeus herbstii* H. Milne Edwards, 1834 inhabiting degraded reefs are bolder than those present on healthy reefs [17]. However, in some cases boldness is not associated to higher success in resource acquisition and reproduction, as found in the common hermit crab *P. bernhardus* which individuals heavily investing in the production of large spermatophores were less bold and less likely to take risks, thus safeguarding in this way their investment for future reproduction (see in [15]).

Personality is crucial when identifying traits associated with the invasive behaviour of some alien crustaceans [2,15]. As stated above, alien species are usually more aggressive than their native counterparts, thus likely to disrupt the invaded communities. For example, positive correlations among aggression, activity and boldness were found in populations of the signal crayfish *Pacifastacus leniusculus* (Dana, 1852) (see in [2]), meaning that it was able to outcompete the native endangered Shasta crayfish *P. fortis* in highly productive habitats and establish itself in low productivity streams. Personality related to agonistic behaviour and aggressiveness is also important in confined environment, such as aquaculture. Studies found that cannibalism and high levels of aggressiveness are common in commercially cultured crustaceans [4], and that bold individuals have higher feeding efficiencies than shy individuals, with higher glycogen and glucose content providing the corresponding energy basis for individual boldness [16]. An experimental reduction in food was shown to lead to a decrease in boldness and fighting frequency in the swimming crab *P. trituberculatus* [*16*]. However, starving animals is clearly not a solution for crustaceans being reared for aquaculture. Reducing stocking densities, providing abundant food and refuges, as well as favouring habitat enrichment and even manipulating molt-inhibiting hormones are more sustainable options that are shown to support both animal welfare and yield and quality of crustacean production [*4*].

Assessing agonistic behaviours in crustaceans is thus valuable for both explaining the ecology of this diversified taxon and providing useful inputs for applied sciences, such as aquaculture production and species conservation and containment.

9.3 Mating behaviour

Crustaceans are a remarkably successful group showing a wide diversity of life-history patterns with various sexual systems that range from gonochorism to various hermaphroditic conditions (simultaneous or sequential), with parthenogenesis occurring among several groups with very distant phylogenetic relationships (reviewed in [*18,19*]). The evolution of these several sexual systems is linked to the reproductive adaptation of different species to a diversity of ecological niches [*20*]. Moreover, due to the exceptional variety of habitats, body plans, and lifestyles, crustaceans exhibit a wide spectrum of mating systems (reviewed in [*19,21,22*]), a fundamental aspect of the reproductive biology of these animals.

Mating system refers to the general pattern by which males and females mate within a population or species. Mating systems, as well as the factors that determine them (e.g., parental care, territory, spatial distribution of mate, sexual selection), are diverse among crustaceans. The basic types of mating systems in Crustacea are monogamy and polygamy (reviewed in [*18,21,22*]). Although different subtypes of these two mating systems are widespread among crustaceans, caridean shrimps inhabiting coral reef niches exhibit a large diversity of subtypes of mating systems. Social monogamy (i.e., adult males and females forming exclusive pairs shares resources for an extended period of time) is more prevalent among the shrimps living in symbiotic association with sedentary-living invertebrates such as sponges. In the sponge-dwelling synalpheid shrimps, social monogamy has even attained the peak of complexity, leading to community living and eusociality (reviewed in [*23*]).

From an evolutionary perspective, the mating behaviour of a species is a product of a selection to increase reproductive success, with differences being seen between males and females. Sexual dimorphism is one of the most striking examples of sexual selection. Crustaceans exhibit pronounced sexual dimorphism, involving morphological traits (e.g., body size, large chelae) as well as behavioural traits (e.g., courtship displays, levels of aggression) (e.g., [*21*]). Usually, males compete for mates, possess exaggerated traits (e.g., big claws), and exhibit larger morphological and behavioural variation. Indeed, large dominant males are not only more successful in contests over females but are also preferred by females (e.g., [*18*]). Females have a higher parental investment compared to males, typically being the limiting sex and thus the resource that triggers the competition in males. Generally, in a continuum of competitive interactions, there are two extreme modes of competition, the "contest" and the "scramble". In contest competition, one individual gains access to the resource, thus limiting the ability of other competitors, while in the scramble mode (mainly found in lower crustaceans of the Anostraca), each individual tries to maximize its share in the limited resource, without direct interference from competitors (reviewed in [21]). Ecological and temporal constrictions strongly influence the spatial and temporal distribution of receptive females, which in turn influences the number of potential mates for males and the ability to monopolize them [24]. Based on the mating system classification described by Emlen and Oring [24], three general categories can be applied for most of the crustacean groups: (1) males that may search for or attract receptive females, defending them from other males; (2) males that defend and mate with the females associated with resources required for breeding and survival; and (3) males that never defend females and associated resources competing in order to maximize the rate of reproduction.

Overall, the contest for mating opportunities in crustaceans may be resource or female centred and it may be manifested in resource defence, direct competition for females, mate guarding, sperm competition, and alternative male mating strategies. For example, in many crustacean species (e.g., cancrid crabs), copulation occurs only when the female is soft shortly after moulting, with multiple copulations occurring in a relatively short period of time and sperm competition being the rule. Thus, this combination of soft-shell mating and sperm competition leads to pre- and post-copulatory mate guarding of females (e.g., [22]). Mate guarding, defined as temporary male-female association exceeding the insemination time, in crustaceans varies among taxa and can include a pre-copulatory mate guarding, to assess the female status or a post-copulatory mate guarding, considered as a behavioural adaptation in males to avoid sperm competition (reviewed in [21]). In contrast, resource defence in crustaceans usually involves a refuge such as crevices, burrows, or cavities. For example, in fiddler crabs of the genus *Uca* spp. males attract and guide receptive females into their burrows by waving their big claw, and when females enter in the burrows the males plug the entrance. Copulation takes place underground and after oviposition, males leave the burrow searching for more females with which to copulate. Thus, the possession of a good burrow is crucial for the reproductive success of the male, with the males competing for breeding burrows and defending them from other intruding males ([25]; Figure 2).

Finally, in several crustaceans, males exhibit alternative mating strategies and tactics, i.e., discontinuous variations in mating behaviour and morphology, mainly in species in which sexual selection through male contest competition is strong (see in [18]). As well as small "sneaker" males and large fighter males as seen in the marine amphipod *Jassa marmorata* Holmes, 1905, there are crustacean species that can change sex (reviewed in [19]). For example, in the partially protandrous alpheid shrimp *Athanas kominatoensis* Kubo, 1942 all the mature individuals are males and are capable of changing sex, and smaller and subordinate males become females [26].

However, in the context of mate choice, both male traits and female preferences co-evolve, and thus sexual selection may be stronger through mate choice than the male-male competition [21]. Females can choose a variety of male traits indicating a good partner and in crustaceans, female mate choice takes various forms. For example, chemical substances released with urine, visual and tactile signals and defended resources are used to distinguish a "good male" (e.g., see in [1]). Female mate choice is not limited to selecting a male for copulation, but also occurs at the

■ Resident wins ☐ Intruder wins

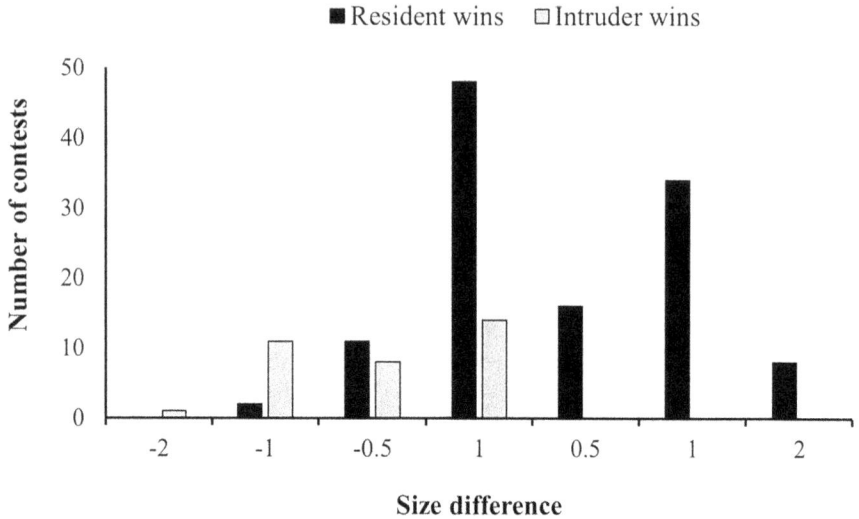

Figure 2. Number of contests won as a function of the size difference resident size category–intruder size category in males of sand fiddler crab *Uca pugilator* (0 = same size category, no obvious difference in size; ± 0.5 = same size category, one contestant obviously larger; ± 1 = one contestant medium in size, the other either large or small; ± 2 = one contestant small, the other large). Modified after Pratt et al. [*25*].

sperm level, in a process called cryptic female choice. For example, in the rock shrimp *Rhynchocinetes typus* H. Milne Edwards, 1837, the female can delay spawning when inseminated by subordinate males, reducing the fertilization success of their sperm [*27*]. Although mate choice is usually exercised by females, in crustaceans there is also evidence for the occurrence of male mate choice, as in the case of the swimming blue crab *Callinectes sapidus* (Rathbun, 1896), where the red marking size and brightness is correlated with body size, fecundity, and reproductive status of the females (reviewed in [*21*]).

Despite several studies investigating the factors influencing mating behaviour, strategies, and systems in crustaceans, the role of the interplay between multiple factors is still understudied. Crustaceans have colonized almost all conceivable habitats on the planet. Their diversity of lifestyles and colonized habitats have led to a huge diversity of reproductive adaptations, giving rise to multiple sexual and social systems, not yet fully understood.

9.4 Anti-predatory behaviour

Interactions between predators and prey play an important role in structuring communities and shaping evolution. Predation is an important component of most ecosystems and predators are known to control the populations, distributions, and behaviour of their prey, with indirect effects of predation on both prey and ecosystems having been widely demonstrated (e.g., [*28*]). In response to the threat of predation, natural selection has resulted in the evolution of a wide variety of morphological, physiological, chemical, and behavioural defensive adaptations in prey species. Crustaceans offer several examples of these adaptations.

Among terrestrial isopods, runners, spiny, rollers and clingers forms show behavioural and morphological adaptations that reduce predatory attacks for their survival. For example, spiny isopods show numerous protuberances located dorsally in the tergites and protect themselves mainly through morphological adaptation from predators [29]. Rollers such as *Armadillidium vulgare* (Latreille, 1804) can conglobate into a tight ball as a mechanism of protection, and clingers such as *Porcellio scaber* Latreille, 1804 either cling to a solid substrate or run to avoid predation.

Early detection of predator presence is key factor for the survival and the ability to distinguish predatory stimuli by direct tactile or visual cues, by non-contact chemical cues, or by mechanical (vibrational) stimuli is crucial for prey. Terrestrial decapods, for example, are highly dependent on visual signals and some land crabs rely exclusively on these cues to avoid predation [30]. In contrast, aquatic organisms often use chemical signals for communication (see [1]). Kairomones, substances released by living organisms, which are perceived by, and benefit individuals of other prey species are commonly involved in prey-predator recognition systems [1]. Disturbance cues, for example, are released before prey injury and have been found in crustaceans, and experimental evidence showed that urinary ammonia is a likely candidate for the disturbance cue in crayfish (see in [1]). There are many advantages in being able to detect a predator using non-contact cues, as they enable prey to avoid predators before direct interaction can occur.

After predator detection, risk assessment is performed before applying the decision to react. Aggregation, hiding, and crypsis are three of the most common anti-predatory strategies in aquatic species found also in crustaceans (see in [31]). Behavioural plasticity is among the most common anti-predator responses. Responses can be diverse and predator specific, even among the same taxa. For example, freshwater amphipods can decrease levels of activity, exploit structural refuge for hiding, or display grouping behaviour in the presence of kairomones from predators [31]. The most frequent benefit of aggregation, the act of gathering with conspecific individuals, is to lower predation risk through dilution or confusion effects (e.g., [31]). When discovered by a predator, many animals respond by adopting a characteristic threatening posture which appears to intimidate and startle the enemy, thus providing an opportunity for escape. Crayfish, for example, in response to predators, reduce their activity, increase the use of refuges, and can display meral spread with claws, used also in intraspecific encounters (see in [1]). Moreover, crayfish can employ one of several escape responses, including freezing (i.e., completely stop of activity), and swimming using the characteristic "tail-flip" (see in [1]).

Autotomy and regeneration of limbs have been observed for a variety of organisms and there is evidence that cheliped autotomy can be an effective mechanism to escape predation also in crustaceans (e.g., porcelain crabs; [32]). Indeed, in predator-prey interactions between different species of crabs, approximately 85% of the crab predators handled and consumed the autotomized claw of their prey instead of continuing search [33]. Despite its apparent immediate advantages for survivorship, autotomy can have important negative consequences for the prey. The costs can include impaired locomotion, foraging, survivorship, and/or reduced reproduction, in addition to the long-term energetic costs of replacing the autotomized appendage [33]. Male fiddler crabs of the genus *Uca* possess an enlarged major cheliped, lacking in females, and this cheliped is used in courtship, combat, and sound production (see in [*34*]). For example, in these peculiar crabs, males that have autotomized their major claw search for vacant mating burrows instead of fighting for occupied ones [*34*]. Crustaceans extensively use also alarm cues to escape predation (i.e., cues released from injured conspecifics when attacked by a predator): when detecting this odour, animals respond accordingly, performing antipredatory behaviour such as increasing decreasing display, or increasing/decreasing locomotion (reviewed in [*1*]).

One of the most fascinating behaviours demonstrated by different groups of animals encountering dangerous situations, crustaceans included, is thanatosis or feigning death. Thanatosis is a state of immobility in response to external stimuli that usually suppresses other behaviours. In this way, the probability of being predated is reduced due to the loss of interest by the predator towards the immobile prey and the increase of interest towards other mobile prey [*35*]. During thanatosis, animals exhibit a rigid body posture called tonic immobility with low respiratory activity and cardiac frequency [*36*]. Thanatosis in the freshwater anomurans of the Aeglidae family seems to be an important anti-predator behaviour, which varies according to the developmental stage of the animals and on the surrounding environment [*37*].

The various concealment strategies to avoid detection and predation are striking and sometimes unique among Brachyura. Concealment is observed in adults as well as in larval stages and juveniles. Individuals may also display colour patterns that match those of the background (crypsis). Concealment among brachyurans includes not only crypsis and mimicry, but also two other modalities of camouflage that are unique to Brachyura: carrying behaviour occurs in several families, involving morphological modifications of the last pair or last two pairs of pereiopods, and decoration behaviour, involving specialized hooked setae in the spider crabs (Majoidea) [38]. Decorating behaviour is one of the characteristic behaviours of majid crabs in which crabs attach other organisms, such as algae, sponge, and hydroids, to hooks on their carapace after manipulating them with the mouthparts: the adaptive value of this decoration appears to be anti-predatory. Items used in decoration are often chemically defended plants or sessile animals, and it seems plausible that predators detect the crab but actively avoid attacking because of repellent smell or taste from the decorations. However, not all decorations provide the animal with chemical defence, and, likely, decoration often functions through crypsis via background matching, masquerade, and/or disruption [*39*]. Behaviour and camouflage type can also be linked to habitat use. For example, the prawn *Hippolyte obliquimanus* Dana, 1852 exists in two main morphs—a more sedentary homogenous type that uses background matching to resemble different algal backgrounds, and a more mobile striped form that seems to rely on transparency for concealment ([*40*]; Figure 3). Transparent prawns are mainly males (the homogenous forms are mostly female) and thus are advantaged in their mobile life history and generalist habitat use, including mate-searching behaviour [*40*].

As demonstrated, the anti-predator adaptations of crustaceans are extremely complex, and combinations of the various strategies, described above, are frequent and can be found in Amphipoda, Isopoda, and Decapoda. Selective pressure has frequently produced parallels in anti-predator approached, not only among crustaceans but also with other arthropods and vertebrates (e.g., [*33, 39*]).

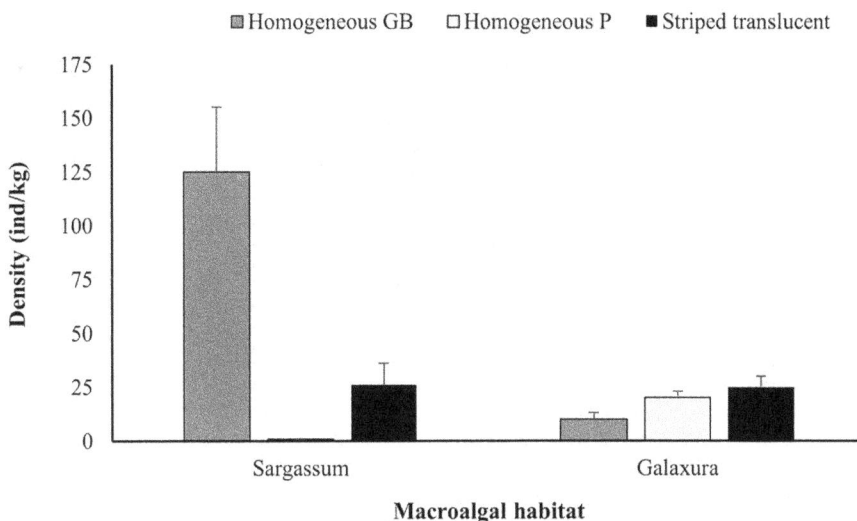

Figure 3. Average density of colour morphs of the prawn *Hippolyte obliquimanus* in main macroalgal habitats *Sargassum furcatum* and *Galaxaura marginata*. Colour morph abbreviations: GB, Greenish-brown; P, Pink. Figure provides mean values +1 standard error. Modified after Duarte and Flores [40].

9.5 Conclusion

Understanding animal behaviour is becoming increasingly important for its contribution to nature and species conservation as well as to sustainable and ethical farming (including aquaculture and mariculture). Crustaceans represent excellent model organisms for ethological studies. The results provided here demonstrate the diversity of the behavioural patterns present within and among the different crustacean groups and show how many aspects (e.g., personality and mating behaviour and the role of the interplay of multiple factors in mating behaviour) require further study. Newly discovered abilities of these animals (e.g., individual recognition, personality), traditionally considered to only be found in vertebrates, highlight their importance in the study of behavioural ecology and applied ethology. It also increases the importance of the debate on their sentience as it is highly likely that decapod crustaceans (crabs, lobsters, prawns, and crayfish) can feel pain [41]. As such, this opens a new avenue of research on the possible presence of emotions in these invertebrates. Acquiring information on new behavioural traits, as well as increasing knowledge on understudied ones is thus particularly critical for the implication on animal welfare legislation. It will also play a key role in improving our knowledge and understanding of crustaceans in order to support conservation activities and develop more sustainable and ethical food production.

Acknowledgments

We are in debt with Dr. Valerio Zupo for his continuous support and patience during the drafting of the chapter. Our warmest thank to Dr. Jodey Peyton (UK) for English and draft revision.

References

1. T. Breithaupt, and M. Thiel [eds.]. (2011). *Chemical communication in crustaceans* (Springer, Berlin), 565 pp.
2. F. Gherardi, L. Aquiloni, and E. Tricarico. (2012). *Curr. Zool.* **58:** 567–579.
3. M. Briffa. (2013). In *Animal contests* (Cambridge University Press, Cambridge), pp. 86–112.
4. N. Romano, and Z. Chaoshu. (2017). *Rev. Fish. Sci. Aquac.* **25(1):** 42–69.
5. J.S. Weis. (2010). *Mar. Freshw. Behav. Physiol.* **43(2):** 83–98.
6. E. Tricarico, and L. Aquiloni. (2016). In *Biological Invasions and Animal Behaviour* (Cambridge University Press, Cambridge) pp. 291–308..
7. L. Ayres-Peres et al. (2015). *J. Zool.* **297(2):** 115–122.
8. R. Caldwell, and H. Dingle. (1975). *Naturwissenschaften* **62:** 214–222.
9. S. M. Lane, and M. Briffa. (2020). *Anim. Behav.* **167:** 111–118.
10. E. Tricarico, and F. Gherardi. (2007). *Behav. Ecol.* **18:** 615–620.
11. G. Innocenti, N. Pinter, and B.S. Galil. (2003). *Can. J. Zool.* **81:** 173–176.
12. J. Kobak, M. Rachalewski, and K. Bącela-Spychalska. (2021). *NeoBiota* **69:** 51–74.
13. L. Aquiloni, and E. Tricarico [eds.]. (2015). *Social recognition in invertebrates: the knowns and the unknowns* (Springer, Berlin, 2015), 266 pp.
14. Ł. Jermacz et al. (2015). *Behav. Ecol.* **26:** 656–664.
15. M. Briffa. (2020). In *The Natural History of the Crustacea. Reproductive Biology* (Oxford University Press, New York) pp. 503–525.
16. X. Su, B. Zhu, and F. Wang. (2022). *Aquac. Res.* **53(2):** 419–430.
17. B.A. Belgrad, J. Karan, and B.D. Griffen. (2017). *Anim. Behav.* **123:** 277–284.
18. T. Subramoniam. (2013). *J. Biosci.* **38(5):** 951–969.
19. C. Benvenuto, and S. Weeks. (2020). In *The Natural History of the Crustacea. Reproductive Biology* (Oxford University Press, New York) pp. 197–241.
20. G. Vogt. (2020). In *The Natural History of the Crustacea: Reproductive Biology* (Oxford University Press, New York), pp. 145–176.
21. A. Barki. (2008). In *Reproductive biology of crustaceans. Case studies of decapod crustaceans* (Science Publishers, Enfield, NH, USA), pp. 223–265.
22. A. Palaoro, and J. Beermann. (2020). In *The Natural History of the Crustacea: Reproductive Biology* (Oxford University Press, New York), pp. 275–304.
23. S. Thanumalaya. (2006). *Sexual biology and reproduction in crustaceans* (Academic press, London).
24. S.T. Emlen, and L.W. Oring. (1977). *Science* **197(4300):** 215–223.
25. A.E. Pratt, D.K. McLain, and G.R. Lathrop. (2003). *Anim. Behav.* **65(5):** 945–955.
26. Y. Nakashima. (1987). *J. Ethol.* **5(2):** 145–159.
27. M. Thiel, and I.A. Hinojosa. (2003). *Behav. Ecol. Sociobiol.* **55(2):** 113–121.
28. A. E. Scherer, and D.L. Smee. (2016). *Chemoecology* **26(3):** 83–100.
29. H. Schmalfuss. (1984). *Symp. Zool. Soc. London* **53:** 49–63.
30. J.M. Hemmi. (2005). *Anim. Behav.* **69:** 615–625.
31. M. Rolla, S. Consuegra, and C.G. de Leaniz. (2020). *Aquat. Invasions* **15(3):** 482–496.
32. M.L. Knope, and R.J. Larson. (2014). *Mar. Ecol.* **35(4):** 471–477.
33. Z. Emberts, I. Escalante, and P.W. Bateman. (2019). *Biol. Rev.* **94(6):** 1881–1896.
34. I. Booksmythe et al. (2010). *Behav. Ecol. Sociobiol.* **64(3):** 485–491.
35. T. Miyatake, S. Nakayama, Y. Nishi, and S. Nakajima. (2009). *Proc. R. Soc. Lond. B* **276:** 2763–2767.
36. T. Miyatake et al. (2004). *Proc. R. Soc. Lond. B* **271:** 2293–2296.
37. A.B. Coutinho et al. (2013). *J. Nat. Hist.* **47(41-42):** 2623–2632.
38. B.C. Guinot, and M.K. Wicksten. (2015). In *Treatise on Zoology – Anatomy, Taxonomy, Biology* (Brill, Leiden and Boston), pp. 583–638.
39. G.D. Ruxton, and M. Stevens. (2015). *Biol. Lett.* **11(6):** 20150325.
40. R.C. Duarte, and A.A. Flores. (2017). *J. Mar. Biol. Assoc. UK* **97(2):** 235–242.
41. R.W. Elwood. (2019). *Phil. Trans. R. Soc. B* **374:** 20190368.

Chapter 10

Crustacean Ecology in a Changing Climate

Jörg D. Hardege* and Nicky Fletcher

10.1 Introduction

One of the greatest challenges facing today's marine ecosystems is the ongoing threat of climate change. The change in seawater chemistry (also known as Ocean Acidification) is driven by the increased presence and accumulation in the atmosphere of the primary greenhouse gas, Carbon Dioxide (CO_2). The main but not sole source of this increase is through anthropogenic activity. The burning of fossil fuels, cement production, agriculture and deforestation are contributing significantly to this increase [1]. The overall concentration of CO_2 has increased at an alarming rate since the pre-industrial age, with levels rising from 280 ppm to in excess of 400 ppm in recent years [2]. As the oceans cover 70% of the world's surface they are a key component to the carbon cycle and absorb approximately 30% of the world's carbon emissions [1]. This ever-increasing rate of CO_2 absorption is resulting in the increased acidity of our oceans through the formation of carbonic acid and H^+ ions, as such altering the seawater pH (Figure 1A). With climate change, sea surface temperature (SST) has also warmed dramatically, resulting in a range of changes seasonally and geographically. Figure 1B highlights how the interactions of physico-chemical changes in seawater interact, since the solubility of CO_2 is temperature and salinity as well as pressure dependent.

The net result of the complexity of these interactions is a geographical diversity of the impacts of climate change globally and even locally, which is reflected in pH prediction maps (Figure 2).

Through the dissolution of CO_2 the average open ocean pH levels have dropped by 0.1 units since the pre-industrial age with a current average of 8.15. The Intergovernmental Panel on Climate Change (IPCC) has predicted that pH levels

School of Biological, Biomedical, Environmental Sciences, University of Hull, Hull, UK.
* Corresponding author: j.d.hardege@hull.ac.uk

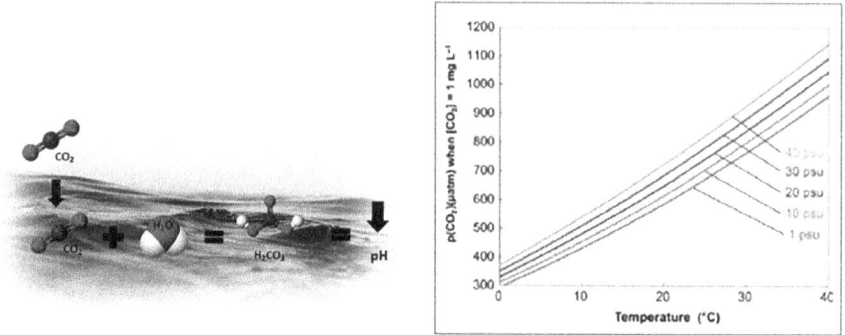

Figure 1. (A) Chemistry of climate change, how CO_2 results in pH change. (B) The effect of temperature and salinity on the amount of dissolved CO_2 in water. [3].

will fall to an average of 7.7 by the year 2100 with global atmospheric temperature increasing by 1.8–4°C (IPCC) and to 7.4 by the year 2300 [4; Figure 2].

Seawater pH varies significantly over the seasons with higher CO_2/lower pH in summer [5], such as in the North Sea due to the North Atlantic Oscillation. In addition, the day/night cycle brings about diurnal pH fluctuations that impact biota's responses to ocean acidification [6]. The variability and complexity of predicting oceanic mean pH and CO_2 concentration (pCO_2) is highlighted in near shore, estuarine ecosystems where latitude, freshwater influx, local upwellings tidal changes and macrophyte growth [7] influence water chemistry [8].

Climate change is predicted to further increase the variance and consequently, extreme lows of oceanic and coastal seawater pH globally [8]. The increase in pH amplitudes (–0.56 pH) will exceed the mean seawater pH decrease (–0.42 pH) in open oceanic waters [7; Figure 3A) and even more so in estuarine systems. Salinity distributions are shifted in estuarine environments due to sea level rise pushing seawater further up the estuary but also fluctuations appear with seasons and rainfall. Here, low salinity determines a lower pH buffering capacity [9; Figure 3B].

The decrease in dissolved oxygen or de-oxygenation in the world's oceans is a serious consequence of our warming oceans. Oxygen is crucial to all aerobic life and this reduced availability would have long lasting and disastrous effects on marine organisms. Reduced oxygen will also impact the biogeochemical cycling of the ocean elements (in particular the nitrogen cycle), which will have global impacts and could ultimately enhance global warming further [10]. Marine organisms appear to be insensitive to fluctuating oxygen levels as long as concentration levels remain high enough; if they reduce too far this leads to organism stress and over an extended period would result in mortality and the area becoming hypoxic or a 'dead zone' [10].

As described, through increasing research we now have a good understanding of the physico-chemical implications of climate change on the environment, and we are starting to understand some of the physiological and behavioural challenges marine life is exposed to in changing oceans (Table 1).

Crustaceans inhabit a wide variety of ecosystems; marine, freshwater, brackish and are known to be able to regulate or conform to fluctuating external environmental variables like salinity, temperature and pH [27]. As described (Figure 3A, B),

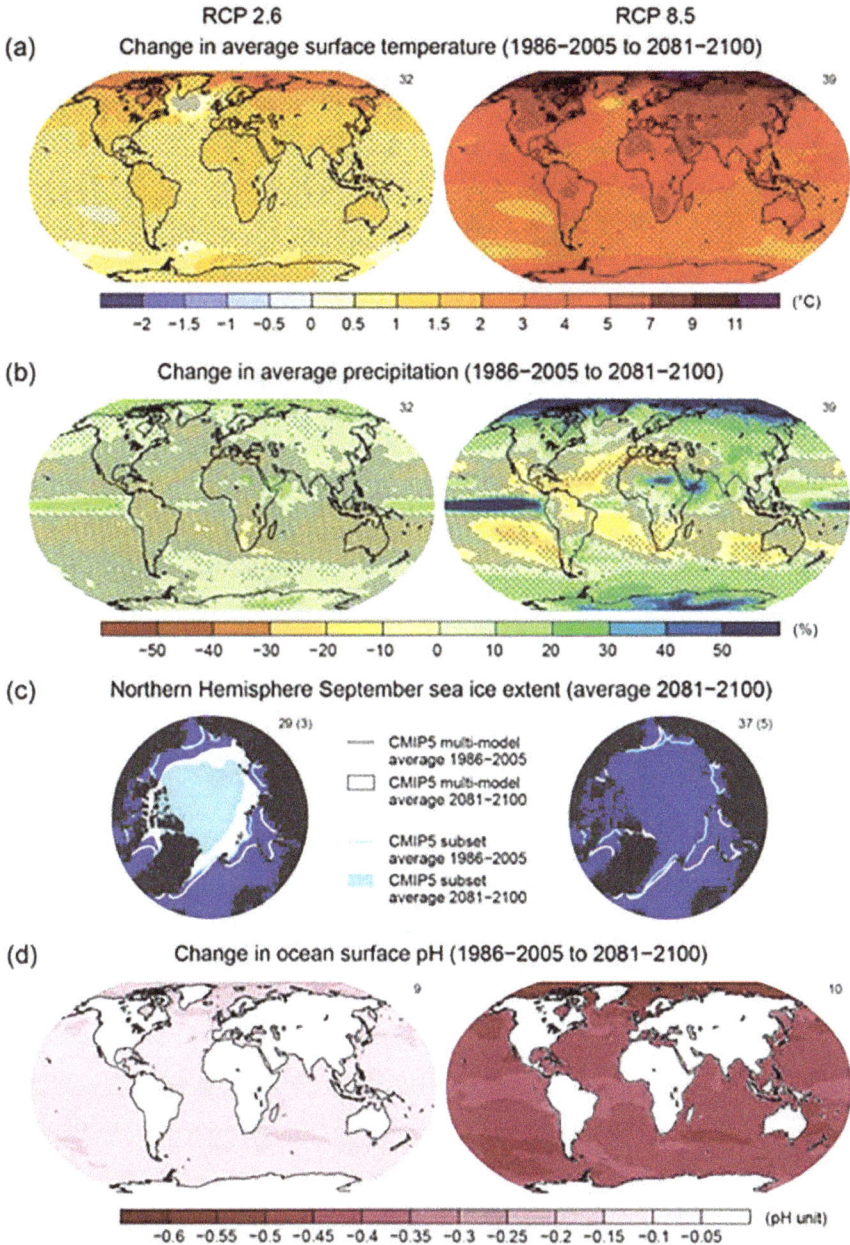

Figure 2. Maps of projected late 21st century annual mean surface temperature change, annual mean precipitation change, Northern Hemisphere September sea ice extent, and change in ocean surface pH (from: IPCC report on climate change 2013); see: http://www.climatechange2013.org/images/figures/WGI_AR5_FigSPM-8.jpg.

intertidal habitats are already exposed to fluctuations in both temperature and pH with organisms adapting to the changing environment. However, with ocean acidification

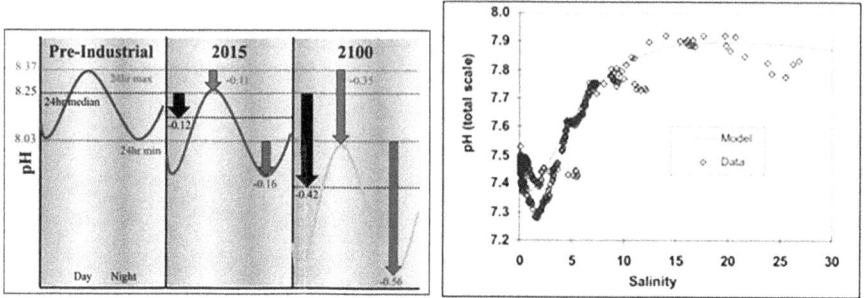

Figure 3. (A) Variance of daily (diurnal, nocturnal) pH fluctuation in pre-industrial times, 2015 and 2100. The median (black), maximum and minimum values are shown [7]. (B) Relation between water pH levels and increasing of salinity in estuaries [9].

Table 1. Examples of challenges incurred by marine crustaceans due to Climate Change.

Species	Impacts	References
Metapenaeus joyneri	Reduced swimming ability	[11]
Callinectes danae	Osmoregulatory function impaired	[12]
Chionoecetes bairdi	Structural integrity of the carapace decreases	[13]
Elasmopus rapax Cyathura carinata	Reduction in fitness, unable to reproduce	[14]
Eriphia verrucosa	Decreased growth rate	[15]
Homarus americanus	Increase in disease	[16]
Homarus americanus	Calcification processes impaired	[17]
Homarus americanus	Increase in disease susceptibility and cardiac performance	[18]
Leptuca uruguayensis; Leptuca leptodactyla	Increased use of energy reserve and burrow duration	[19]
Nephrops norvegicus	Immune suppression and protein damage	[20]
Pagurus bernhardus	Effects chemosensory responses to food odour	[21]
Palaemon intermedius and *Palaemon serenus*	Increased foraging in presence of predator, a shift in growth/survival trade off	[22]
Panopeus herbstii	Impaired foraging behaviour	[23]
Panulirus argus	Increased mortality	[24]
Petrolisthes cinctipes	Juvenile survival decreased and heart rate increased.	[25]
Petrolisthes cinctipes	Respiration rate decreased	[26]

and the warming of the waters, these fluctuations will become more extreme causing physiological stress on the organisms that inhabit these zones. To date, few studies have addressed the impact of diurnal, variable pH on marine life [28].

Crustaceans are species rich and show a wide range of morphology and life history traits, they are primary and secondary consumers and therefore have extensive ecological importance [27]. They are commercially valuable and highly invasive due to their resilient nature. Despite this, crustacean physiology (including growth rate, respiration rate, immunology, chemoreception and neuro-physiology) is impacted by the climate change driven alterations in the physico-chemical properties of the aquatic environment [*16,27*]. In the last decade research has increased looking at the physiological and behavioural implications of climate change to assess how this will impact marine ecosystems in years to come (Table 1).

10.2 Physiological implications of climate change on crustaceans

Maturation, moult and respiration rate of crustaceans are impacted by temperature [*29*]. Organisms that use calcium carbonate ($CaCO_3$) to create shells are at particular risk of physiological constraints with the changes of water chemistry incurred through climate change. These organisms include many crustaceans that use calcite in the production of their shells. After the $CaCO_3$ has formed, it dissolves back into the ocean unless there is high enough concentrations of carbonate ions. Due to this and other variables (including temperature, CO_2 and enriched water), there is a distinct saturation boundary whereby all $CaCO_3$ below this line dissolves and all above does not [*1*]. All organisms that have shells reside above this saturation line, and the increase in CO_2/resulting decrease in seawater pH affects this fine balance, raising the saturation line closer to the surface. Altering pH also makes calcification harder and weakens any existing structures [*1*]. This is the case in the commercially valuable Dungeness crab, *Metacarcinus magister*, known to have severe carapace dissolution (due to increased acidification of US coastal waters), potentially impacting mechanoreceptors that have an important function in the crab's sensory and behavioural capabilities [*30*]. Negative impacts to the exoskeletal properties of the Tanner crab, *Chionecetes bairdi* have also been observed, with post-terminal moult females exhibiting shell thinning and degradation to structural integrity [*13*].

Changes in environmental conditions have been found to have numerous impacts on crustacean physiology [*16*]. This is partially due to crustaceans being ectotherms and therefore relying on external sources for temperature control, thus changes in temperature affect their physiological and immunological responses making them more prone to disease, which in turn affects morbidity and mortality rates [*16*]. Ocean acidification and warming has shown to affect both disease susceptibility and cardiac performance in the American Lobster, *Homarus americanus* increasing death rates and lowering hemocyte counts [*18*] Shields [*16*] has concluded that global warming through anthropogenic stresses is causing such a major impact on the disease framework (resulting in changes to both the host and the pathogen), that in years to come there will be many new emerging pathogens causing detrimental impact to crustacean hosts.

Over the past decades there has been a measurable decline in dissolved oxygen in our oceans, this is resulting in oxygen deficiency and hypoxia in some regions. Although this is known to be a naturally occurring phenomenon on the short-term relating to seasonal change, over time the gradual decrease in dissolved oxygen concentrations is a direct result of climate change and the warming of our oceans [*31*]. This decline in oxygen availability and the impacts temperature and pH has on the respiration capabilities of our marine organisms could have catastrophic implications.

The respiration rate in the porcelain crab, *Petrolisthes cinctipes*, is affected by both the combination of altered temperature and pH, and by temperature as an independent factor, however is less impacted by pH change alone [26]. Reduced metabolism of embryonic development and heart rate in juvenile *P. cinctipes* were observed under constant reduced pH conditions, negatively impacting survival rates [25]. It is also possible that the observed 30% decrease in the swimming ability of the shallow water coastal decapod, *Metapenaeus joyneri*, is the result of reduced metabolic scope brought about through exposure to CO_2 enriched water [11].

The increased metabolic demand on organisms incurred through global warming is resulting in greater food consumption and increased risk taking by prey species [22]. Two different species of fiddler crab have shown differing responses to global warming depending on their habitat. Here those that reside in a vegetated area (*Leptuca uruguayensis*) are more vulnerable to increased temperatures than those that inhabit non vegetated (*Leptuca leptodactyla*), the latter was shown to be able to adjust its metabolic rate and reduce its ammonia excretion in response to rising temperatures. However, both species used increased energy reserves [19]. Alteration in seawater pH can be at least partially buffered internally through hemolymph proteins (mainly haemocyanin) and bicarbonate ions, therefore strong iono- and osmoregulators will be more able to withstand fluctuating pH levels (at least in the short term—there is a comprehensive review [27]). Immune suppression and protein damage has been reported in the Norway lobster, *Nephrops norvegicus* [20]. Osmoregulatory function and capacity in *Callinectes danae* was affected by reduced pH and altered salinities [12] showing physiological costs of coping with altered water chemistry. Evidence has shown the impacts these oceanic condition changes have on both predator and prey species and how one physiological change can alter the behaviour or fitness of the next, suggesting that this chain reaction will occur further up the food chain. An example of this can be found in the decrease in the condition index of the mussel, *Brachidontes pharaonis*; when held at reduced pH this resulted in the decreased growth rate of its predator *Eriphia verrucosa,* probably due to the decrease in food quality [15].

The increase in ultraviolet radiation brought about by the reduction in the Earth's stratospheric ozone layer due to climate change is impacting aquatic ecosystems. This was demonstrated by MacFadyen [32], who showed how the increase in UV exposure rates to *Daphnia pulicaria* impacted their DNA repair processes and suggested how crustaceans that are reliant on this process—specifically in low temperature environments- may be less able to survive.

10.3 Behavioural Implications of climate change on crustaceans

The warming of the oceans and the change in oceanic chemistry has wide reaching implications on animal behaviour, which not only affects individual behaviours, but potentially affects community structure and food web dynamics, through altering the state and availability of different trophic levels [33]. As many aquatic organisms to communicate use chemoreception, a change in abiotic conditions has been shown to decrease this sensory function, affecting the organisms' ability to locate the odour source and reducing prey consumption [23]. De la Haye [21] investigated the source of this reduced responsiveness to chemical stimuli, using the hermit crab *Pagurus bernhardus* and observed a reduction in response to food stimuli as well as showing a decline in antennule flicking (a clear detection method in crustacea). Demonstrating there was no degradation to the cue clearly indicated that both physiology and direct chemosensory function is impaired by altered pH. Likewise, post-larval spiny lobsters have shown to be less responsive to settlement cues due to reduced pH seawater affecting olfactory capabilities, resulting in increased mortality [24].

Many crustaceans carry their developing young under their abdomen with the female providing brood care in the form of ventilation over the eggs/young until hatching. This behaviour is induced by the developing young releasing a cocktail of small peptides with synthetic a tripeptides analogues glycyl-L-histidyl-Llysine (GHK) and glycyl-glycyl-L-arginine (GGR) as well as the dipeptide L-leucyl-L-arginine (LR) inducing this behaviour also in shore crabs, *Carcinus maenas* females [34]. A reduction in seawater pH from 8.2 to 7.7 alters the protonation of these peptides and with it their overall charge and impaired female responses to the cues [34]. This highlights that peptide sensory signalling molecules are pH sensitive and small pH changes alter the bioavailability of the active form of the signals. With crustaceans relying heavily upon olfactory cues for key behaviours including reproduction and predator-prey interaction, this dependence on environmental pH can have significant impacts upon fitness and survival of crustaceans [35]. Interestingly, this pH dependence also includes toxic compounds including those from microalgae that accumulate in mussels, key prey organisms for many crustaceans [36]. This potentially impacts animal fitness and survival.

How climate change will impact community structure and food webs is dependent on the differing abilities of the organisms to adapt to the physical, biological and chemical changes in the environment. Associated extreme events such as ocean heat waves will have contrasting impacts on fitness and physiology of different species, altering community dynamics [37]. How species within a community respond to these pressures will depict how the composition of marine community structures will change. In the *Posidonia oceanica* beds, decapod crustacean communities are altered through the presence of a CO_2 vent, decreasing the pH to similar conditions expected over the coming century. This lowered pH has caused a shift in the assemblages and has reduced diversity of species [38]. Predators shape the prey communities but rely on prey for energy resources, therefore the impact on any part of the food chain would have a huge effect on food webs and influence potential negative effects on marine habitats and ecosystems. Ocean acidification and temperature elevation combined has the potential to increase biomass in the primary trophic level but this biomass is turning to detritus rather than transferring up the secondary and tertiary levels, resulting in constraining the energy in the primary producers and weakening or potentially collapsing food webs [33]. Wernberg [39] provides evidence of a shift in community structure after an extreme oceanic heating event on the west coast of Australia, resulting in the change of biodiversity patterns of seaweeds, invertebrates and fish. These unpredictable extreme weather events also cause problems with modelling impacts on ecosystems, which are primarily based on a gradual warming scenario and impact local biodiversity change and community structures. Therefore, this will affect the reliability of climate indicators for global biodiversity and ecosystem change.

Whilst much research has been undertaken on the effects of climate change on organisms over a specific phase of a life cycle, less focus has been paid to species plasticity over various life stages/multiple generations, though this research has increased over recent years [40]. These cross-generation impacts will provide much needed insight into species adaptation over time and whether parental conditioning to stressors caused by global warming can prepare the following generation for altered climates. Research on a variety of organisms including Annelida, Mollusca, Arthropoda and Cnidaria have resulted in a wide variety of outcomes [40]. The focus on multigenerational studies in crustacea has primarily been on copepods due to their importance in marine trophic energy transfer [40]. Functional traits such as growth rate, fertility and hatching success and mortality rate have been researched in a variety of copepod species eliciting a wide range of responses, some cross generation species responses being affected more negatively than others. Madiera [37] have concluded that certain species of copepods can adapt physiologically to temperature stress allowing them to survive with minimal impact on fitness and develop a thermal tolerance to oceanic heating events. Analysis of copepods as a whole showed no significant impacts of ocean acidification in particular pH change, suggesting that they have good phenotypic plasticity [40]. Brachyuran crabs are well known to have their life stages controlled by temperature, and a change in this can affect their reproduction, development and metabolism. They do however seem to have developed physiological, morphological, behavioural and biochemical responses that allow the animal to adapt to compensate for these changes in the short term [41].

Primary consumer like zooplankton, small fish and crustaceans are at a greater risk from predation due to the increased risk taking in response to elevated temperature; this however is counterbalanced by the increase in consumer biomass brought about as a result of the elevated CO_2 availability [33]. Adaptations can be observed in the porcelain crab, who despite showing varied changes in their metabolism, appear to be able to adapt their physiology to changing conditions and therefore to future declines in oceanic pH [42]. This implies that nature in all its complexities can perhaps compensate for the negative impacts of climate change but will nevertheless result in a change in our ocean ecosystems that would take decades to recover. Although, marine ecosystems as a whole may adapt to global oceanic changes this would undoubtedly result in significant physiological costs and potentially even the loss of more vulnerable species. The simplification of ecosystems and the reduction in niche diversity is then likely to result in more dominant species reducing species richness through the loss of more uncommon and subordinate species [43]. This highlights the complexity by which the physico-chemical changes associated with climate change impact crustaceans at all levels of organisation, (from cellular to whole ecosystems) and the significant challenges associated with understanding such complexities. Such data is an urgent requirement when decisions upon climate change risk limitations are taken, and crustaceans (with their prominent role in the marine food webs as well as for food security and many coastal human communities) play a central role here.

References

1. Royal Society. (2005). Ocean acidification due to increasing atmospheric carbon dioxide. Policy Document 12/05. London: *The Royal Society*, pp. 1–60.
2. R.A. Betts, C.D. Jones, J.R. Knight, R. Keeling, and J.J. Kennedy. (2016). *Nat. Clim.* **6:** 806–810.
3. R.P. Ellis, M.A. Urbina, and R.W. Wilson. (2016). *Glob. Chang. Biol.* **23:** 2141–2148.
4. K. Caldeira, and M.E. Wickett. (2003). *Nature.* **425:** 365.
5. T. Takahashi et al. (2014). *Mar. Chem.* **164:** 95–125.
6. J. Godbold, and M. Solan. (2013). *Philos. Trans. R Soc.* **368(1627):** 20130186.
7. S. Pacella et al. (2018). Seasonal patterns of estuarine acidification in seagrass beds of the Snohomish Estuary, WA. Salish Sea Ecosystem Conference. 400.
8. P. Landschützer, N. Gruber, D.C.E. Bakker, I. Stemmler, and K.D. Six. (2018). *Nat. Clim.* **8:** 146–150.
9. L.M. Mosley, B.M. Peake, and K.A. Hunter. (2010). *Environ. Model Softw.* **25:** 1658–1663.
10. R.F. Keeling, A. Körtzinger, and N. Gruber. (2010). *Annu. Rev. Mar. Sci.* **2:** 199–229.
11. A. Dissanayaka, and A. Ishimatsu. (2011). *ICES J. Mar. Sci.* **68:** 1147–1154.
12. A.C. Remaglia, L. Mantovani de Castro, and A. Augusto. (2018). *J. Comp. Physiol.* **188:** 729–738.
13. G.H. Dickinson et al. (2021). *J. Exp. Biol.* **224(3):** jeb232819.
14. M. Conradi et al. (2019). *Mar. Pollut. Bull.* **143:** 33–41.
15. S.T. Dupont et al. (2015). PeerJ PrePrints. https://peerj.com/preprints/1438/3:e1438v1.
16. J. Shields. (2019). *J. Crustac. Biol.* **9(6):** 673–683.
17. L. Nagle, S. Brown, A. Krinos, and G.A. Ahearn. (2018). *J. Comp. Physiol.* **188:** 739–747.
18. A.M. Harrington, R.J. Harrington, D.A. Bouchard, and H.J. Hamlin. (2020). *J. Crustac Biol.* **40(5):** 634–646.
19. B.S. da Vianna, C.A. Miyai, A. Augusto, and T.M. Costa. (2020). *Physiol. Behav.* **215:** 112765.
20. B. Hernroth, H. Nilsson Sköld, K. Wiklander, F. Jutfelt, and S. Baden. (2012). *Fish Shellfish Immunol.* **33:** 1095–1101.
21. K.L. De la Haye, J.I. Spicer, S. Widdicombe, and M. Briffa. (2012). *J. Exp. Mar. Biol. Ecol.* **412:** 134–140.
22. E. Marangon, S.U. Goldenberg, and I. Nagelkerken. (2020). *Behav. Ecol.* **31:** 287–291.
23. L.F. Dodd, J.H. Grabowski, M.F. Piehler, I. Westfield, and J.B. Ries. (2015). *Proc. Royal Soc.* **282:** 20150333.
24. P.M. Gravinese. (2020). *Sci. Rep.* **10(1):** 18092.
25. L. Ceballos-Osuna, H.A. Carter, N.A. Miller, and J.H. Stillman. (2013). *J. Exp. Biol.* **216:** 1405–1411.
26. A.W. Paganini, N.A. Miller, and J.H. Stillman. (2014). *J. Exp. Biol.* **217:** 3974–3980.
27. N.M. Whiteley. (2011). *Mar. Ecol. Prog. Ser.* **430:** 257–271.
28. Y. Shang et al. (2020). *Sci. Total Environ.* **722:** 138001.
29. A.A. Kuhn, and M.Z. Darnell. (2019). *J. Crustac. Biol.* **39:** 22–27.
30. N. Bednaršek et al. (2020). *Sci. Total Environ.* **716:** 136610.
31. IPCC (Intergovernmental Panel on Climate Change). (2021). Climate Change 2021; Science, Solutions, Solidarity. Cambridge University Press.
32. E.J. MacFayden et al. (2004). *Glob. Chang. Biol.* **10(4):** 408–416.
33. S.U. Goldenberg. (2018). *Nat. Clim.* **8:** 229–233.
34. C.C. Roggatz, J.D. Hardege, M. Lorch, and D. Benoit. (2016). *Glob. Chang Biol.* **22:** 3914–3926.
35. C. Porteus, C.C. Roggatz, Z. Velez, P. Hubbard, and J.D. Hardege. (2021). *J. Exp. Biol. in press.*
36. C.C. Roggatz et al. (2019). *Nat. Clim.* **9:** 840–844.
37. C. Madiera, M.C. Leal, M.S. Diniz, H.N. Cabral, and C. Vinagre. (2018). *Mar. Environ. Res.* **141:** 148–158.
38. V. Zupo, and T. Viel. (2021). *Bull. Reg. Nat. His.* **1:** 19–27.
39. T. Wernberg et al. (2012). *Nat. Clim.* **3:** 78–82.
40. M. Byrne, S.A. Foo, P.M. Ross, and H.M. Putman. (2019). *Glob. Chang. Biol.* **26(1):** 80–102.
41. A.M. Azra et al. (2019). *Rev. Aquac.* **12:** 1211–1216.
42. H. Carter, L. Cebellos-Osuna, N.A. Miller, and J.H. Stillman. (2013). *J. Exp. Biol.* **216:** 1412–1422.
43. I. Nagelkerken, S.U. Goldenberg, E.O.C. Coni, and S.D. Connell. (2018). *Sci. Total Environ.* **645:** 615–622.

CHAPTER 11

The Biodiversity of Freshwater Crustaceans Revealed by Taxonomy and Mitochondrial DNA Barcodes

Adrian A. Vasquez,[1,2,]* *Brittany L. Bonnici,*[1,2] *Donna R. Kashian,*[3]
Jorge Trejo-Martinez,[4] *Carol J. Miller*[1] *and Jeffrey L. Ram*[2]

11.1 Introduction

Crustaceans are a biodiverse and ubiquitous group found in the frigid waters of the Antarctic to the completely light deprived subterranean caves of the tropics. They are vital members of aquatic ecosystems often providing important ecological and economical contributions. For example, copepods are the most abundant group of multicellular animals and have been observed in atypical aquatic habitats such as leaf litter and phytolemata [*1*]. Healthy fish communities are dependent on a continuous supply of food of which planktonic crustaceans make up a large percentage. In order to predict fish stock declines it is imperative to assess and monitor current zooplankton communities. In some habitats, like the Great Lakes, invasive crustaceans may become problematic to its fish stocks. Due to their small size, species identification is difficult; however, DNA barcoding using the mitochondrial cytochrome oxidase I (COI) gene can provide a powerful application in identifying species in many animal groups and sometimes in suggesting whether cryptic species may be present.

In this chapter we outline current knowledge about molecular barcodes for copepods, cladocerans, and ostracods. We include recently determined new barcodes in each group, thus adding sequences to various databases. Our data review here will enable future researchers to more readily identify planktonic crustaceans

---

[1] Healthy Urban Waters, Department of Civil and Environmental Engineering, Wayne State University, Detroit, MI 48202 USA.
[2] Department of Physiology, Wayne State University, Detroit, MI 48201 USA.
[3] Department of Biological Sciences, Wayne State University, Detroit, MI 48202 USA.
[4] Science Department, Corozal Junior College, San Andres Road, P.O. Box 63 Corozal, Belize, Central America.
* Corresponding author: avasquez@wayne.edu

from freshwater habitats and to compare specimens to one another from distant biogeographic regions, especially comparisons to North American fauna that are a major focus of this chapter. This can lead to insights into zooplankton population dynamics by metabarcoding and other emerging technologies and may contribute to modeling future fish stocks as a vital resource for many North American communities.

11.2 Crustacean species in the Laurentian Great Lakes

Numerous planktonic crustaceans are found in the Great Lakes, and several surveys have characterized them to species level. A relatively detailed example compared crustacean zooplankton collected annually in all five Great Lakes, in spring and summer, by the US Environmental Protection Agency from 1997 to 2016 [2]. This study of crustacean zooplankton in the Great Lakes summarized findings going back to at least the late 1800s [2]. In brief, the largest proportion of spring collections in all five Great Lakes were dominated by calanoid copepods; frequently named species encompassed *Limnocalanus macrurus* and *Leptodiaptomus sicilis*. The list of copepods included an invasive brackish water species identified as *Eurytemora affinis*; however, our studies (reviewed below) indicate that the designation *E. affinis* is probably incorrect, as barcode analyses have demonstrated that most historically identified *E. affinis* in the Great Lakes are likely a more recently described species *E. carolleae* [3]. Cyclopoid copepods in spring collections of the Great Lakes included *Diacyclops thomasi*. Cladocerans were also found although not in large numbers in the spring except in Lake Erie, where collections of *Bosmina longirostris*, *Eubosmina coregoni*, and various species of *Daphnia* were recorded. Work in the Great Lakes by our research group has also identified these cladocerans, for which new barcode records of these diverse organisms are described below. Cladocerans also included the invasive species *Bythotrephes longimanus*. In summer collections (usually in August of each year), many of the same species were found with additional species increasing in biomass, including the copepod *Skistodiaptomus oregonensis*, especially in Lake Erie; and cladocerans increasing substantially in proportion, especially the invasive *B. longimanus* in the eastern and central basins of Lake Erie, the invasive *Cercopagis pengoi* in Lake Ontario, and a huge growth in the biomass in Lake Erie of *Daphnia mendotae* and *Daphnia retrocurva*. Several other invasive species that have been noted in recent years in western Lake Erie are *Daphnia lumholtzi* and *Thermocyclops crassus*.

During the same period in which the above observations were made, the Ram laboratory conducted plankton collections in 2012 and 2013 in the Maumee River and Maumee Bay region of western Lake Erie (see map in Vasquez, Hudson [3]), with the objective of determining mitochondrial barcodes of various planktonic species and discovering new invaders. The Maumee River and Maumee Bay were thought to be among the most likely sites for discovery of new invaders as the Maumee River port in Toledo, OH is the second largest port in the Great Lakes in terms of the amount of ballast water from outside the Great Lakes being discharged as ships entered the port, or took on cargo. The plankton collection methods, morphological taxonomic analysis, DNA extraction, cytochrome oxidase I gene amplification, and bioinformatics methods have been described previously [*3*] with additional methods explained throughout this chapter.

Plankton collections in western Lake Erie by the Ram laboratory during 2012 and 2013 identified multiple examples of crustaceans that had not been reported by Barbiero, Rudstam [*2*]. Figure 1 identify all taxa that we found at frequencies above 0.3% of the total in each year.

In addition, the category of "Other" includes in 2012 *Mesocyclops edax*, Calanoida, *Skistodiaptomus pygmaeus*, *Epischura lacustris*, *Megacyclops viridis*, *Chydoridae*, Cyclopoida, Harpacticoida, *Eubosmina longispina*, *Gammarus* sp., *Ceriodaphnia* sp., *Ergasilus* sp., *Leptodiaptomus minutus*, *Argulus* sp., *Daphnia lumholtzi*, Diaptomidae, *Tropocyclops prasinus mexicanus*, *Chydorus sphaericus*, *Diacyclops thomasi*, *Microcyclops* sp. and in 2013 *Chydoridae*, *Daphnia lumholtzi*, Diaptomidae, *Leptodiaptomus minutus*, *Diacyclops thomasi*, *Alona* sp., *Epischura*

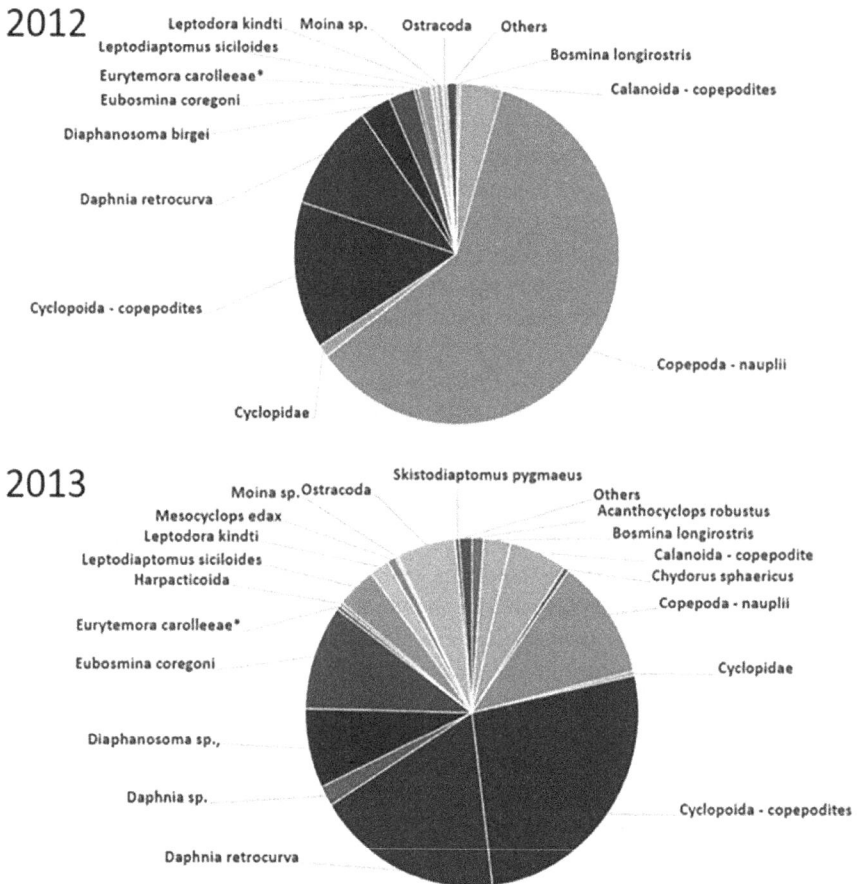

Figure 1. Biodiversity of crustacean taxa from the Western basin of Lake Erie during 2012 and 2013. Each pie chart section represents percent abundance out of 33,185 specimens collected in 2012 (May, July, and August) and 19,162 collected in 2013 (May, June, July, and August). * indicates where we changed the taxonomic identification from *Eurytemora affinis* to *Eurytemora carolleeae* due to subsequent DNA barcoding [1].

lacustris, Ceriodaphnia sp., *Cyclops strenuus, Ilyocryptus* sp., *Sida cristalina,* Cyclopoida, *Leptodiaptomus ashlandi, Leptodiaptomus sicilis, Eucyclops* sp., Cladocera, Macrothricidae, *Ergasilus* sp., *Eucyclops agilis, Eubosmina longispina, Ergasilus* sp., *Microcyclops rubellus, Skistodiaptomus pallidus, Skistodiaptomus* sp., *Alona setulosa, Scapholeberis* sp., *Tropocyclops prasinus mexicanus,* and *Eubosmina logisrostris.* Among the taxa that we found that were not identified in Barbiero, Rudstam [2] were *Moina* (Figure 2A), Harpacticoida (Figure 2B), *Diaphanosoma birgei, Skistodiaptomus pygmaeus, Megacyclops viridis, Ergasilus, Argulus, Eubosmina longispina, Scapholeberis, Eubosmina longirostris* and *Acanthocyclops robustus* and *Gammarus* and Ostracoda which are crustaceans that were not a focus of their work. *Moina* is a cosmopolitan freshwater cladoceran genus reported as presenting taxonomic difficulties where DNA barcoding has been useful to clarify the systematics of this group [4].

Keeler, Tucker [5] reported *Acanthocyclops robustus,* Harpacticoida, *Moina micrura, Diaphanosoma brachyurum, Scapholeberis* and Ostracoda. The work by Keeler, Tucker [5] was not included in the assessment by Barbiero, Rudstam [2] which may be the reason for the lack of some taxa. It is also possible that some taxa such as *Gammarus,* Ostracoda and Harpacticoida are generally not seen in planktonic studies but rather in benthic studies.

Some crustacean groups require revision for both taxonomy and clarification of cryptic species. DNA barcoding has strengthened crustacean taxonomy and has been shown to be a useful tool in unraveling missed Cladocera taxa in groups such as *Diaphanosoma birgei* and *Ceriodaphnia* in Mexico and Guatemala [6]. Application of COI DNA barcoding to planktonic crustaceans has shown great potential in resolving species identities [7]. DNA barcoding was used to update copepod and cladoceran species lists by identifying cryptic species and improving the knowledge of zooplankton distribution especially in little known areas like the tropics [6]. Table 1 in this review and supplementary online information (https://digitalcommons. wayne.edu/physio_frp/3) summarizes the DNA barcode confirmations, lists additional taxonomic references, lists new species-level identifications from our group based on specimens collected from Toledo Harbor in 2012 and 2013, and those

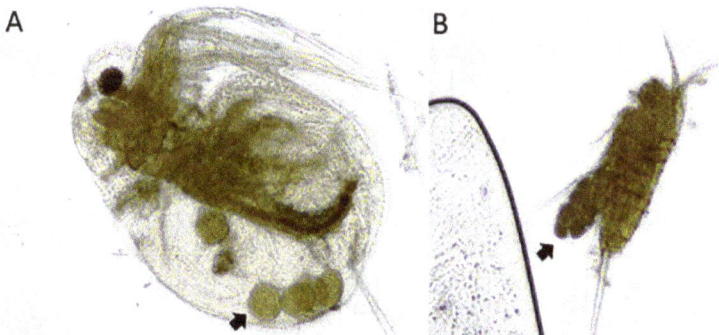

Figure 2. Crustaceans collected in Lake Erie. (A). *Moina* sp. with eggs in the brood chamber indicated by arrow. (B). Harpacticoida with eggs indicated by arrow.

Table 1. New species and genus level COI barcodes and confirmations of previous species level barcodes.

RamLab ID[1]	Morphospecies identification	Barcode sequence length	Nearest previous GenBank accession ID	Query percent	Percent identity	GenBank taxon Species if > 96.5% ID; Genus if > 90.5% ID.	Confirmation or new barcode
LE090914S14	*Acanthocyclops americanus*	641	KC016141.1	97	99.1	*Acanthocyclops americanus*	Confirmation
1-4LHLE05302017	*Acanthocyclops americanus*	658	KC016171.1	100	99.7	*Acanthocyclops americanus*	Confirmation
2-6HLE05302017	*Acanthocyclops americanus*	658	KC617189.1	100	99.4	*Acanthocyclops americanus*	Confirmation
5-10MM07142017	*Acanthocyclops brevispinosus*	658	MG449165.1	100	99.4	*Acanthocyclops* sp.	New for species
LH100614SK	*Acanthocyclops robustus*	515	KC016182.1	100	86.5	--	New for species
LP041615SJ	*Acanthocyclops* sp.	628	KC016192.1	98	93.2	*Acanthocyclops* sp.	Confirmation of genus
LP041615SH	*Acanthocyclops venustoides*	641	KU613358.1	99	80.2	--	New for species
HR032015SB	*Acanthocyclops vernalis*	639	MG448835.1	97	80	--	New for species
LCL090914S10	*Acanthocyclops vernalis*	641	MG320342.1	98	97.4	*Acanthocyclops* sp.	New for species[2]
1BZEB1P22215	*Apocyclops dimorphus*	658	MK558266.1	97	81.6	*Apocyclops spartinus*	New for species
1PMV070212S1	*Bosmina* sp.	660	LS991483.1	100	99.1	*Eubosmina coregoni*	Contradiction[3]
1PMV070212S2	*Bosmina* sp.	660	LS991483.1	100	99.1	*Eubosmina coregoni*	Contradiction[3]
2PMU053012S6	*Calanoida* sp.	643	EU770472.1	98	97.7	*Leptodiaptomus siciloides*	Confirmation
2PMU053012S8	*Calanoida* sp.	643	MG449882.1	97	99.7	*Leptodiaptomus minutus*	Confirmation
2PMD070212S2	*Calanoida* sp.	643	HM883988.1	95	99.7	*Leptodiaptomus siciloides*	Confirmation

2PMD080912S2	Ceriodaphnia dubia	651	MG450106.1	99	98.6	Ceriodaphnia dubia	Confirmation
2BZEB1P22215	Cyclopoida sp.	658	MK558266.1	98	81.6	Apocyclops spartinus	New for family
1PMW060212S10	Cyclopoida sp.	642	KC016180.1	97	99.1	Acanthocyclops americanus	Confirmation
2PMD080912S1	Daphnia lumholtzi	652	LS991497.1	100	99.5	Daphnia lumholtzi	Confirmation
2PMU053012S4	Diaphanosoma sp.	639	MG448880.1	98	98.9	Diaphanosoma sp.	Confirmation of genus
2PMB060212S6	Diaphanosoma sp.	653	KC617625.1	98	98.0	Diaphanosoma sp.	Confirmation of genus
2PMN060212S2	Diaphanosoma sp.	649	KC617625.1	98	98.7	Diaphanosoma sp.	Confirmation of genus
LE082614	Homocyclops ater	642	MG449912.1	98	99.4	Cyclopidae sp.	New for species[4]
SCL061614	Leptodiaptomus sicilis	641	KP213174.1	98	98.2	Leptodiaptomus cf. sicilis	Confirmation
2PMN081313S1	Leptodiaptomus siciloides	597	HM883988.1	97	99.7	Leptodiaptomus siciloides	Confirmation
2PMS060212	Leptodiaptomus siciloides	642	HM883988.1	97	99.8	Leptodiaptomus siciloides	Confirmation
LE082614	Macrocyclops albidus	641	MG317646.1	98	98.4	Cyclopidae sp.	New for species
2PMS061212	Mesocyclops edax	641	MG449552.1	97	99.7	Cyclops sp.	New for species[5]
LE080317-6-26	Neoergasilus japonicus	658	KR049037.1	88	93.3	Neoergasilus japonicus	New for species
LE080317-727	Neoergasilus japonicus	658	KR049037.1	88	93.1	Neoergasilus japonicus	New for species
LE080317-828	Neoergasilus japonicus	658	KR049037.1	88	93.3	Neoergasilus japonicus	New for species
LE080317-929	Neoergasilus japonicus	657	KR049037.1	88	93.1	Neoergasilus japonicus	New for species
MMLE081817-1035	Neoergasilus sp.	658	KR049037.1	88	92.5	Neoergasilus japonicus	Confirmation of genus
LSC091317-1137	Neoergasilus sp.	658	KR049037.1	88	93.1	Neoergasilus japonicus	Confirmation of genus

Table 1 contd. ...

...Table 1 contd.

RamLab ID[1]	Morphospecies identification	Barcode sequence length	Nearest previous GenBank accession ID	Query percent	Percent identity	GenBank taxon Species if > 96.5% ID; Genus if > 90.5% ID.	Confirmation or new barcode
LSC091317 1238	*Neoergasilus* sp.	658	KR049037.1	88	93.3	*Neoergasilus japonicus*	Confirmation of genus
2PMV060212S1	Ostracoda sp.	661	HM883981.1	98	95.7	Candonidae sp.	Confirmation of Class
1LMUSK061914	*Skistodiaptomus pallidus*	631	MG450145	97	99.8	*Skistodiaptomus pallidus*	Confirmation
3LMUSK061914	*Skistodiaptomus pallidus*	626	MG450145.1	97	99.8	*Skistodiaptomus pallidus*	Confirmation
BHL060614S2	*Skistodiaptomus pallidus*	635	MG450145.1	96	99.8	*Skistodiaptomus pallidus*	Confirmation
2PMP052113S2	*Skistodiaptomus reighardi*	636	HM045395.1	100	83.2	--	New for species
2PMN081313S2	*Skistodiaptomus reighardi*	633	HM045395.1	99	82.6	--	New for species
2LMUSK061914	*Skistodiaptomus* sp.	616	HM045397	98	83.6	--	New for genus
3-7MM07142017	*Thermocyclops crassus*	658	MN641910.1	99	82.8	--	New for species
4-8MM07142017	*Thermocyclops crassus*	658	MN641910.1	99	82.9	--	New for species
LE082614	*Tropocyclops* sp.	642	KC617424.1	98	82.6	--	New for genus

[1] Collection location abbreviations in the RamLab ID include the following: All sequences starting with PM, Toledo Harbor in western Lake Erie; LMUSK, Lake Muskoday, Belle Isle, Detroit; SCL, Saint Clair River; BHL, Blue Heron Lagoon, Belle Isle; LE, Lake Erie; LSC, Lake St. Clair; MMLE; Metzgers Marsh, Lake Erie; MM, Metzgers Marsh; LP, Leonard Preserve, Manchester, Michigan; HR, Huron River Drive, Ypsilanti, Michigan; LCL, Little Cedar Lake, Orion, MI; HLE, Harbor Lake Erie; LHLE, Lorain Harbor Lake Erie; BZEB1P, Cenote in Shipstern Reserve, Corozal, Belize, Central America. Collection dates are mostly shown in the RamLab ID in six characters as MMDDYY.

2 Confirmation of previous genus identification, but new for the species level identification of the barcode.

3 Best matches (top 5, with > 99% identities) include both *Bosmina* (e.g., MG449616.1) and *Eubosmina* as genus. We endorse the identification of this barcode as *Bosmina*.

4 The "New barcode for species" is also a confirmation of a previous family identification (*Cyclopidae*)

5 Contradicts the previous *Cyclops* identification; however, the previous identification may be using a taxon that some have said is synonymized between *Mesocyclops* and *Cyclops*. Identified by us as *Mesocyclops edax*.

subsequently obtained from other freshwater environments in the Great Lakes region and in the Shipstern Peninsula.

DNA barcoding has been particularly useful in assisting taxonomic studies of freshwater zooplankton because "specialists" in this area are few and overburdened with taxonomic work [6]. Moreover, species level identification of organisms that are generally less than 1 mm in size (Figure 3) often requires detailed dissections and high power microscopy to study fine structures of taxonomic value (like the 5th leg in copepods) [6] and the length/width ratios for the left basipod in the male fifth leg in *Eurytemora* [3].

Freshwater copepods are thought to number up to 2,814 species globally distributed throughout almost all freshwater habitats including lakes, subterranean habitats and phytolemata [8]. COI barcodes of the copepods *Skistodiaptomus reighardi*, shown in neighbor-joining tree constructed using MEGA X [9] (see Table 1 and Figure 4) and multiple specimens from the cyclopoid complex of *Acanthocyclops* (Figure 3D, Figure 4 and Table 1) in this review include several new species-level barcodes to be added to public databases.

COI barcodes could help resolve taxonomic inconsistencies among copepods, for example, *Eurytemora affinis*, an invasive brackish water copepod, was reported as present in the Great Lakes since the 1960's (as reviewed by Vasquez, Hudson [3]). However, DNA barcoding revealed multiple barcodes, suggesting invasions by multiple representatives of a species complex of divergent populations going back at least 20 years [8]. With the use of DNA barcodes, we were able to identify the species in the Great Lakes as *Eurytemora carolleeae* [3] (Figure 3A). Morphospecies comparisons to archived specimens by our group revealed that *E. carolleeae* has been present in the Great Lakes since at least 1962 but was frequently identified as

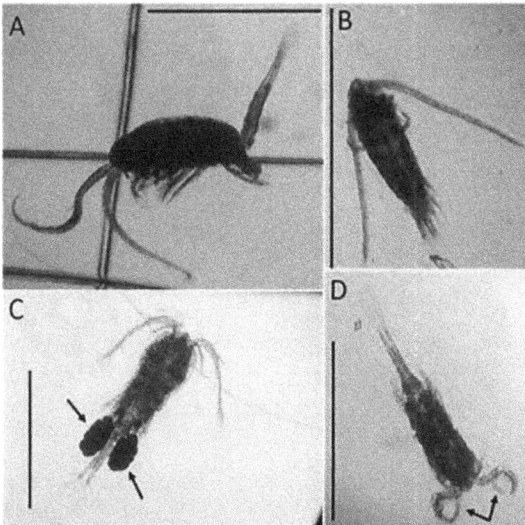

Figure 3. Copepods of the Great Lakes that were barcoded. (A) *Eurytemora carolleeae* (B) *Skistodiaptimus pallidus* (C) *Cyclopoid* sp. (arrows show egg sacs) (D) *Acanthocyclops* sp. (arrows show male with modified antennules used for grasping female). Scale bar = 1 mm.

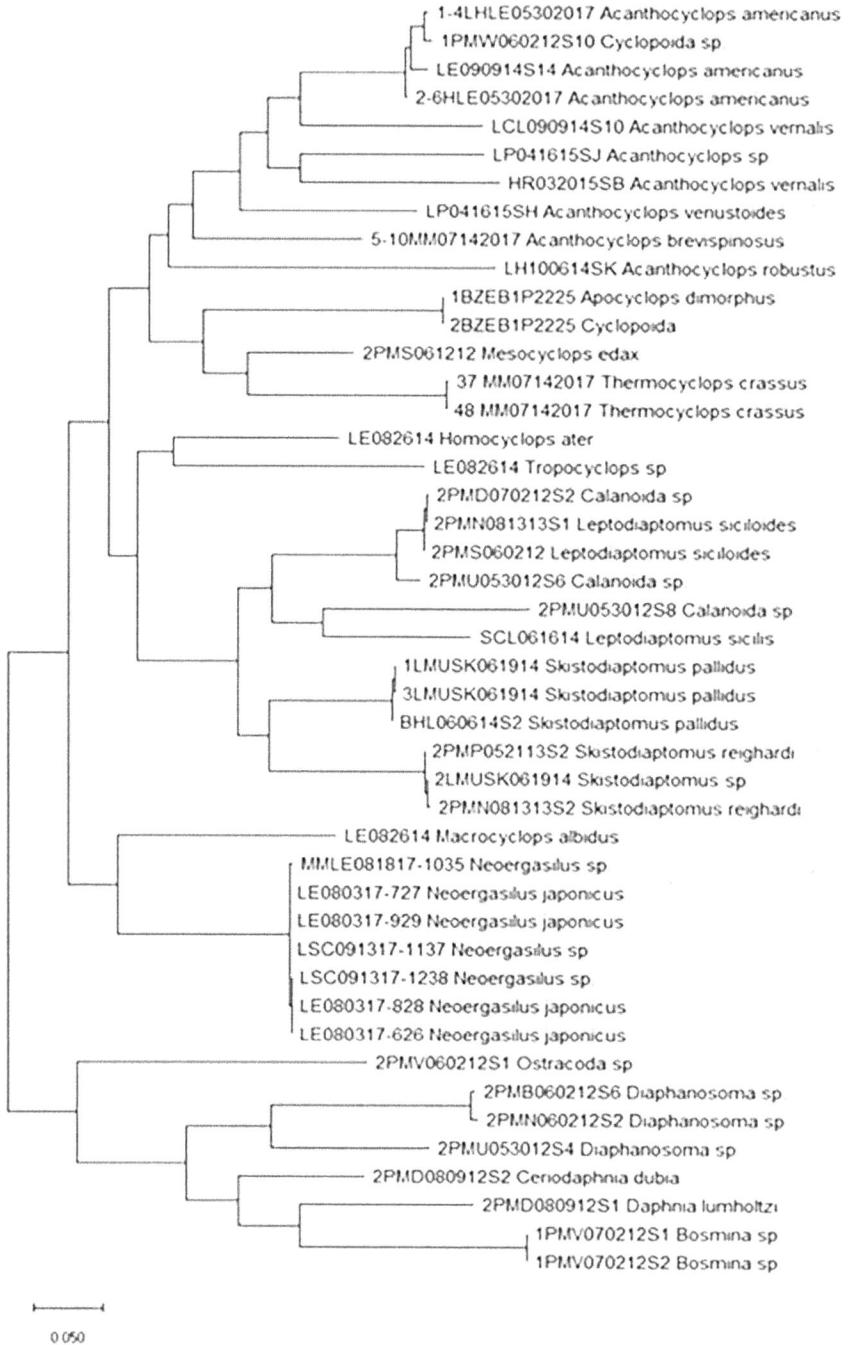

Figure 4. Neighbor-joining COI DNA barcode tree of crustaceans. The tree is constructed from 45 nucleotide sequences, and the scale bar represents 5 percent difference. The distances were computed with the Maximum Likelihood method using MEGA X [9]. Sequence names utilize the abbreviations described in footnote 1 of Table 1.

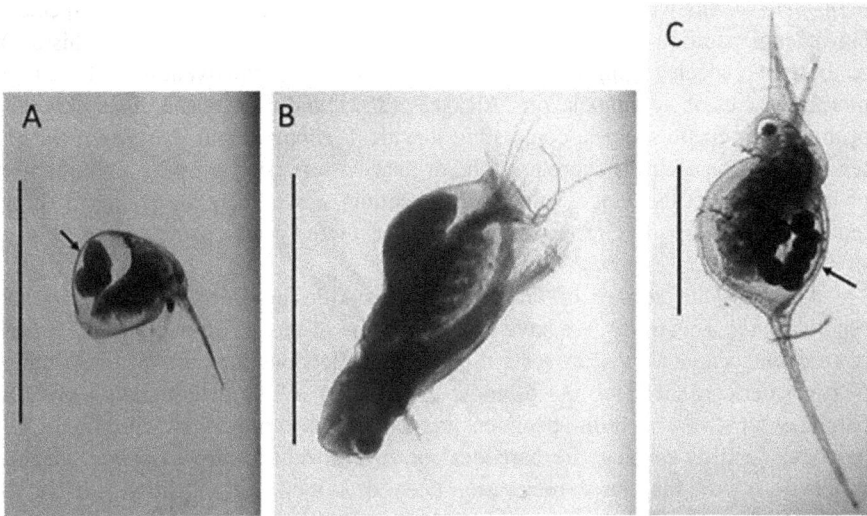

Figure 5. Crustaceans from the Great Lakes that were barcoded. (A) *Bosmina* sp. with arrow showing eggs in brood chamber (B) *Diaphanosoma* sp. (C) *Daphnia lumholtzi* (invasive species of the Great Lakes) with arrow showing eggs in brood chamber. Scale bar = 1 mm.

E. affinis, which is understandable since *E. carolleeae* had not been described as a distinct species until 2011 [*10*].

Cladocerans are primarily freshwater organisms with an estimated 620 species globally; the actual number is likely 2–4 times greater [*11*]. COI DNA barcodes in public databases reveal potential controversies such as the correct morphospecies identification of COI barcodes for the cladoceran *Bosmina* (water flea; see Figure 5A). We identified our specimens as *Bosmina* and obtained COI barcodes that were > 99% identical to GenBank sequences variously identified as *Eubosmina* and *Bosmina* (see Table 1). COI DNA barcodes for *Diaphanosoma* sp. (see Figure 5B) and *Daphnia lumholtzi* (see Figure 5C) may be useful for early detection of these potentially harmful invasive crustaceans (see Table 1).

Forro, Korovchinsky [*11*] noted that molecular technologies can help to describe the diversity of cladocerans, and this is important since cladoceran eggs can undergo desiccation and remain viable, becoming good candidates for invasion in new habitats.

11.3 Cyclopidae: several invasive species events

11.3.1 Acanthocyclops complex

The copepod family of Cyclopidae has especially gotten attention due to numerous instances of invasive species moving from North America to Europe and others arriving recently in the Great Lakes. Alekseev [*12*] recounts a complicated history of *Acanthocyclops americanus* (Marsh 1892) which was first described in Wisconsin (USA) and subsequently reported by Lowndes in England in 1926 and later in other European countries. For a while taxonomists combined the identification of *A. americanus* with *Acanthocyclops robustus* (Sars 1863), a European species. However, separation (and therefore, in effect, confirmation of this invasive history) of these two species using COI barcodes confirms the distinctiveness of these two morphospecies of *Acanthocyclops*. Alekseev (2021) indicates that *A. americanus* is primarily a pelagic species, contrasting it with *A. robustus* and *A. vernalis* which tend to occur in the near-shore area and/or near-bottom zone in lakes. With the use of morphological techniques and DNA barcoding, researchers were able to show that the *Acanthocyclops* group is comprised of three species namely *A. robustus*, *A. americanus* and *A. vernalis*.

Although the survey of Great Lakes plankton by Barbiero, Rudstam [2] reported only *A. vernalis*, we have found all three of the above-mentioned species of *Acanthocyclops* as well as several others in collections for Lake Erie and other North American sites (see the footnote to Table 1). While determination of COI barcodes for *A. americanus* specimens matched 99% to previously described COI barcodes for this species, the barcodes for several other *Acanthocylcops* species differ from any that have previously been described (i.e., identity < 97% to previously described barcodes) and thus may represent new species level barcodes. These include COI barcodes for *A. vernalis*, *A. robustus*, and *A. venustoides*, all of which are less than 90% identical to any previous GenBank records (see Table 1). The resolution of whether these barcodes represent new invaders to the Great Lakes may depend on additional genetic studies of the *A. robustus* (*robustus-vernalis*) group. Our barcode on *A. robustus*, a European species, suggests it is different from the other *Acanthocyclops* and therefore might be an invader in North America. The neighbor-joining tree in Figure 4 shows that the DNA barcodes of our North American specimens of *A. robustus*, *A. americanus* and *A. vernalis* differ by more than 3% from one other, in agreement with previous work on these taxa. Other novel *Acanthocyclops* COI barcodes include sequences for *A. brevispinosus* and *A. venustoides* (Figure 4 and Table 1). An online taxonomic key for specimens of *Acanthocyclops* and other copepods from the Great Lakes can be found at https://www.glsc.usgs.gov/greatlakescopepods/MainMenu.php?TITLE=Introduction%20*%20Keys%20to%20Major%20Groups&SHOW=Introduction%20*%20Keys .

11.3.2 Invasive thermocyclops in North America

Thermocyclops is another group of species for which invasions have recently been reported. *Thermocyclops crassus*, a Eurasian member of this widely distributed group has now been reported as introduced in Mexico and other countries in the Americas. *Thermocyclops crassus* was first reported in the Great Lakes by Connolly, Watkins [13]. To our knowledge no DNA barcodes for *Thermocyclops crassus* have previously been uploaded to public databases, as of June 2021.

Several specimens of *Thermocyclops crassus* were collected from Metzgers Marsh, Lake Erie in 2017, for which we report their COI DNA barcode (see Table 1 and our on-line sequence database). The COI barcodes of these *T. crassus* specimens are only 83% identical to other *Thermocyclops* COI species level barcodes that have been uploaded to GenBank (with one match at 99% to *Thermocyclops* sp. but at a low query percent of 45%). This invasive genus may comprise several diverse cryptic species not yet identified as distinct morphospecies.

11.3.3 Cyclopoids from Belize

To contrast the genetic distance of cyclopoids from tropical freshwater habitats with those from the Great Lakes, Figure 4 shows the placement of two cyclopoid copepod DNA barcodes collected from a cenote (collapsed freshwater sinkhole) referred to as the "Orchid Pond" at the Shipstern Reserve located in Corozal, Belize, Central America. Their barcodes are > 10% distant from the nearest branches of North American cyclopoids (see Figure 4 barcodes for BZEB1PS122215 *Apocyclops dimorphus* and BZEB1PS222215 *Cyclopoid* sp.). The nearest identification is an 82% match to *Apocyclops spartinus* whose native habitat is salt marshes which differs from the habitat sampled in Belize. This suggests a novel barcode for *Apocyclops dimorphus* which was keyed based on the taxonomic work of Elias-Gutierrez et al. "Copepods and Cladocerans of the continental waters of Mexico" (2008). *Apocyclops dimorphus* was collected only once in Cuatro Ciénegas, Coahuila which is a collection of freshwater habitats in northern Mexico similar to the Cenote in the Shipstern Reserve. Our search in the GenBank depository of DNA barcode sequences did not find any *A. dimorphus* COI sequences. *A. dimorphus* was reportedly found in Florida, USA in a roadside swale ditch and grassy swale with methods of capture that included half pint dippers and turkey basters. Shipstern Reserve, although recognized as a national protected area under the National Protected Areas Systems Act of 2015 of Belize, is still rapidly being encroached by agricultural use which could be detrimental to this unique biodiverse region. This highlights the urgency for continuous work that includes a combination of morphology and DNA barcodes, much needed to assess biodiversity in these aquatic habitats.

DNA barcoding has been hailed as a potential game changer for invasion ecology where early and correct identification of potentially invasive species can make a difference in proper management of invasive species. Further work is encouraged in this area, including applying new technologies like metabarcoding that could increase the ability to detect invasive species. We have shown that in a well-studied aquatic ecosystem like the Great Lakes, where significant investments of research effort and dollars have been made, many new species can potentially be discovered, therefore we should expect even more discoveries from less studied habitats in places like Belize, Central America.

11.4 Parasitic copepods of fish

Several invasive parasitic copepods have found their way into the Great Lakes in the past decades. Some of these include the invasive copepods *Argulus japonicus* and *Neoergasilus japonicus*. *Neoergasilus japonicus*, of Asian origin, was first reported in the Great Lakes two decades ago and recently found to have expanded its range to Lake Oneida, New York [*14*]. Although Barbiero, Rudstam [*2*] did not report these genera, we found them in our study of Lake Erie. Invasive parasitic copepods of fish are widely thought to be spread by aquaculture and in our studies we found

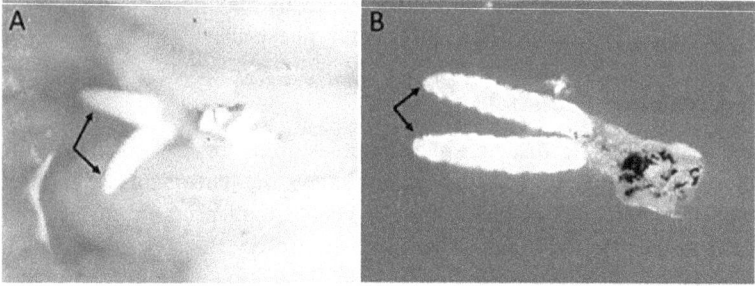

Figure 6. (A) Parasitic copepod (*Neoergasilus japonicus*) attached to the gill raker of a Yellow Perch (Lake St. Clair Metropark – Lake St. Clair) with arrows showing egg sacs. (B) *Ergasilus* parasitic copepod dissected from the gill raker of *Poecilia* fish sampled in the Shipstern Peninsula, Corozal district, Belize, Central America with arrows showing egg sacs.

N. japonicus in yellow perch (see Figure 6A), blacknose shiner, redear sunfish, and longear sunfish.

The DNA barcodes generated from these specimens, when compared to DNA barcodes of *N. japonicus* from Korea, were found to be only 93% identical, suggesting a possible subspecies of *N. japonicus* (see Figure 6 and Table 1). Although this invasive copepod has been found as far south as Nicaragua it is not known whether it exists in Belize. *Ergasilus*, another parasitic genus of copepods (one of 6 genera known to be parasitic in North America), was found in the gill rakers of *Poecilia mexicana/orri* from Belize (colloquially known in Belize as "poopsie" or "boglis" for which the results of our DNA barcodes indicate that it could be either species). These fish were collected from "chorros" (colloquial name probably originating from the Spanish word chorro meaning "a jet of water" used to identify small engineered spring-fed streams) in the Corozal District, Belize near the residential Orchid Bay Resort (Figure 6B). Poeciliinae is represented by 14 species in Belize with both *Poecilia mexicana* and *P. orri* present. *Poecilia* fish are important in the aquarium industry and parasitic copepods have been found in some species. Given the rise in aquaculture (fish farming) and the crowded conditions of these farms, these parasitic copepods could cause infection in fish leading to death and warrant further study of their potential spread using taxonomy and DNA barcodes.

11.5 Future needs of freshwater crustacea biodiversity research

Freshwater aquatic habitats are the most threatened ecosystems on earth. Limnological research has revealed a great loss of biodiversity in freshwater habitats and increased resources should be spent on restoring aquatic habitats. Esselman,Boles [15] remarked on the lack of limnological research in Belize and the need to invest more in this area, which remains true today, 20 years later. Belize is considered a biodiversity hotspot but not much is known about its freshwater invertebrates. In general, crustaceans are vital for any ecosystem in which they are found, often being prey items or of positive or negative economic value in aquaculture as food or parasites, respectively. The findings reviewed here show the great need to invigorate basic research on the biodiversity of crustaceans and other organisms in aquatic habitats: to discover new species, to clarify the status of known taxa, to detect invasive species, and to understand the trophic and pathogenic relationships among the organisms in these still largely unstudied environments. This can be useful for many areas of science including invasion biology and studying the effect of changing global weather patterns on aquatic organisms.

The development of new molecular technologies that can be used to assess the genetic diversity of a given habitat gives much promise to advancing knowledge of crustacean biodiversity. As mentioned in this review crustaceans are a suitable group to develop methods of molecular identification coupled with morphological studies. The US federal government spent $1.4 billion on Great Lakes Restoration Initiative (GLRI) projects during the years 2010 to 2016 and supported part of the work presented here (https://www.glc.org/wp-content/uploads/GLRI-Project-Summary-Report-20180924.pdf). Despite this enormous research funding and other similar investments made in the Great Lakes habitats, much remains to be learned about the biodiversity of crustaceans in the Great Lakes. As for Belize, comparable funding has not even been available, but the need is at least as great if not more so: on a positive note, Belize still has large tracts of undisturbed forests, streams and other rare and threatened habitats for freshwater crustaceans, although there are still challenges such as agricultural encroachment on these enclaves. Therefore, funding agencies must be duly informed on the current and future threats that freshwater habitats are facing.

. The United Nations declared the last decade (2011–2020) a Decade on Biodiversity (https://www.cbd.int/undb/home/undb-strategy-en.pdf) and have now declared this decade (2021–2030) a Decade on Ecosystem Restoration (https://www.decadeonrestoration.org/). This is a good direction, and the way forward to bring awareness to the loss of freshwater ecosystem health and the need to restore these habitats. Within this scope, the study of biodiversity is critical as we have shown, but there is still much work needed. Finally, as in study regions such as Belize, the active involvement and collaboration with local Belizean researchers (in this study, Dr. Jorge Trejo-Martinez planned and participated in field and laboratory preparation of samples obtained in Belize) has been gainful. This facilitates research in understudied environments and encourages Belize researchers to continue collaboration in the process of designing and participating in studies of their environments, reporting results, and creating a database that all stakeholders can access. This vital data can be used in the efforts toward supporting best management practices of sensitive environments of the host country and people. This approach establishes stronger scientific and educational links between countries, that is necessary if any attempt is made to restore threatened freshwater ecosystems as declared by the United Nations for this decade.

The future of crustacean biodiversity research in regions like Belize is promising if agencies/departments support biodiversity research and scientists genuinely collaborate with their counterparts in regions of the world where there is a great potential for discovering new species. Further research in freshwater habitats such as the Great Lakes (where more studies have been done on its fauna compared to studies in the tropics) is also apparent, and much emphasis should be placed on clarifying and resolving inconsistencies concerning the biodiversity found in these freshwater habitats, such as those presented in this chapter.

Acknowledgements

This work could not have been done without the help of many people including family and friends, and we would like to extend our thanks to them. We apologize to any whom we did not mention. We thank Patrick Hudson of the United States Geological Service, Ann Arbor, Michigan, for his taxonomic expertise and identifying many of the organisms newly barcoded in the present work. We thank the Belle Isle Aquarium Research Field Station, Lake St. Clair Metropark Laboratory field station, and Jose Mai, Dean of Corozal Junior College, for granting us laboratory space to conduct our studies in Belize and facilitating travel to visit El Colegio de la Frontera Sur (ECOSUR) in Chetumal, Quintana Roo, Mexico. We thank Dr. Manuel Elias-Gutierrez, Senior Investigator at ECOSUR, for many conversations on DNA barcoding of aquatic invertebrates, and for a copy of his comprehensive work: "Cladocera y copepoda de las aguas continentales de Mexico." We thank Janiel Cruz for her help in collecting invasive parasitic copepods and Justin Vasquez, Nathan Vasquez, Hugo Vasquez, Solangel Vasquez, Victor Vasquez, Daniel Vasquez, Michal Ram, Ricardo Cowo, Rachel Leads and students from Corozal Junior College for their help in collecting specimens in Belize. We thank Heron Moreno of the Shipstern Conservation and Management Area for granting us access to the reserve and Mauro Gongora of the Belize Fisheries Department for assisting us in obtaining a research permit to conduct our work in Belize. We also thank many students and staff from the Ram Lab that have assisted us in our research with aquatic invertebrates.

Funding Sources

The work in Toledo Harbor was supported by grants from EPA (GL-00E00808-00) to JLR and an NIH fellowship to AAV (NIH R25 GM058905-17). Work in Belize was partially supported by the Office of the Vice President of Research to JLR and AAV and Stellar Links (a local business and stakeholder of the environment in Belize) to AAV. Corozal Junior College (a tertiary level educational institution in Belize) provided support to JTM. AAV, BLB and CJM were supported by the Fred A. and Barbara M. Erb Family Foundation.

References

1. J.E. Bron, D. Frisch, E. Goetze, S.C. Johnson, C.E. Lee, and G.A. Wyngaard. (2011). *Front. Zool.* **8**.
2. R.P. Barbiero, L.G. Rudstam, J.M. Watkins, and B.M. Lesht. (2019). *J. Great Lakes Res.* **45(3):** 672–690.
3. A.A. Vasquez, P.L. Hudson, M. Fujimoto, K. Keeler, P.M. Armenio et al. (2016). *J. Great Lakes Res.* **42(4):** 802–811.
4. L. Montoliu-Elena, M. Elias-Gutierrez, and M. Silva-Briano. (2019). *Limnetica.* **38(1):** 253–277.
5. K.M. Keeler, T.R. Tucker, C.M. Mayer, W.W. Taylor, and E.F. Roseman. (2019). *J. Great Lakes Res.* **45(5):** 888–900.
6. M. Elias-Gutierrez, E. Suarez-Morales, M. Gutierrez-Aguirre, M. Silva-Briano, J.G. Granados Ramirez et al. (2008). *Zootaxa,* (1839): 1–42.
7. F.O. Costa, J.R. deWaard, J. Boutillier, S. Ratnasingham, R.T. Dooh, M. Hajibabaei et al. (2007). *Can. J. Fish. Aquat. Sci.* **64(2):** 272–295.
8. Boxshall, G.A. and D. Defaye. (2008). *Hydrobiologia.* **595:** 195–207.
9. S. Kumar, G. Stecher, M. Li, C. Knyaz, and K. Tamura. (2018). *Mol. Biol. Evol.* **35(6):** 1547–1549.
10. V.R. Alekseev, and A. Souissi. (2011). *Zootaxa.* **2767:** 41–56.

11. L. Forro, N.M. Korovchinsky, A.A. Kotov, and A. Petrusek. (2008). *Hydrobiologia.* **595:** 177–184.
12. V.R. Alekseev. (2021). *Water.* **13(10):** 16.
13. J.K. Connolly, J.M. Watkins, C.C. Marshall, J.M. Adams, L.G. Rudstam et al. (2017). *J. Great Lakes Res.* **43(3):** 198–203.
14. C.C. Marshall, P.L. Hudson, J.R. Jackson, J.K. Connolly, J.M. Watkins et al. (2019). *J. Great Lakes Res.* **45(6):** 1348–1353.
15. P.C. Esselman, and E. Boles. (2001). pp. 35–68. *In:* Wetzel, R.G., and B. Gopal [eds.]. *Status and future needs of limnological research in Belize, Limnology in Developing Countries,* Vol. 3.

PART 3
Aquaculture and Biotech

Chapter 12

Crustaceans as Pathogens and Most Common Pathogens of Crustaceans

Francesca Carella

12.1 Introduction

In the last years, numerous pathogens have been reported in crustaceans belonging to different families, especially penaeids, comprehending bacterial, viral, mycotic and parasitic diseases [1,2,3,4]. Some of them had a serious impact on wild populations but also on aquaculture production. Considerable financial losses occurred through the on growing phase of animal hatcheries and in many cases associated with poor diet, low water quality or pond bottom characteristics [5]. Viral diseases represent the most significant in this group of animals, and many are listed by the World Animal Health Organization (OIE, Office International des Epizooties) that keeps them under constant evaluation to avoiding the risk of spreading diseases and facilitate a safe international trade.

Many groups of crustaceans are also important parasites of fish, shellfish, and other aquatic animals with devastating effects on aquaculture, in many cases showing important ecological and economic relevance [6]. Most known parasitic crustaceans include copepods, isopods, brachyuran, tantulocarids, and cirripedes [7]. Related reports describe highly important lesions at muscles, gills, and skin level [8].

This chapter is related to those pathogens that cause significant disease in economically important crustaceans that are used for either aquaculture or commercial harvest and to the most common parasitosis due to different group of crustaceans.

12.2 Crustaceans' viral diseases

Marine viruses are abundant biological entities in the ocean. Studies on most marine viruses have focused on pelagic systems, but the study of virology it-self from the

Department of Biology, University of Naples Federico II, Complesso Universitario di Monte S. Angelo, Via Cinthia, Ed. 7, 80136 Naples, Italy.
Email: francesca.carella@unina.it

perspective of host-pathogen interaction in the marine environment has grown only in the last 20 years. Viruses are obligate intracellular pathogens. Viral infection counts on the entry into the host cell, the most critical step, which requires the interaction between cell receptor and viral proteins, in both enveloped and non-enveloped virus.

Viral infections are common in crustaceans. Since their initial discovery in this group in the 1960's [9], over 50 viruses have been described from different crustacean families. Twenty or more are reported in penaeids [10], 6–10 in the group of crayfish, 4 in *Macrobrachium* sp. [11] and 3 in some crab species. The molecular mechanisms that underlie the crustacean antiviral immune responses are barely known but are gaining more importance in the scientific research due to the highly negative effects on the economy in several countries and on shrimp's culture industries. DNA viruses have been described in most host groups studied [12]. Crustacean viruses belong to various families like the *Baculoviridae, Bunyaviridae, Herpesviridae, Picornaviridae, Parvoviridae, Reoviridae, Rhabdoviridae, Togaviridae, Iridoviridae* or a new discovered viral family, the *Nimaviridae* [13]. Among these, the most intensively studied have been characterized from cultured penaeids such as Infectious hypodermal and hematopoietic necrosis virus (IHHNV), the white spot syndrome virus (WSSV), yellow head virus (YHV), and Taura syndrome virus (TSV) [14]. Because of animal movements in aquaculture industry, pathogens were transferred from the regions where they initially appeared to new regions, introducing pathogens that encountered totally naive hosts with little or no innate resistance. The molecular mechanisms and the sensed recognition receptors that trigger crustacean antiviral immune responses have been in some case recognized, as well as the antiviral responses, including inflammatory cytokines and adaptive immune molecules [14]. Several recent reports using crustacean models have suggested that the invertebrate innate immune system may be capable of some form of immune memory, and few attempts of animal's vaccination have been reported [15].

About disease diagnostic, as for all invertebrates, no established cell lines are available so light and electron microscopy are the elective methods. Other methods can include immunoassay or nucleic acid probes [10].

12.2.1 Infectious hypodermal and hematopoietic necrosis virus (IHHNV)

IHHNV is an icosahedral, non-enveloped virus with a single-stranded DNA genome of 4.1 kb, a virion of 22 nm diameter [16] and considered the smallest of the known penaeid shrimp viruses. Because of its characteristics, IHHNV has been classified as a member of the *Parvoviridae* and a probable member genus *Brevidensovirus* [16,17]. After its initial discovery in cultured shrimp in Hawaii in 1981, IHHNV was subsequently described in America and in wild shrimps collected along the Pacific coast from Peru to Mexico. IHHNV often causes an acute disease with mortalities up to 90% in juveniles but also in subadult individuals the shrimp *Paeneus stylirostris*. Gross signs are represented by a reduction in food consumption, signs of weakness and immobility (Table 1).

The virus replicates at ectodermal and mesodermal tissue from the early phase of animal life cycle. Main targeted organs are cuticle, gills, tissues involved in haematopoiesis, antennal gland, and nervous system (Table 2).

Table 1. OIE listed crustacean diseases as of 2020 and those being considered for listing.

Name of the disease	Aetiological agent description
Acute hepatopancreatic necrosis disease	Infection with *Vibrio parahaemolyticus* (Vp AHPND)
Crayfish plague	Infection with *Aphanomyces astaci* Oomycote of the family *Leptolegniaceae*
Necrotising hepatopancreatitis	Infection with *Hepatobacter penaei*
Infectious hypodermal and haematopoietic necrosis virus	Infection with a virus of the family *Parvoviridae*, genus *Penstyldensovirus*
Infectious myonecrosis virus (IMNV)	Infection with Putative totivirus closely related to *Giardia lamblia* virus, family *Totiviridae*
White tail disease (WTD)	Due two viral pathogens, *Macrobrachium rosenbergii nodavirus* (MrNV), a nodavirus (primary) and extra small virus (XSV) (associate), satellite virus
Taura syndrome (TS)	Infection with Taura syndrome virus (TSV), Genus *Aparavirus*, Family *Dicistroviridae*, Order *Picornavirales*
White spot syndrome virus (WSSV)	Infection with genus *Whispovirus,* Nimaviridae family
Yellow head (YHD1)	Infection with yellowhead virus genotype 1, genus *Okavirus*, family *Roniviridae*, order *Nidovirales*

Table 2. Crustacean common viral diseases, OIE listed and not, and distinctive characteristics.

Disease Name	Agent	Classification & Type
Taura syndrome (TS)	Taura syndrome virus (TSV)	Dicistroviridae; ssRNA
White spot disease (WSD)	White spot syndrome virus (WSSV)	Nimaviridae; dsDNA
Yellowhead disease (YHD)	Yellowhead virus (YHV) & gill associated virus (GAV)	*Roniviridae*; ssRNA
Infectious hypodermal & hematopoietic necrosis (IHHN)	IHHN virus (IHHNV)	*Parvoviridae*; ssDNA
Infectious myonecrosis (IMN)	IMN virus (IMNV)	*Totiviridae*; dsRNA
White Tail Disease (WTD)	WTD virus (MrNV)	*Nodaviridae*; ssRNA
Tetrahedral baculovirosis	*Baculovirus penaei* (BP)	*Baculoviridae*; dsDNA
Spherical baculovirosis	*Monodon baculovirus* (MBV)	*Baculoviridae*; dsDNA

Virus spread on eggs, larvae and post larval animals, with reduced hatching success and farming performance. In *P. stylirostris* and *P. vannamei* experimental studies showed both vertical and horizontal transmission of the virus to the offspring or through cannibalism, respectively.

Susceptible species belong to penaeids like the yellow leg shrimp (*P. californiensis*), white leg shrimp (*P. vannamei*), blue shrimp (*P. stylirostris*), giant tiger prawn (*P. monodon*) and the northern white shrimp (*P. setiferus*). Different crustacean's species have been reported PCR positive to the virus, but an active action wasn't always demonstrated, like in other penaeids species (*P. duorarum, P. occidentalis, P. japonicus, P. semisulcatus*), or the crab Cuata swimcrab (*Callinectes arcuatus*), and the teleosts mazatlan sole (*Archirus mazatlanus*), yellowfin mojarra (*Gerres cinereus*), tilapias (*Oreochromis* sp.), Pacific piquitinga (*Lile stolifera*) and blackfin snook (*Centropomus medius*).

Three genetic variants of the virus have been documented and described as Type 1, from the Americas and East Asia (mainly Philippines); Type 2, from South-East Asia; Type 3A, East Africa, India and Australia; Type 3B, the western Indo-Pacific region including Madagascar, Mauritius and Tanzania [18]. The first two genotypes are infectious to the representative penaeids, *P. vannamei* and *P. monodon*, while the latter two genetic variants are not infectious to these species [19,20].

Selection of resistant population to IHHNV were produced for the species *P. stylirostris* [20]. Unfortunately, to date, IHHNV resistance do not have increased survival rates for the other common viral diseases [21].

12.2.2 The white spot syndrome virus (WSSV)

Nowadays WSSV is considered one of the most dangerous viral pathogens of shrimp in the world [22]. The WSV belongs to the family *Nimaviridae* and it is a notifiable disease caused by the White Spot Syndrome Virus (WSSV), an enveloped, rod-shaped virus containing a double-stranded DNA responsible of high mortality prevalence all around the World. The virus has been referred to in the literature by a variety of different names over the years and infects a wide range of decapod crustaceans. In the acute form of the disease, infected prawns are lethargic and anorexic, displaying a numerous white spot at the cuticle and a pink to red discolouration. The lesions are associated with systemic destruction of the ectoderm and mesoderm. Infected nuclei are hypertrophic, display *Cowdry A-type* inclusions, lightly to deeply basophilic with marginated chromatin containing the viral particle at transmission electron microscopy (TEM) [23]. Clinical signs are not easily correlated to infection intensity. Variable susceptibility between different hosts has been reported comprehending both freshwater and marine species like crayfish *Austopotomobius pallipes* and *Pascifastacus leniusculus*, crabs and lobsters and different species of penaeid shrimps. Other species show reduced sensitivity but may remain vectors of the virus like some crab [24], rotifers [25], bivalves, polychaete worms [26] and *Artemia* sp. [23].

The infection can be transmitted horizontally through cannibalism or predation and dead of moribund animals. Outbreaks can be related to environmental stressors of different nature, like osmotic stress and increased water temperature (18 to 30°C) with reduces mortality at 32°C.

Viral pathogenesis and virus enter host cells remains largely undetermined. It seems that a shrimp small GTPase protein binds to viral VP28 involved in the infection, a common cellular factor for other viral entry mechanisms [27] (Figure 1).

Considering the high virulence of the pathogen, the disease control was approached in different ways over the years. The methods comprehend tools of the innate immune response, management practices, pre-exposure to the inactivated pathogens and phytotherapy. Lately, the development of a vaccine for the WSSV has become an intense focus for the researcher in the fields of aquaculture. Most of these reports have concentrated on the characterizations of WSSV proteins. To date, nearly 40 different envelope proteins of WSSV have been identified, and few of them

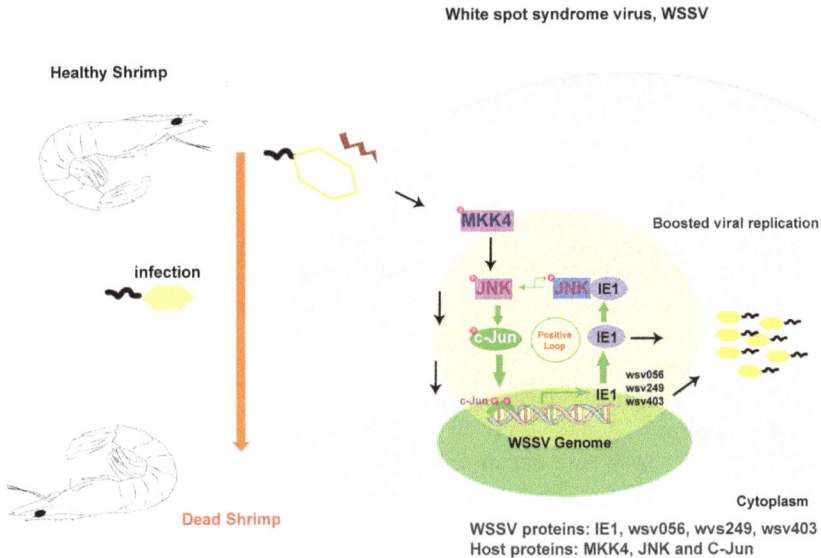

Figure 1. WSSV pathogenesis adapted from Young 2019. WSSV steals the host JNK pathway via its immediate-early 1 (IE1) protein that can drive to its replication, enhancing JNK autoactivation by autophosphorylation, resulting to a positive feedback loop for the virus. Produced viral proteins are wsv056, wsv249, and wsv403, beneficial to virus genome replication.

explored for their possible use in developing recombinant WSSV subunit vaccines. Vaccine efficacy relies on the administration method, oral or injected, the amount of virus used in the challenge, vaccine amount, VP28 subunit protein structure, and vaccine buffer.

In the end, the application of vaccines to prevent WSSV infection is considered possible, with clear results. At the same time, WSSV vaccines still have some problem that need to be solved before to be commercialized.

12.2.3 Yellow Head Virus (YHV)

Yellowhead disease (YHD) is a viral listed exotic disease (EC Directive 2006/88). Virus are enveloped, rod-shaped particles (40–60 nm × 150–200 nm); virions contain a 20–22 kDa nucleoprotein (p20) that forms the nucleocapsid and two envelope glycoproteins of 110–135 kDa (gp116) and 63–67 kDa (gp64) that form the prominent peplomers on the virion surface, classified by the International Committee on Taxonomy of Viruses to the genus *Okavirus*, family *Roniviridae*, order *Nidovirales* [28]. The first description was reported during mortality events in black tiger prawn (*P. monodon*) in Thailand in 1990 [29]. YHV has since been reported widely in Asia, while other genotypes are reported as the Yellowhead disease complex (including Gill Associated Virus—GAV). The virions are located within vesicles in the cytoplasm of infected cells and in intercellular spaces. Yellow head virus is considered as genotype 1, one of six known genotypes in the yellow head complex of viruses and is the only known agent of YHD. Gill-associated virus (GAV) is designated as genotype 2. GAV and four other known genotypes in the complex (genotypes 3–6) occur commonly in healthy *Penaeus monodon* in certain geographic areas like East of Africa, Asia, and Australia. Gross appearance of the disease is represented by yellow discolouration of the dorsal cephalothorax and bleached appearance of muscles [30] caused by the underlying yellow hepatopancreas who appears soft and necrotic when compared with healthy shrimps. In many cases, total harvest loss can happen within a few days of the first appearance of gross signs. Gross signs of GAV disease include swimming near the surface, feeding interruption, a blushing of body and appendages, and pink to yellow coloration of the gills for the regressive phenomena damage. On the other side, in some case, shrimp with chronic infection with YHV or GAV, may not exhibit any sign of disease.

The virions accumulate in the cytoplasm and obtain an envelope by budding at the endoplasmic reticulum into intracellular vesicles. YHV infection can be transmitted horizontally to susceptible crustacean species, by direct injection or ingestion of infected tissue or tissue extracts and by co-habitation [31].

YHD outbreaks were reported only in the black tiger prawn (*P. monodon*) and the white Pacific shrimp (*P. vannamei*). Nevertheless, natural infections have also been detected in other peneids shimps like the kuruma prawn (*P. japonicus*), Pacific blue prawn (*P. stylirostris*), white prawn (*P. setiferus*), red endeavour prawn (*Metapenaeus ensis*), mysid shrimp (*Palaemon styliferus*) and in the krill (*Euphasia superba*).

Environmental factors involved in disease outcome were changes in pH or dissolved oxygen, or temperature modification. YHD can reach up to 100% mortality of shrimps in 3–5 days from the appearance of clinical signs. GAV instead has been associated up to 80% in *P. monodon* in Australia.

12.2.4 Taura Syndrome Virus (TSV)

Taura syndrome is caused by Taura syndrome virus (TSV), a small, simple RNA virus. Based on its characteristics, TSV has been assigned by the International Committee on Taxonomy of Viruses (ICTV) to the newly created genus *Cripavirus* in the new family *Dicistroviridae* (in the "superfamily" of *Picornaviruses*) [32]. The virion is a 32 nm diameter, non-enveloped and the genome consists of a linear, positive-sense single-stranded RNA of 10, 205 nucleotides and the virus replicates in the cytoplasm of host cells.

At least four genotypes have been documented based on the gene sequence encoding for the VP1, a structural protein of the virus. The described genotypes are: (1) the Americas group; (2) the South-East Asian group; (3) the Belize group; (4) the Venezuelan group [33]. Susceptible species are the giant river prawn (*Macrobrachium rosenbergii*), fleshy prawn (*P. chinensis*), some copepod (*Ergasilus manicatus*) and the two barnacles *Chelonibia patula* and *Octolasmis muelleri*. Other species harbouring the virus, but not manifesting the disease comprehend kuruma prawn (*P. japonicus*), southern white shrimp (*P. schmitti*), the blue crab (*Callinectes sapidus*).

Animal with acute disease can display clinical signs with consequent behavioural change. Typically, severely affected shrimps have breading problems, move to the water surface of to the shore where dissolved oxygen levels are higher. TSV infects and mainly replicates in the cuticular epithelium (or hypodermis) of the exoskeleton, foregut, hindgut, gills and appendages, and connective tissues, while enteric organs and muscle do not show any sign of colonization [*34*]. Gross signs of the disease can be displayed by juveniles, subadult and adult shrimps, and in some case pathognomonic. In the acute phase moribund animals have expanded red chromatophores that gives to the animal exoskeleton a roseate coloration who becomes bright red at the tail and pleopods, so initially defined as red tail disease. Epithelial cuticles show focal necrosis and softer exoskeleton also associated to a whole disease condition with watery digestive tissue. Generally, the animal dyes during the highly stressful moment of the moult process. In the chronic phase the animal shows no obvious signs of disease. However, individuals are more susceptible to environmental stressors, like change in temperature and salinity. Between the two phases, a transition phase can be observed, where animals display melanic areas at exoskeleton and limb level due to haemocytes' melanin deposition. At light microscopy, absence of inflammatory response, help in distinguish the acute phase of infection from the transitional phase.

12.3 Most common Bacteria and mycotic infection in Crustaceans

Numerous bacterial and fungal diseases have been spread in the last years due to movements of crustacean's specimens all around the world. Fungi and fungal-like organisms easily occur in waters, acting as saprobes and colonizing decomposing organic matter, or as pathogens leading to disease outbreaks. Bacteria belonging to the genus *Vibrio* are common in the culture conditions in aquaculture systems, but also other bacteria have been isolated like *Photobacterium*, *Aeromonas*, *Chlamydia like* organism, Rickettsia intracellular organisms (RLOs) *Alteromonas*, *Pseudoalteromonas*, *Clostridium*, *Cytophaga* and *Chromobacteria* and many of them related to the so-called *shell disease* of crustaceans [*35*].

The Crayfish plague is a mycotic epidemic disease, OIE listed, caused by the pathogenic oomycete *Aphanomyces astaci. A. astaci* is part of the water moulds recently reclassified as protists with diatoms and brown algae, in a group called the Stramenopiles or Chromista. Crayfish plague has decimated stocks of native crayfish in Europe, but also Australia and Asia. However, only three North American species, *Orconectes limosus*, *Pacifastacus leniusculus* and *Procambarus clarki* are known vectors to European species [*36*]. The pathogen introduction to Europe is reported in 1850, quickly spread pandemic through ships travelling from the USA or Canada to the north of Italy, carrying the infected crayfish species *P. leniusculus* and *P. clarki* [*37*]. The pathogen is considered highly invasive and listed in the Global Invasive Species Database (http://issg.org/database/welcome/).

Transmission among animals occurred through the release of zoospores attaches to the naive crayfish population. Experimental studies showed that the zoospores can swim actively in the water displaying a positive chemotaxis towards the host [*38*].

The tissue initially infected is the outer cuticle of the exoskeleton, in particular the softer parts of the ventral abdomen and joints, resulting then in the areas especially affected. Infectious bacterial diseases are among the world's leading causes of death. The speed of progression of our understanding of bacteria interaction with the host at the molecular level is providing novel insights and perspectives on pathogens and pathogenicity. The elementary steps for the bacterial infection are represented by pathogen adhesion to tissue host, evasion of host defences, pathogen multiplication with host tissue damage (with production of endotoxin or exotoxin) and transmission from the infected animal to other susceptible animals. Bacterial diseases are common in crustaceans, in all the phase of development, from larvae to adults. An overview of the reported bacterial diseases, with the geographic distribution and tissue tropism is presented in Table 3. The *Shell disease syndrome*, also called black spot disease, is one of the most described disease in crustaceans [*39*]. Chitinolytic bacteria are believed to be the primary aetiological agents. Chitin it-self is an abundant polymer in the marine environment, thus bacteria involved in its degradation are common [*40*]. Several bacteria are capable of degradation of the chitin component; when breach of the cuticle occurs, bacteria can enter the lesion, infect the body cavity and leading to the death of the animal. Destruction of the epicuticle can also be due to predatory or cannibalistic attacks, chemical attack or scrapes from sediment. In the American Lobster (*Homarus americanus*), poor water quality overcrowding and improper diet have also been related to disease pathogenesis [*41*]. Various bacteria and fungi are involved in the disease outcome, in some case specific for the host species and linked to geographical location. Bacteria belong to the *Vibrionaceae* family are the most common isolated pathogens, along with *Pseudomonas, Alteromonas Flavobacterium, Photobacterium, Moraxella* and *Pasteurella* [*42*]. In the spiny lobster (genus *Jasus* and *Panulirus*), *Vibrio* sp. like *V. vulnificus, V. parahaemolyticus, V. alginolyticus* are the most common bacteria associated, but also other species have also been isolated including *Shewanella* spp. and *Aeromonas hydrophila*. The second most important group of disease in crustaceans is represented by *Vibriosis*. Many *Vibrio* sp. resulting in septicaemic disease have been reported like the hepatopancreatic necrosis disease (AHPND) in *P. monodon* and *P. vannamei*, responsible for the production loss of US$44 billion between 2010 and 2016 in China, Malaysia, Mexico, Thailand, and Vietnam and related with different *Vibrio* sp. AHPND is currently listed on the OIE list of notifiable terrestrial and aquatic animal diseases [*43,44,45*] (Table 3).

Aquaculture is a sustainable solution for ensuring global food security. With the development and intensification of aquaculture, new bacterial diseases have recently appeared in commercially exploited crustacean species. The threat of bacterial diseases on production can have devasting effects on achieving this goal. The understanding of bacterial infections mechanisms gaps is needed in future research, along with the understanding of disease management for hatchery productivity.

12.4 Crustaceans Microsporidia

The phylum Microsporidia is a group of parasites infecting all major taxa in all environments. They are closely related to fungi, but their nature has been revised constantly [*46*]. Many microsporidian parasites have been described as infecting

Table 3. Overview of the main bacterial disease recorded from different Crustaceans.

Bacteria/associated disease	Global location	Host	Tissue
Septic Hepatopancreatic Necrosis Disease		All farmed shrimps	haepathopancreas
V. alginolyticus, *V. campbellii*	China, Malaysia, Vietnam Thailand, Mexico, e Philippines	(Hatchery, and grow-out, post larvae and adults)	
V. harveyi, *V. parahaemolyticus,* *V. penaeicida,* *V. vulnificus,* *V. owensii*			
Lobster hatchery vibriosis (phyllosome)	ubiquitous	lobster like *Jasus edwardsii*	All the body
V. anguillarum, V. harveyi *V. owensii* *V. parahaemolyticus*		*Palinurus* sp.	Tail and body
Tail necrosis *V. atlanticus,* *V. crassostreae,* *V. cyclitrophicus,* *V. gigantis, V. splendidus*	New Zealand, Australia	lobster *Jasus* sp. *Palinurus* sp. *Thenus* sp. *Jasus* sp. *Panulirus* sp. *Thenus* sp	
Red body disease *V. alginolyticus*		*P. homarus, P. longipes,* *P. ornatus, P. polyphagus*	Systemic
Common Vibriosis			
Vibrio parahemolyticus	Chesapeake Bay	Blue crab *Callinectes sapidus*	Haemolymph
Vibrio parahemolyticus	Chesapeake Bay		
Vibrio spp.		Hoseshoe crab, *Limulus polyoemus*	Haemolymph
	Puerto Rico	Callinectes bocourti,	
	Connecticut USA	Rock crab, *Cancer irroratus*	
		Rock crab, *Cancer irroratus*; Shore crabs *Carcinus maenas*	
Vibrio harveyi	Australia	*J. edwardsi*	Systemic
Vibrio algynoliticus	Australia	*Panulirus homarus*	Systemic
Vibrio sp.	Australia	*P. argus, P. laevicauda*	Systemic
Limp Lobster Disease *Vibrio fluvialis* or **Others** *Hyphomicrobium indicum*	Gulf of Maine	*Homarus americanus*	Haemolymph

Table 3 contd. ...

...Table 3 contd.

Bacteria/associated disease	Global location	Host	Tissue
Shell Disease *chitinolytic bacteria*	New York Bight	Blue crab *Callinectes sapidus*	Shell, internal organs
(See the text for more details)			
	Nova Scotia	*Homarus americanus*	Shell, internal organs
	Rowan Bay, Alaska	Dungeness crab, *Cancer magister*	
		Shore crab, *C. pagurus*	
	Brittany, France		
	Langland Bay, Swansea, UK.		
	New York Bight	Rock crab, *C. irroratus*	Shell, internal organs
	Mid-Atlantic Bight	Jonah crab, *C. borealis*	
	South Florida, USA	Stone crab, Menippe mercenaria	
	New York Bight	Red crab, Geryon quinquedens	
Aerococcus viridans	UK and USA	*H. americanus, H. gammarus*	Systemic
Leucothrix mucor		*Palinurus*	crustacean eggs and limbs
Rickettsia-like organism	Mediterranean coast of France	Carcinus mediterraneus	Connective tissue of hepatopancreas, gut, gills and gonads
	Alaska, USA	Blue king crab Paralithodes	Hepatopancreatic
		platypus	epithelium
	Jiangsu, China	Cheses mitten crab Eriocheir	
		sinensis	
Chlamydia-like organism	Willapa Bay and northern Puget	Dungeness crab *Cancer magister*	Connective tissue and
	Sound, Washington, USA		connective tissue cells
	Laboratory-reared	Rock crab Cancer irroratus jonah	Haemocytes and
		crab Cancer borealis	haematopoietic tissue
Rhodobacteriales-like	Swansea, Wales, UK	European shore crab, Carcinus	Connective tissue and the
organism		maenas	blood vessels

decapod crustaceans [*47*]. Life cycle stages involves three main phases distinguished in proliferative, sporogonic, and environmental [*48*]. The infective phase is represented by a microspore contain a sporoplasm and a polar filament involved in the infection and enters the host cell. In the following phases merogon matures, followed by sporogonic development. Spores can then infect adjacent cells or other individuals [*49*]. To date, at least 35 genera aquatic arthropod-infecting taxa have been reported, 17 fish-infecting taxa, and six taxa associated with non-arthropod invertebrates [*50*]. Within the Microsporidia, to date almost fifty genera were reported infecting major classes of crustacean comprehending Malacostraca, Maxillipoda, Ostracoda, and Branchiopoda [*51*]. Animals infected comprehends a wide group, from crustaceans belonging to the plankton to other of the benthos (Table 4).

The most severe epizootic due to microsporidians was reported in the Gulf of Mexico in 1929 in white shrimp *Litopenaeus setiferus* due to *Agmasoma penaei*. The infection prevalence was 90% and associated with almost 100% mass mortality and a resulting damage of fish industry for several years [*52*]. About disease impact on the host, many tissues and organs have been described, and infection can be restricted and organ-specific while in other cases, multiple organs and tissues can be affected. In certain species, microsporidian may exclusively infect germ cells (male or female gametes) and be transmitted vertically to offspring. Some of the initial observations of microsporidian infections in aquatic arthropods involved taxa colonizing the skeletal musculature [*53*].

A rarely reported phenomenon of intranuclear microsporidian was described from the hepatopancreas of the edible crab *Cancer pagurus* and of the hermit crab (*Eupagurus bernhardus*) [*54*] in the new genus *Enterospora*, family *Enterocytozoonidae*, same genus of the humans' parasites. Of the genus *Enterocytozoon* belongs *Enterocytozoon hepatopenaei* (EHP) that infect the black tiger shrimp *Penaeus monodon* in Asia. EHP infects the hepatopancreas, responsible of the hepatopancreatic microsporidiosis (HPM), a condition that has been associated with animal weakness and slow growth. In EHP infection happens through polar tube into the host cell cytoplasm. There, with its plasma membrane in direct contact with the host cell cytoplasm, the sporoplasm matures resulting in a distinctive branched plasmodium, unique to the *Enterocytozoon* group. During this process, infected HP epithelial cells swell with a rupture allow spores to be released, bringing to auto-infection and diffusion into the environment [*55*] (Figure 2).

Schematics illustration of the life cycle and transmission of EHP with infecting spore into host cell with the polar tube; the sporoplasm creates a branched plasmodium with the formation of extrusion and generation of new infecting spores.

12.5 Crustaceans as Parasites

Three main groups of parasitic crustaceans are reported in the literature as responsible of major disease of commercially important aquaculture species and belonging to the subclass of Copepoda, Branchiura and Isopoda [*56*]. Literature account about 2000 species of arthropods as parasite, the majority of which belong to the group of Copepods [*57*]. Copepods are small to microscopic with free living and parasitic stages, and they can be both ectoparasites and endoparasites. Members of the

Table 4. Microsporidia infecting aquatic crustaceans.

Crustacean Class	Genus of Microsporidia	Crustacean host
Malacostraca	*Enterocytozoon*	*Penaeus monodon*
Malacostraca	*Enterospora*	*Cancer pagurus*
Malacostraca	*Agmasoma*	*Penaeus setiferus*
Malacostraca	Perezia	*Litopenaeus setiferus*
Malacostraca	*Thelohania*	*Crangon vulgaris*
Malacostraca	*Vairimorpha*	*Cheraz destructor*
Malacostraca	*Ameson*	*Callinectes sapidus*
Malacostraca	*Cucumispora*	*Dikerogammarus villosus*
Malacostraca	Triwangia	*Caridina formosae*
Malacostraca	*Abelspora*	*Carcinus maenas*
Malacostraca	Pleistophora	*Crangon* spp.
Malacostraca	*Fibrillanosema*	*Crangonyx pseudogracilis*
Malacostraca	*Hepatospora*	*Eriocheir sinensis*
Malacostraca	*Inodosporus*	*Palaemonetes* spp.
Malacostraca	*Mrazekia*	*Asellus aquaticus*
Malacostraca	*Myospora*	*Metanephrops challengeri*
Malacostraca	*Nadelspora*	*Cancer magister*
Malacostraca	*Ormieresia*	*Carcinus mediterraneus*
Malacostraca	*Dictyocoela*	*Gammarus* spp.
Malacostraca	*Vavraia*	*Crangon crangon*
Maxillopoda	*Alfvenia*	*Acanthocyclops vernalis*
Maxillopoda	*Ambylospora*	*Mesocyclops albicans*
Maxillopoda	*Cougourdella*	*Megacyclops viridis*
Maxillopoda	*Desomozoon*	*Lepeophteirus salmonis*
Maxillopoda	*Encephalitozoon*	*Macrocylops distinctus*
Maxillopoda	*Facilispora*	*Lepeophteirus* spp.
Maxillopoda	*Holobispora*	*Thermocyclops oithonoides*
Maxillopoda	*Marssoniella*	*Cyclops strennus*
Maxillopoda	*Lanatospora*	*Macrocyclops albidus*
Maxillopoda	*Nelliemelba*	*Boeckella triarticulata*
Maxillopoda	*Parathelohania*	*Microcyclops varicans*
Maxillopoda	*Paranucleospora*	*Lepeophteirus salmonis*
Maxillopoda	*Tuzetia*	*Macrocyclops albidus*
Maxillopoda	*Pyrotheca*	*Cyclops albidus*
Maxillopoda	*Trichotuzetia*	*Cyclops vicinus*
Branchiopoda	*Larssonia*	*Daphnia pulex*
Branchiopoda	*Norlevinea*	*Daphnia longispina*

Table 4 contd. ...

...Table 4 contd.

Crustacean Class	Genus of Microsporidia	Crustacean host
Branchiopoda	*Agglomerata*	*Sida crystallina*
Branchiopoda	*Glugoides*	*Daphnia magna/pulex*
Branchiopoda	*Ordospora*	*Daphnia magna*
Branchiopoda	*Gurleya*	*Daphnia maxima*
Branchiopoda	*Nosema*	*Artemia salina*
Branchiopoda	*Gurleyides*	*Ceriodaphnia reticulata*
Ostracoda	*Flabelliforma*	*Candona* spp.
Ostracoda	Binucleospora	*Candona* spp.

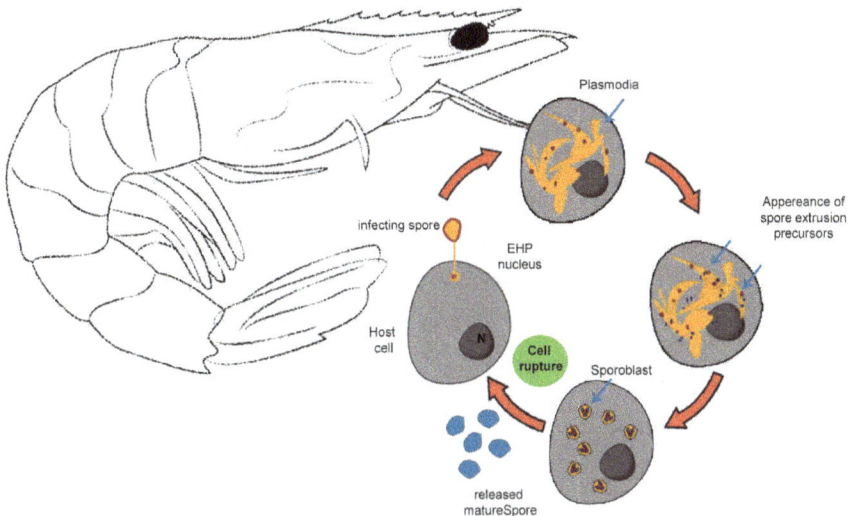

Figure 2. *Enterocytozoon hepatopenaei* (EHP) life cycle, adapted from Chaijarasphong et al. 2020.

Branchiura and Isopoda are instead generally bigger with both sexes parasitic. All the three groups have progressively caused serious problem in cultured fish and wild animal population. They have been described worldwide in different environment from marine to freshwater ecosystems (Table 5).

12.5.1 Copepods as parasites

Copepods are an abundant group of crustaceans belongs to the zooplankton in the water column, mostly considered to create symbiotic associations with different aquatic species. As parasites, they have an extraordinary range of hosts in the marine animals, in fish, but also in invertebrates like echinoderms and cnidarians. In marine invertebrates the morphology of parasitic copepods is different from vertebrates, with an adaptation bring to the reduction of body segmentation and loss of limbs. Literature descriptions report a multifaceted and complex life cycle that comprehends different larval stages, interspersed with a moult. From the eggs hatch

Table 5. Major crustacean parasites.

Subclass	Family	Morphology related parasite
Copepoda		
	Ergasilidae	Second antennae modified for hanging to the host.
	Myticolidae	First antennae have four joints, the second have three, and the last are forming a hook-like appendages.
	Caligidae	Presence of suckers on the frontal edge of the body
	Pennelidae	They show highly specific morphological adaptation linked to host spectrum;
	Phylichthyidae	Parasitize the lateral line and other sensory canals of teleosts. Buccal area with buccal appendages exposed or displaying a siphon.
Isopoda		The mouth forms a cone with maxillipeds that tear at the flesh and tiny pointed mandibles that pierce into the tissue to penetrate blood vessels or blood sinuses.
Brachiura		They have claws as organs of attachment to the host and a tubular sucking mouth equipped with rasping mandibles located at the tip of the mouth tube.

a free-swimming nauplius larvae after which they moult to form copepodid stages. In certain case, copepodids are free swimming. The family *Caligidae*, the most species-rich of the copepods, is mainly represented by ectoparasitic forms and have a direct life cycle, consisting of a free living planktonic nauplius stage, free swimming infective copepodid stages, attached chalimus, pre-adults and adult stages.

The family *Ergasilidae* comprehends mainly the genus *Ergasilus*. it comprises 24 different genera and more than 140 species, but taxonomic keys that define a species are constantly debated.

This group of parasites combine both freshwater and marine species, parasites of teleost fish. Morphological features resemble free-living copepods. Parasitic phases are represented by fertilized females. Males and developmental stages live both in the water column. The main characteristic of Ergasilids is a large, recurved, second antennae, which are used for attachment to the host along with feeding appendages which permit the removal of host tissue or blood. Tissue tropism and lesions associates are at the level of skin, gills, nasal cavities where the parasite perform his feeding activity. Affected fish species are seabass, seabreams and grouper, and bring to skin haemorrhage, impaired respiration, gill hypertrophy/hyperplasia, anaemia and in certain cases to retardation of the growth. Although known to be pathogenic to the host under certain conditions, the presence of small number of parasites is generally thought to be tolerated by the host, unless additional factors contribute to the pathogenicity. In the freshwater environment of Danube River in Hungary, an intense copepod infestation was recorded from a reservoir visibly diseased catfish, *Silurus glanis*. The copepods were confirmed to belong to the Ergasilids *Sinergasilus major* [58]. The species *S. major* has been recorded from seven host species other than *Cyprinus idella*, but the hosts belongs all from different genera and to separate families: four of *Cyprinidae* (*Elopichthys bambusa*, *Scardinius erythrophthalmus*, *Mylopharyngodon piceus*, *Squaliobarbus curriculus*, one *Siluridae* (*Silurus asotus*), one of *Bagridae* (*Tachysurus fulvidraco*) and one *Sinipercidae* (*Siniperca chuatsi*) [*59*].

In the marine environment *Ergasilus sieboldi* is widespread in temperate parts of Europe and Asia.

It was firstly described by von Nordmann in 1832 infecting various freshwater fish in Europe and is the type species of the genus. In the life cycle, the male dies soon after copulation and is thus rarely encountered in fish. Once attached to the host, the female adult parasite grows due to the production of eggs. Following this morphological change, the effective swimming capability of the parasite is reduced. *Ergasilus lizae* parasite is important in mullet culture in the Mediterranean and Middle East and connected to the morbidity and mortality episodes of largemouth bass (*Micropterus salmoides*).

In invertebrates', the family *Myticolidae*, affect bivalves belonging to the genus *Mytilus* sp., like *Mytilus edulis* and *M. galloprovincialis* in the Mediterranean and the Atlantic Sea, but also reported in oysters (*Ostrea edulis* and *Crassostrea gigas*), clams (*Tapes decussatus*) and cockcles (*Cerastoderma edule*). Reported wrongly as "Red Worm Disease", *Mytilicola intestinalis* is a parasitic copepod that occurs in the intestinal tract of the affected individuals. The Adult phase of the parasite is red, with an elongated body and displays a sexual dimorphism. Male individuals are sexually mature at 3 mm and reaching a maximum length of 4.5 mm. The female becomes sexually mature at about 4.6 mm in length, reaches a maximum of about 9.0 mm (Figure 3). The second antennae are modified as hooks used as anchors for resisting expulsion from the host.

The parasite attachment can result in metaplastic changes including replacement of ciliated columnar cells by non-ciliated cuboidal cells [*60*]. Although not typical, a marked haemocytic response may be observed. Parasite burden tends to increase with the host size as larger hosts have a higher filtration rate and high bivalve population density can increase parasite diffusion and spreading. Moreover, bivalves from lower shore levels and sheltered areas display heavier infestations.

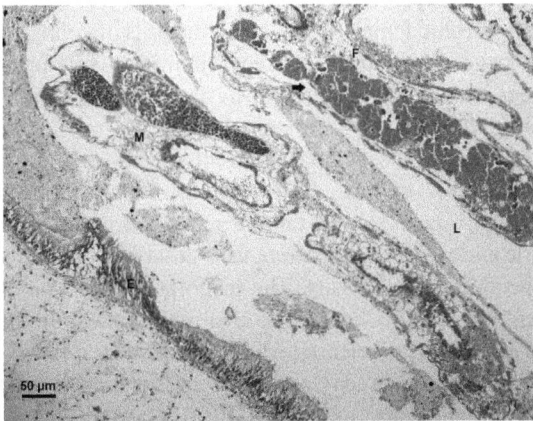

Figure 3. *Mytilicola intestinalis* male (M) and female (F) individuals in the intestine lumen (L) of the mussel *M. galloprovincialis* from Gulf of Pozzuoli (Campania, Naples, Italy). Note the presence of eosinophilic large eggs (arrow).

Another *Mytilicola* species exist, named *M. orientalis* firstly documented in Pacific oysters *C. gigas* from Japan. It was introduced in British Columbia then found in Europe transported to France in the 1970s and spreading to the Dutch Delta. At present, the two copepod species have overlapping host ranges, both mussels and clams.

Members of the family *Caligidae*, with the species of *Caligus* and *Lepeophtheirus* represent a group of siphonostomatoid copepods commonly known as *sea lice* deeply impacted finfish aquaculture worldwide [*61*]. *Caligidae* family is a family of parasitic copepods consists of 31 genera and more than 450 species. A total of 115 sea lice species are reported in 217 fish species, comprising 58 species of *Caligus* and 22 of *Lepeophtheirus* [*62*]. In particular, *Caligus amblygenitalis* and *C. chiastos* are associated with of visible eye damage of the bluefin tuna (*Thunnus thynnus*), from South Australia, with lice grazing on the cornea and bringing to keratitis, panopthalmitis and cataracts with associated blindness, leading to significant production losses. *C. epidemicus* is also a parasite responsible of the mortality of more than 10 marine cultured fish species. *C. rogercresseyi* is a host-dependent parasite that affects rainbow trout and Atlantic salmon in Chile [*63*]. The parasite was transmitted to the farmed fish by the native rock cod *Eleginops maclovinus* and *Odonthestes regia*. About disease management of *C. rogercresseyi* it results really complex. Several treatments used to control sea lice have been attempted, but the majority are administered as baths, only affect the life stages of lice on fish. The residual effect of bath treatments is minimal, not preventing reinfection.

The family *Pennellidae* comprehends mesoparasites of fish, cetacean, and cephalopods [*64*]. The genus *Pennella* includes nine species (*P. instructa, P. sagitta, P. filosa, P. balaenoptera, P. exocoeti, P. diodontis, P. benzi, P. hawaiensis, P. makaira*), and other six species are still not taxonomically defined. Parasites of this genus are present at muscle level. *P. balaenoptera* can be found in different marine mammals, like whales, dolphins, porpoises, or pinnipeds and additionally, *P. filosa* is easily described in pelagic fish like the swordfish (*Xiphias gladius*), Atlantic bluefin tuna (*Thunnus thynnus*), striped marlin (*Kajikia audax*), blue marlin (*Makaira nigricans*). *Pennelidae* displays a life cycle with two naupliar stages, a copepodid stage in which males and pre-metamorphic females develop on the gills of cephalopods, a chalimus stage and an adult stage. Adult males are free swimming, while females are parasitic and hematophagous and after metamorphosis they attach to and enter the body surface of the host (a fish or a marine mammal).

The family *Phylichthyidae* include different species of the genus *Sarcotaces*. They include *Sarcotaces arcticus, S. japonicus, S. komaii, S. namibiensis, S. pacificus, S. shiinoi, S. verrucosus*. So far, the genus was reported from the northern parts of the Atlantic and the Pacific. *S. arcticus* is described on the red snapper found encysted under the skin muscle tissue. The juvenile copepod inserts the head into the flesh and when dies, and the host tissue enclosed it in a solid cyst. Female body typically contains a black fluid, released when the parasitized fish is filleted, and the cyst broken. The copepodids larvae released into environment from the adult female. The juveniles look for a new host to continue the life cycle.

12.5.2 Brachyura as parasites

This group cause mortalities of fish in captivity and in ponds, lakes and estuaries. They are primarily ectoparasites of fishes but have occasionally been reported in amphibians. Defined as fish lice, the group comprise 175 species and 4 genera placed in a single family, the *Argulidae*. Brachyurans are parasitic from the second larval stage and abandon the host looking for a new one throughout development. Changes during the larval phase are gradual, undergoes a profound metamorphosis around the fifth larval stage until changing from a long limb to a short but powerful circular sucker. Typical of the genus, *Argulus* sp. displays a stylet inject a secretion contain digestive enzymes able to break into host tissues. Produce secretion have also a pre-digestive function.

12.5.3 Isopods as parasites

This group comprehends blood-feeding species settle on different part of the body, like fish buccal cavity, gill chamber or body skin and fins, bringing to fish morbidity and mortality [65] They are mostly ectoparasites. Gill's damage comprehends erosions of lamellae, anaemia, gill fusion and granulomas. Important families include *Gnathiidae* and *Cymothoidae* [66]. The firsts are blood feeders on gills and skin, while the second is more common in body cavities like mouth or forms their own cavities in the skin. Gnathiids are larval parasites of fish, the adults being free living and non-feeding, on the contrary Cymothoids are parasitic in the adult phase. Over parasite reproduction, parasite increases in size, associated with the development of marsupium full of juveniles.

12.6 Concluding remarks

Disease is a natural component of animal populations. In the context of animal farming, the potential transfer of disease agents around the globe is fully demonstrated by the epidemic nature of several important diseases of marine animals, including crustaceans. Moreover, mortality events can be widespread and can even damage the socioeconomics of impacted fishing communities. Crustacean parasites and pathogens of crustacean are an important component of aquatic animal disease. Disease management and biotechnology needs to approach the disease outcome.

Different innate immune molecules and mechanisms have been reported in shrimps and other crustaceans that plays an important potential role against invading bacterial, fungal and viral pathogens. The so-called *immune priming*, after prior antigenic exposure, is a potential key determinant to elucidate crustacean's immunity and to develop operative vaccines. In the last years growing interest in the study of invertebrate's immunity was performed to fight resistant agents with the discovery of new bioactive compounds for biotechnological applications. This approach opens the way for the design of practical promising strategies to control crustacean disease also for other invertebrate pathogens.

References

1. T.W. Flegel. (1997). *World J. Microbiol. Biotechnol.* **13:** 433–442.
2. T.W. Flegel. (2006). *Aquaculture* **258:** 1–33.
3. D.V. Lightner, and R.M. Redman. (1998). *Aquaculture* **164:** 201–220.
4. D.V. Lightner. (1999). *J. App. Aquac.* **9:** 27–52.
5. J.F. Wickins, and C. Lee. (2002). DO' Crustacean Farming: Ranching and Culture. 2nd ed. *Aquac. Res.* **34:** 269–270.
6. S.C. Johnson, J.W. Treasurer, S. Bravo, K. Nagasawa, and Z. Kabata. (2004). *Zool. Stud.* **43:** 229–243.
7. A.P. Thamban, S. Kappalli, H.A. Kottarathil, A. Gopinathan, and T.J. Paul. (2015). *Zool. Stud.* **54:** 42.
8. J.D. Williams, and C.B. Boyko. (2012). *PLoS ONE* **7:** e35350.
9. C. Vago. (1966). *Nature* **209:** 1290.
10. D.V. Lightner, FAO. (2000). Fisheries Technical Paper **395:** 38.
11. S.K. Johnson, and S.L.S. Bueno. (2000). Freshwater Prawn Culture: the farming of *Macrobrachium rosenbergii*. Oxford, UK: Blackwell Science; pp. 239–258.
12. D.C. Behringer, M.J.I.V. Butler, J.D. Shields, and J. Moss. (2011). *Diseases Aquat. Org.* **94:** 153–160.
13. P. Jiravanichpaisal. (2005). Digital comprehensive summaries of Uppsala dissertations from the Faculty of Science and Technology **47:** 1–56.
14. H. Liu, K. Soderhall, and P. Jiravanichpaisal. (2009). Fish Shellfish Immunol. **27:** 79–88.
15. Y.H. Chang, R. Kumar, T.H. Ng, and H.C. Wang. (2018). Dev. Comp. Immunol. **80:** 53–66.
16. H. Shike, A.K. Dhar, J.C. Burns, C. Shimizu, F.X. Jousset et al. (2000). *Virology* **277:** 167–177.
17. L.M. Nunan, B.T. Poulos, and D.V. Lightner. (1998). *Aquaculture* **160:** 19–30.
18. T.F. Jr. Duda, and S.R. Palumbi. (1999). *Mar. Biol.* **134:** 705–710.
19. K.F.J. Tang, B.T. Poulos, J. Wang, R.M. Redman, H.H. Shih et al. (2003). *Dis. Aquat. Org.* **53:** 91–99.
20. K.F.J. Tang, S.V. Durand, B.L. White, R.M. Redman, C.R. Pantoja et al. (2000). *Aquaculture* **190:** 203–210.
21. A. Alcivar-Warren, R.M. Overstreet, A.K. Dhar, K. Astrofsky, W.H. Carr et al. (1997). *J. Invertebr. Pathol.* **70:** 190–197.
22. J.G. Sánchez-Martínez, G. Aguirre-Guzmán, and H. Mejía-Ruíz. (2007). *Aquac. Res.* **38:** 1339–1354.
23. G.D. Stentiford, J.R. Bonami, and P. Alday-Sanz. (2009). *Aquaculture* **291:** 1–17.
24. A.S. Sahul Hameed, M.X. Charles, and M. Anilkumar. (2000). *Aquaculture* **183:** 207–213.
25. D.C. Yan, S.L. Dong, J. Huang, X.M. Yu, M.Y. Feng, and X.Y. Liu. (2004). *Dis. Aquat. Org.* **59:** 69–73.
26. K.K. Vijayan, V.S. Raj, C.P. Balasubramanian et al. (2005). *Dis. Aquat. Org.* **63:** 107–111.
27. K. Sritunyalucksana, W. Wannapapho, C.F. Lo, and T.W. Flegel. (2006). *J. Virol.* **80:** 10734–42.
28. P.J. Walker, J.R. Bonami, V. Boonsaeng et al. 2004. Virus Taxonomy, VIIIth Report of the ICTV. London, UK: Elsevier/Academic Press; p. 973–977.
29. C. Limuswan. (1991). Handbook for Cultivation of Black Tiger Prawns. Bangkok, Thailand: Tansetakit; pp. 202.
30. C. Chantanachookin, S. Boonyaratpalin, J. Kasornchandra et al. (1993). *Dis. Aquat. Organ.* **17:** 145–157.
31. D.V. Lightner. (1996). A Handbook of Shrimp Pathology and Diagnostic Procedures for Diseases of Cultured Penaeid Shrimp. Baton Rouge, Louisiana, USA: World Aquaculture Society; p. 304.
32. M.A. Mayo. (2002a). *Arch. Virol.* **147/5:** 1071–1076.
33. J.O. Wertheim, K.F.J. Tang, S.A. Navarro, and D.V. Lightner. (2009). *Virology* **390:** 324–329.
34. M.G. Bondad-Reantaso, S.E. McGladdery, I. East, and R.P. Subasinghe FAO. (2001). Fisheries Technical Paper **402/2:** 240.
35. L. Nyhlén, and V. Unestam. (1975). *J. Invertebr. Pathol.* **26:** 353–366.
36. T. Unestam. (1972). *Rep. Inst. Freshwat. Res. Drottingholm* **52:** 192–198.
37. D.J. Alderman. (1996). Rev. Sci. Tec. **15:** 603–632.
38. L. Cerenius, and K. Soderhall. (1984). *J. Invertebr. Pathol.* **43:** 278–281.
39. C.L. Vogan, A. Powell, and A.F. Rowley. (2008). *Environ. Microbiol.* **10(4):** 826–835.
40. N.O. Keyhani, and S. Roseman. (1999). *Biochim. Biophys. Acta* **1473:** 108–122.

41. R. Smolowitz, A.Y. Chistoserdov, and A. Hsu. (2005). *J. Shellfish Res.* **24:** 749–756.
42. R.G. Getchell. (1989). *J. Shellfish Res.* **8:** 1–6.
43. [OIE] Office International des Epizooties, Office International des Epizooties, Paris, France (2020).
44. [OIE] Office International des Epizooties, Office International des Epizooties, Paris, France (2019).
45. [OIE] Office International des Epizooties, Office International des Epizooties, Paris, France (2019).
46. E.U. (1998). Canning Evolutionary relationships among protozoa. London, UK: Chapman and Hall; pp. 77–90.
47. J.A. Couch. (1983). The Biology of Crustacea. New York: Academic Press; pp. 79–111.
48. E.U. Canning, and J. Lom. (1986). The Microsporidia of Vertebrates. New York: Academic Press; pp. 1–16.
49. P.J. Keeling, and N.M. Fast. (2009). *Annu. Rev. Microbiol.* **56:** 93–116.
50. C. Azevedo. (1987). *J. Invertebr. Pathol.* **49:** 83–92.
51. J. Vavra, and J. Lukes. (2013). *Adv. Parasitol.* **82:** 254–319.
52. G. Gunter Estuaries. (1967). Washington, DC: *Am. Assoc. Adv. Sci.*; pp. 621–638.
53. G.D. Stentiford, and A.M. Dunn. (2014). Microsporidia: Pathogens of Opportunity, First Edit. John Wiley & Sons, Inc., Iowa, pp. 579–603.
54. G.D. Stentiford, and K.S. Bateman. (2007). *Dis. Aquat. Organ.* **75:** 73–78.
55. T. Chaijarasphong, N. Munkongwongsiri, G.D. Stentiford et al. (2020). *J. Invertebr. Pathol.* **1:** 107458.
56. R. Huys, and G.A. Boxshall. (1991). Copepod Evolution. London: Ray Society; p. 468.
57. Z. Kabata. (1979). Parasitic Copepoda of British fishes. London: Ray Society; pp. 1–468.
58. Q.M. Dos Santos, and A. Avenant-Oldewage. (2021). *Int. J. Parasitol. Parasites Wildl* **15:** 127–131.
59. I.E. Bykhovskaya-Pavlovskaya, A.V. Gusev, M.N. Dubinina et al. (1964). Key to Parasites of Freshwater Fishes of the U.S.S.R. Israel Program for Scientific Translation, Jerusalem.
60. M.N. Moore, D.M. Lowe, and J.M. Gee. (1978). *J. Conseil.* **38(1):** 6–11.
61. C.J. Hayward, N.J. Bott, and B.F. Nowak. (2009). *J. Fish Dis.* **32:** 101–106.
62. M.J. Costello. (2006). *Trends Parasitol.* **22(10):** 475–83.
63. C. Hamilton-West, and G. Arriagada. (2012). *Prev. Vet. Med.* **104:** 341–345.
64. S. Ohtsuka, K. Harada, K. Miyahara et al. (2007). *Fish Sci.* **73:** 214–216.
65. R.C. Brusca. (1981). *Zool. J. Linn. Soc.* **73:** 117–199.
66. Y. Honma, S. Tsunaki, A. Chiba, and J.S. Ho. (1991). Report of the Sado Marine Biological Station, Niigata University **21:** 37–47.

CHAPTER 13

Biotechnologies Linked to Crustaceans

Antonietta Siciliano, Giovanni Libralato* and Marco Guida

13.1 Introduction: an overview of *Daphnia* spp. as a versatile model organism

Who in ecotoxicology has not heard of *Daphnia* species? Known to many as a fundamental model organism for freshwater habitat, *Daphnia* spp. is the cornerstone of many important findings in ecotoxicology and other areas of ecology [*1*]. Also called water fleas, *Daphnia* spp. are small planktonic crustaceans, belonging to the class Branchiopoda and to the subclass Cladocera. They have a wide distribution occurring in a highly diverse set of freshwater habitats. Microcrustacea are significant grazers of detritus, bacteria, and especially algae species.

The genus *Daphnia* includes more than 100 species of freshwater planktonic crustaceans, of which the most representative and studied are *D. magna* and *D. pulex* [*2*]. For over a century, freshwater branchiopods are recognised model organisms in various fields of research, including evolutionary, ecological, toxicological and physiological studies. The microcrustaceans are ecologically important in freshwater habitats, occupying a key position in the aquatic food chain as the intermediate link between primary and secondary productivity. *Daphnia* is a very efficient filter feeder of algae, bacteria and protozoans, and a major food source for a whole range of aquatic invertebrates or vertebrates [*1,2*].

Besides, the crustaceans can be easily cultivated in large numbers in natural or artificial freshwater media supplemented with unicellular algae. This simple multicellular organism is cyclic parthenogenetic with sex determination ruled by environmental factors. Males are morphologically distinct from females in length, size of the first antennae and first legs armed with a hook [*2*]. Under favorable conditions, *Daphnia* spp. reproduce clonally during most of the year (Figure 1A). Yet, typically before the stressful conditions, such as a change in temperature, food levels or population density, they produce males and consequently undergo sexual

Department of Biology, University of Naples "Federico II", Via Cintia Complesso Monte Sant'Angelo, 80126 Naples, Italy.
* Corresponding author: giovanni.libralato@unina.it

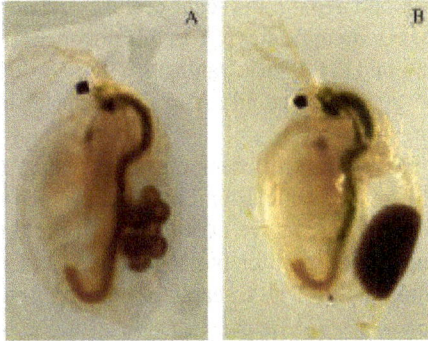

Figure 1. Female of *Daphnia magna* carrying asexual (parthenogenetic) eggs (A). Female of *Daphnia magna* carrying a resting egg (ephippium) (B).

reproduction during which dormant fertilized eggs (called ephippia) are produced (Figure 1B). The fertilized eggs remain in a diapause stage being enclosed by several protective membranes, and can survive these conditions for many decades before hatching [1–3].

Daphnia has a short life cycle of ~ 12 days at 20°C from egg to juvenile and through 4 molts and instars to egg-laying adult and lives up to 6–8 weeks under favorable conditions. A female can generate eggs every four days and the number of offspring produced can be from 1–100, with an average of 6–10 eggs per brood [2]. With its transparent body at all stages of its life cycle and its small size, water flea lends itself to non-invasive optical monitoring and manipulation methodologies. Such approaches can help to investigate the physiologic and molecular mechanisms underlying normal function and dysfunction at all levels from cellular organelles to the whole organism [1].

The small body size and the parthenogenetic life cycle enable experimental analysis on great populations and consequently on fitness changes under multiple environmental stressors. It is also for these reasons that *Daphnia* species are renowned models in evolutionary biology and ecotoxicology widely used as indicators for environmental health and environmental changes. *Daphnia* are generally used as the focal species in ecotoxicology studies, often acting as prescreening tests before vertebrate testing (for both practical and ethical reasons) when testing is desired according to the 3Rs idea: replace, reduce and refine [3,4].

Among freshwater organisms, daphnids are relatively highly sensitive to environmental stressors [5]. Contaminant exposure can induce a shift of behavioral, morphological, or physiological traits in these organisms showing phenoplasticity [1,5]. The abiotic and biotic stressors include everything from chemical substances to synthetic hormones, acidity, salinity, calcium levels, hypoxia, radiation and bacterial pathogens [5]. Water flea responds to stress exhibiting morphological alterations such as enlarged caudal spines, neck teeth or protective helmet development in presence of predator. The completely sequenced *Daphnia* genome, which is approximately 200 Mb, has estimated a lot of genes with homologues in humans [6]. These advantages together with the development of molecular biology and genetic approaches enable the study of classical signaling pathways that underlie development, cell death and ageing. In recent years, *Daphnia* has also emerged as a hopeful candidate for epigenetic studies [7]. In this post-genomic era, the availability of DNA sequence combined *Daphnia* advantages will promote the understanding the complex regulation of genes and cellular and molecular processes that respond to environmental stressors [1].

In this chapter, we focus on recent advances using *Daphnia* models that contribute to shed light on the cellular and molecular mechanisms and we review what we consider the modern tools for *D. magna*, and the future for this organism that can be considered the toolbox for biotechnologies and may begin integrating studies of environmental quality with research of human diseases.

13.2 Integrative multi-omic analyses

The U.S. National Institutes of Health recognize *Daphnia* as a model organism for biomedical research (http://www.nih.gov/science/models/) for its advantages in comparison to established biomedical models and its extreme ecoresponsive genome. Methods that use forward and reverse genetics, with mutants, transgenics, and knockouts can be easily created with *Daphnia* [8]. The range of resources that are available to *Daphnia* species, have contributing significantly to the rapid adoption of *Daphnia* spp. as a model system also for biomedical research [9]. One such specific resource is the Wfleabase (http://wfleabase.org/), an organized repository of *Daphnia* specific sequences developed by the *Daphnia* Genomics Consortium (DGC). WfleaBase contains a wealth of information about gene structures, microsatellites, cDNA, expressed sequenced tags (ESTs) and microarrays [3,8]. The first *Daphnia* genome sequencing project showed a large number of genes (31,000) which exceeds the number found in most other organisms, including human and including a large number of recent gene duplicates. The genome is made compact by an unusually high rate of intron and the high genes number is due to numerous tandem duplications with rapid divergence of expression patterns [6].

In 2017, a new reference genome assembly for the microcrustacean was proposed and demonstrated a ~ 7000 excess genes of previous study appearing to be false positives for the high GC content, lack of introns, and short length of these gene [10]. The complete sequence of the *Daphnia* genome opening to new research fields has increased number of scientific publications on *Daphnia* including research into mechanisms of ageing and sex differentiation in lifespan. Constantinou et al. [11] found significant differences between sexes in physiological and molecular markers such as growth rate, heart rate and lipid peroxidation product accumulation, thiol content decline and age-dependent decline in DNA damage repair efficiency. Another study of Constantinou et al. [11] demonstrated changes in the main lipid groups recorded in human such as triglycerides (TG), diglycerides (DG), phosphatidylcholine, phosphatidylethanolamine, ceramide and sphingomyelin lipid groups as a function of age in male and female of *Daphnia* (Figure 2). These approaches help to investigate the different factors at the basis of the ageing process and to translate the results for human applications.

Water flea genome encodes three DNA methyltransferases (DNMT) enzymes with all of the protein sequences identified as orthologous to the human DNA DNMT

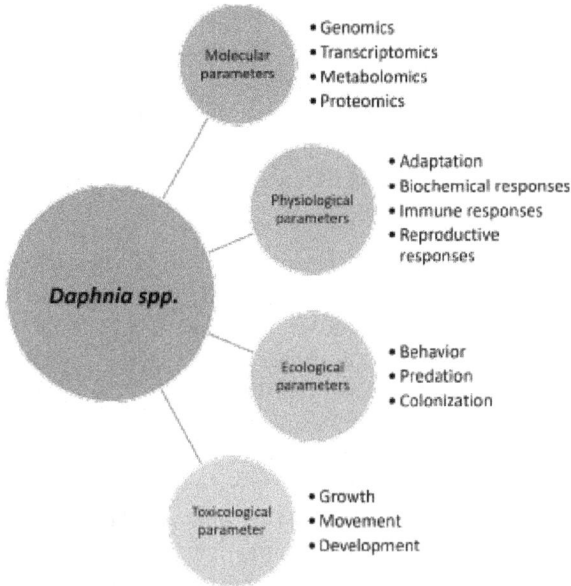

Figure 2. Multilevel-framework for *Daphnia* studies at multiple levels ranging from molecular, physiological, ecological and toxicological parameters.

genes and although the DNA methylation levels and the CpG site frequency across the genome are different, there are significant similarities [*12*]. DNA methylation serves a wide variety of biological functions. Consequently, daphnids are becoming the focus of studies linking methylation, caloric restriction, and ageing due for the understanding the regulatory effect of methylation on disease progression in humans.

Knockdown technology for DNMT genes, helps to assessing the role of DNA methylation and related epigenetic processes correlated with phenotypes that derive from epigenetic alterations [*11,12*].

Maxwell et al. [*13*] with an orthologue identification on 2,727 human disease-associated genes derived from OMIM (Onlin Mendelian Inheritance in Man), identified 1,810 ortholog clusters in *Daphnia*, in particular *D. pulex*, corresponding to 66.4% of observed disease gene orthologs. This value is higher compared to that found for *Caenorhabditis elegans* and *Drosophila melanogaster*, other two invertebrates that have long been valid model organisms in biomedical research. Analyses from assembled *Daphnia* genomes defined xenobiotic detoxification and metabolism genes, identifying several defense related genes such as cytochrome P450 genes, a protein family involved in xenobiotics detoxification, ABC transporter genes, ATPbinding cassette membrane transport-proteins, and glutathione S-transferase (GST) genes, metabolic isozymes [*14*].

Daphnia provides a great opportunity to unveil the role of miRNAs in response to stressor. miRNAs are a family of short non-coding RNAs that act as post-transcriptional regulators and modulate gene expression mainly through suppression of their target gene expression. Unlu et al. identified 205 putative mature miRNA sequences belonging to 188 distinct miRNA families and 36% of these miRNAs identified are predicted to have a role in regulating cellular signaling cascades implicated in controlling developmental pathways in humans (e.g., calcium, Hedgehog, WNT, Notch and chemokine signaling pathways) [7]. Last but not least, *Daphnia* was also suitable for the study of human hypoxic injury as a model organism supported by high homology of HIF (Hypoxia-inducible factors) related genes among *Daphnia* and human. Positive Reactive oxygen species (ROS) effects on HIF mRNA expression, have been shown in *Daphnia* and in humans [*15*].

13.3 Ecotoxigenomic

Since early 1900s, when geneticist Ernest Warren described the effects of environmental stressor on genus *Daphnia*, the research community has gathered an enormous body of knowledge regarding to the ecology and evolution of these keystone freshwater species [*1*]. The extensive literature on the use of *Daphnia* as bioindicator is due to the fact that is considered as high sensitivity analytical tools to screening toxicity of common xenobiotics and monitoring of effluents, contaminated waters and other environmental matrix [*5*]. In the past two decades, due to the rapid advent in omics technologies and interest in environmental risk assessment, *Daphnia* has been incorporated in ecotoxicogenomic that integrate omics studies (i.e., transcriptomics, proteomics, metabolomics, and epigenomics) into ecotoxicological fields [*14*] (Table 1).

Table 1. Overview on ecotoxigenomic studies published in *Daphnia* spp.

Stressors	Method	Species	Reference
Temperature	Protein 2D-Gel Electrophoresis	*D. pulex*	Schwerin et al. 2009
Reproductive stages	qPCR	*D. carinata*	Kong et al. 2015
Copper, cadmium, and zinc	Microarray	*D. magna*	Poynton et al. 2007
Chemicals	RNA-seq	*D. magna*	Lee et al. 2019
Quantum dots-indolicidin	NMR spectroscopy	*D. magna*	Falanga et al. 2018
Cibacron Blue 3GA	MALDI-ToF-MS	*D. magna*	Bayramoglu et al. 2019
Antidepressant	Immunofluorescence microscopy	*D. magna*	Campos et al. 2012, 2016
Tributyltin, pyriproxyfen and bisphenol A	Microarray	*D. magna*	Fuertes et al. 2019
Cadmium	Real-time PCR	*D. magna*	Haap et al. 2016
Cerium and Erbium	Real-time PCR	*D. magna*	Galdiero et al. 2019
Malathion	qPCR	*D. magna*	Trac et al. 2016
Propicanazole; cadmium; copper; zinc; Ibuprofen; bisphenol-A; silver nanoparticles; Copper sulphate; Hydrogen peroxide; pentaclorophenol; naphthaflavone	Comparative genomics	*D. magna*	Kim et al. 2015
Midazolam; pentobarbital; ketamine; Flumazenil	qPCR	*D. pulex*	Dong et al. 2013

There have been various technologies developed to investigate gene expression at a genome wide level (e.g., hybridisation-based and high-density DNA array-based techniques) and metabolite profiles at the proteome (e.g., 2-D gel electrophoresis and MALDI-TOF mass spectrometry) and metabolome level (e.g., NMR based technologies). To date, these strategies have also been applied to study the response of *Daphnia* to environmental chemicals [5]. One basic concept in ecotoxicogenomic is that gene expression profiling change depending on its environment, and the profiling can be used to identify and confirm mechanisms of action of different toxicants and indicate cell health state [5,16]. Thus, a water flea swimming in contaminated water will express a set of genes differently than daphnids accustomed to water without contaminants.

Genome research on the responses of *Daphnia* to stress has important implications for assessing the possible effects of environmental agents on cellular and molecular processes. Consequently, the environmental risk can be more easily linked to similar processes in other organisms, including humans [16]. Many interspecies comparative studies have suggested a commonality in toxicity of different compound across taxa and, thus, cross-species extrapolations for future toxicity testing [5,16,17]. Perhaps the greatest potential of ecotoxicogenomic is the possibility to identifying chemicals related to gene expression profiles, thus helping to understand their mode of action. Daphnids have also been used extensively to analyze the effects and modes of action for a large variety of anthropogenic chemicals singly or in mixture [18]. Since these compounds are designed to target specific pathways, when introduced into the environment they may affect the same or comparable pathways in organisms having the drug target homologies [1].

Heckmann et al. [17] studied ibuprofen (a non-steroidal anti-inflammatory drug) toxicity and potential mechanisms of action in *Daphnia* and demonstrated that this drug inhibited eicosanoid biosynthesis thus suggesting a similar mode of action like mammals. Based on genomic analyses, human targets of antidepressants, anxiolytic, and neuropathic drugs and antihypertensive compounds, are highly conserved among the vertebrates, and 61% of them are also found in *Daphnia* with 49% predicted homology with humans between the serotonin transporter [19]. Moreover, it has been suggested that the primary mechanism of action of antidepressants in *Daphnia*, highlighted by adverse reproductive responses, was similar to the mode of action in humans and like in these the drugs enhanced the accumulation of serotonin in the central nervous [20].

It was demonstrated that the pesticide exposure, one of risk factors for diabetes and obesity in humans, has promoted the accumulation of storage lipids such as triacylglycerols and cholesterol in *Daphnia* upregulating genes related to neuroactive ligand receptor interaction signaling pathways [18]. Other results indicated response to sublethal heavy metals exposure, mediated by upregulation of genes encoding digestive enzymes and associated processes (e.g., oxidative stress and immune suppression), similarly to those described in *D. melanogaster*, plants and fungi under heavy metal treatments. Gene expression profiling, focused on inter-clonal differences in expression of a few genes, have been used before and after multigenerational acquired tolerance to abiotic stressors [5]. By simulating the human-impacted environments, a recent study investigated the early transcriptional response of *Daphnia* to a set of multiple (biotic and abiotic) environmental stressors and found activation of a genotype and/or condition-specific response underlying the tolerance to environmental pertubations. In 2019, taking advantage of an extensive literature for the differentially expressed *Daphnia* genes in response to a stressor, was developed a gene expression database, the *Daphnia* stressor database (http://www.daphnia-stressordb.uni-hamburg.de/dsdbstart.php). This database, accessible from March 2020, will also be critical to interpretation the results of studies on adaptation in natural populations as well as ecological experiments. The integration of ecotoxicological daphnid assays and molecular studies allowed the assessment of xenobiotic effects at environmentally relevant concentrations identifying early warning markers to signal the exposure to stressors [5,18].

13.4 Other environmental current approaches and future directions

Daphnids possess a variety of interesting functions and some of them have been studied from biotechnology point of view. This organism may seem having many interesting features for biotechnology, such as the capacity to remove abiotic or biotic compounds or to produce polymers [1,21–24]. In terms of environmental and human health, the use of *Daphnia* to remediate watersheds and drinking water supplies affected by pathogen or algal bloom is a novel possibility for freshwater bioremediation [1]. In literature, there are several reports where *Daphnia*-based biosensors have been successful for remove different microbial communities: cyst walls degradation and infectivity loss of *Cryptosporidium parvum* and *Giardia lamblia* (zoonotic parasites); reduction populations of *Campylobacter jejuni* (a human pathogenic bacterium); depletion of Avian influenza virus (a potential source for the pandemic emergence); decrease of *Vibrio cholerae* densities [21]. Although some authors reported that the filter-feeders are not able to digest all kinds of microbes, grazing of zooplankton is expected to be an important factor in the reduction and or control of pathogen in natural environments [1].

Due to their role as aquatic grazers, water fleas have been applied to reduce harmful algal bloom. Algal (cyanobacterial) blooms are also a human health concern, as some produce substances (cyanotoxins) that are toxic to the liver and nervous system. The grazing experiments successfully reduced and controlled the wide range of bloom-forming cyanobacterial species at low levels of fish predation proving useful for their potential use in restoration [22].

Several surveillance studies were carried out on *Daphnia* spp. offering potential for the removal of heavy metals or for uptake and biotransformation of organophosphate flame retardants [14] that could control the abundance of these toxic chemicals in natural contaminated waters. More recent applications from daphnids, included the use of crustacea byproducts, such as the exoskeleton for chitin production and characterization. Chitin is a homopolysaccharide of N-acetylglucosamine, and generally a major constituent of the shell of crustacea. After cellulose, it is the most abundant polysaccharide in the biosphere, and its unique attributes, together with its by-products (chitosan, chitotriose, and chitobiose) make this one of the most widely used polymer for industrial, medical and agricultural applications. Application of a biotechnological approach is very promising indeed to both short processing time and use at the industrial scale [*23*].

Another promising application was *Daphnia* stepping into biodiesel production. In order to alleviate the increase demand for fossil fuels, the research is trying to develop novel technologies for biofuel production by the conversion of carbon sources into usable fuel. Ding et al. [*24*] proposed a novel innovative lipid co-extraction strategy using *Daphnia* after digestion of biodiesel-producing microalgae. The optimal lipid extraction and fatty acid methyl esters recovery rates set the stage for potential *Daphnia* utilization as an economical, green, low-energy way for microalgae biodiesel production [*24*].

13.5 Conclusions

With all the research that is currently running on *Daphnia*, a bright future is expected. Gained knowledge and range of available resources to *Daphnia* species are promising for really important findings and novel biotechnological procedures. They can have implications for the water quality in terms of the potential role of *Daphnia* in the biocontrol of biotic and abiotic stressor in recreational waters and drinking water reservoirs and for studies on human diseases and of signaling pathways. The current scientific knowledge will enable us to fully achieve these goals in the near future for boosting the true potential of this wonder organism.

References

1. A. Siciliano, R. Gesuele, G. Pagano, and M. Guida. (2015). *J. Biodivers. Endanger. Species.*
2. D. Ebert. (2020). *Nat. Rev. Genet.* **21(12):** 754–768.
3. C.E. Caceres, and A.J. Tessier. (2004). *Oecologia.* **141(3):** 425–31.
4. M.W. Wojewodzic, and M.J. Beaton. (2017). *Adv. In Insect. Phys.* **53:** 287–312.
5. E. Galdiero, R. Carotenuto, A. Siciliano, G. Libralato, M. Race et al. (2019). *Environ. Pollut.* **254:** 112985.
6. J.K. Colbourne, M.E. Pfrenderdonald, M.E. Gilbertw, W.K. Thomas, T. Abraham et al. (2011). *Science* **331(6017):** 555–561.
7. E.S. Ünlü, D.M. Gordon, and M. Telli. (2015). PLoS One **10(9):** e0137617.
8. T. Watanabe, H. Ochiai, T. Sakuma, H.W. Horch, N. Hamaguchi et al. (2012). *Nat. Commun.* **3(1):** 1–8.
9. C.A. Schumpert, J.L. Dudycha, and R.C. Patel. (2015). *BMC Biotechnol.* 1–13.
10. Z. Ye, S. Xu, K. Spitze, J. Asselman, X. Jiang et al. (2017). *G3 (Bethesda)* **7(5):** 1405–1416.
11. J.K. Constantinou, J. Sullivan, L. Mirbahai et al. (2020). *Sci. Rep.* **10(1):** 1–15.
12. J. Kvist, C. Gonçalves Athanàsio, O. Shams Solari, J.B. Brown, J.K. Colbourne et al. (2018). *Genome Biol. Evol.* **10(8):** 1988–2007.
13. E.K. Maxwell, C.E. Schnitzler, P. Havlak, N.H. Putnam, A.D. Nguyen et al. (2014). *BMC Evol. Biol.* **14(1):** 1–18.
14. B.Y. Lee, B.S. Choi, M.S. Kim, J.C. Park, C.B. Jeong et al. (2019). *Aquat. Toxicol.* **210:** 69–84.
15. H. Li, X. Zhou, J. Fan, S. Long, J. Du et al. (2018). *Sens. Actuators B. Chem.* **254:** 709–718.
16. H.J. Kim, P. Koedrith, and Y.R. Seo. (2015). *Int. J. Mol. Sci.* **16(6):** 12261–12287.
17. L.H. Heckmann, A. Callaghan, H.L. Hooper, R. Connon, T.H. Hutchinson et al. (2007). *Toxicol. Lett.* **172(3):** 137–145.
18. I. Fuertes, B. Campos, C. Rivetti, B. Piña, and C. Barata. (2019). *Environ. Sci. Technol.* **53(20):** 11979–11987.

19. L. Gunnarsson, A. Jauhiainen, E. Kristiansson, O. Nerman, and D.J. Larsson. (2008). *Environ. Sci. Technol.* **42(15):** 5807–5813.
20. B. Campos, C. Rivetti, T. Kress, C. Barata, and H. Dircksen. (2016). *Environ. Sci. Technol.* **50(11):** 6000–6007.
21. P. Ramírez. (2012). *Tecnol. Cienc. Agua.* **3(1):** 69–76.
22. P. Urrutia-Cordero. (2016). *PLoS One* **11(4):** e0153032.
23. M. Kaya. (2013). *Int. J. Biol. Macromol.* **61:** 459–464.
24. W. Ding, Z. Wang, X. Yang, L. Shi, and J. Liu. (2020). *Bioresour. Technol.* **306:** 123162.

CHAPTER 14

Copepods vs. Salmons
Environmental Treats for Crustaceans or Possible Eco-Sustainable Solutions?

Valerio Zupo,[1,*] *Valerio Mazzella,*[2] *Patrick Fink,*[3] *Mahasweta Saha,*[3]
Ylenia Carotenuto[2,4] *and Mirko Mutalipassi*[5]

14.1 Introduction

Crustaceans may be objects of modern aquaculture productions, and in this case, they must be protected by various ectoparasites. However, they may represent, in their turn, threatening pests because some species are obligate parasites of fish [1]. Conversely, multiple anti-parasitic agents used in aquaculture practices may produce heavy impacts on natural populations of planktonic and benthic crustaceans [2]. In particular, the increasing use of drugs against parasitic crustaceans raised concerns about the potential impacts on non-target crustaceans [3], especially when they play key ecological roles, or they are valuable marketable species [4]. In this chapter, we will analyze the effects of anti-parasitic agents and illustrate some new perspectives in the eco-sustainable treatment of fish infestations sustained by parasitic copepods.

Two key genera of sea lice are commonly found as parasites of salmonids in marine and brackish waters: *Lepeophtheirus* and *Caligus*. The so-called "Salmon louse" *Lepeophtheirus salmonis salmonis* (Krøyer 1837) is a copepod (family Caligidae) and it occurs in cold temperate waters of the northern hemisphere [5] as an

[1] Stazione Zoologica Anton Dohrn, Department of Ecosustainable Biotechnology, Villa Comunale. 80121 Naples, Italy.
[2] Stazione Zoologica Anton Dohrn. Integrative Marine Ecology Department. Villa Comunale. Napoli (Italy).
[3] Department River Ecology and Department Aquatic Ecosystem Analysis, Helmholtz Centre for Environmental Research, Magdeburg, Germany.
[4] Plymouth Marine Laboratory. Plymouth, Devon, GB.
[5] Stazione Zoologica Anton Dohrn, Department of Marine Animal Conservation and Public Engagement, Amendolara Excellence Centre, C. da Torre Spaccata, 87071 Amendolara (CS) - Italy.
* Corresponding author: vzupo@szn.it

ectoparasite of all salmonid fish, so representing a major concern to Atlantic salmon (*Salmo salar*) aquaculture. In fact, the economic impacts of *L. salmonis* are estimated to induce a reduction from 6.2% to 8.7% of the productive value of salmonid fish [6]. Taking into account that Norway is the largest producer of salmon fish and that in 2018 a production of 1.28 million tons was declared (FAO, 2020), and including the productions of other nations providing relevant biomasses, the losses for the global salmon farming industry may be evaluated to be more than $1.26 billion. When aquaculture farms are clustered in given regions, as in northern Europe and in Australia, they may further contribute to the outbreak of pathogens and to clinical diseases [7]. As well, along the east coasts of Canada and the USA, *L. salmonis* and *Caligus elongatus* represent emerging concerns both for aquacultural productions and natural conservation [8]. In addition, environmental changes have increased the levels of infestation by *L. salmonis* and a continued increment of losses has been recorded in the last decade [9]. Low-level infestations of sea lice may cause minimal effects on the host; however, high density of parasites and long-term infestations results in progressive worsening of skin damage and, finally, to the death of the host. For this reason, severe control regimes have been established to reduce the release of sea lice larvae from aquaculture facilities into the environment.

Their complete life cycle is well-known [10] and their taxonomical relationships have been described (Figure 1).

The parasites hatch from typical egg strings carried by copepod females, to develop through three planktonic naupliar stages (nauplius 1 and 2, before the infective copepodid stage) nourished at the expense of their yolk sac (non-feeding).

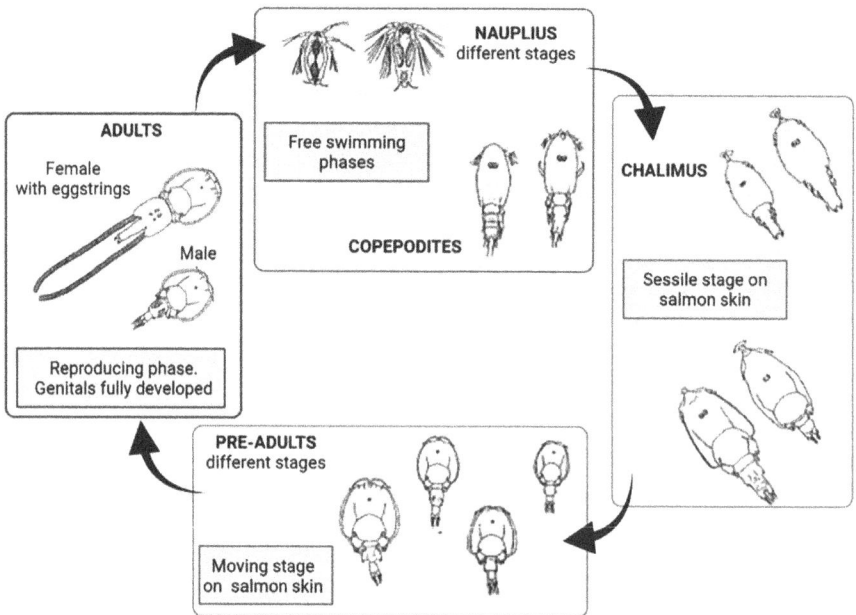

Figure 1. Life cycle of *L. salmonis* with indication of the main planktonic and settled phases. Modified from (Schram and Others 1993).

When a copepodid finds an adequate host, it attaches and develops through five further intermediate stages [*11*] prior to reach the adult stage. However, the number of stages between the infective copepodid and the adult may vary [*12*].

Seventeen species of Caligidae [*1*] are grouped into three genera, i.e., *Caligus* Müller 1785 (with 12 species), *Lepeophtheirus* von Nordmann 1832 (with 4 species) and *Pseudocaligus* A. Scott 1901 (containing a single species). All genera share the same number of stages in the free-living phase: as above mentioned, two nauplii and an infective copepodid stage, normally comprised within eight life stages [*11*]. The copepodid, following the attachment to the host, molts into the first *chalimus* stage. Chalimus stages (in variable number according to species and environmental conditions) are characterized by a frontal filament [*10*] that firmly secures their attachment to the host [*11*]. All known free-living copepods exhibit 6 post-nauplius stages, collectively referred to as *copepodid stages*. In fact, also *Caligus* spp. and most parasitic copepods exhibit 6 copepodid stages during their life cycle. The only genus, among copepods, exhibiting more than 6 copepodid stages is *Lepeophtheirus* [*11*]. Usually, *L. salmonis* exhibits a direct life-cycle completed over a single host. The mature female of a sea louse extrudes a pair of egg strings. The egg stage has a variable time span according to temperature and it may last from 17.5 days (at 5°C) to 5.5 days (at 15°C). Further, the planktonic nauplii stages above mentioned hatch directly into the water. Nauplii and copepodids are planktonic and phototactic and exhibit daily vertical migrations: they reach shallower depths during the day and sink at night. However, their ability to find their host is not light-dependent and interestingly, they may be found on various fish species, as a stop-gap measure, while seeking out their salmonid host (salmons, trouts and chars). They are also responsive to low-frequency water accelerations, indicating the presence of a swimming fish: in this way they manage to find their hosts. Initial attachment of copepodids occurs on the fins of the fish or their scales. Initially, the copepodid clasps the host tissue. Further, it molts to the first sessile stage in the life cycle, up to the parasitic chalimus stage, starting to feed at the expense of the host fish. In fact, their planktonic stages live based on their own fat reserves and cannot feed until a host is found. The chalimus stage molts twice while it is attached to the fish, before becoming a pre-adult or mobile stage. At this stage, it is able to move around on the fish skin and can even swim around in the water column. The generation time for larvae is temperature-dependent too, and it varies from 2 to 4 weeks at 18°C [*13*]. In contrast, adults are very longeval under all temperature conditions and the lifespan of an adult female, under laboratory conditions, was demonstrated to be more than 210 days [*14*]. The adults feed on mucus, tissues and blood of their hosts, immediately inducing immunosuppression [*15*], followed by the appearance of skin sores and lower feeding efficiency.

The parasitic stages present on a host fish can be readily observed and enumerated. In contrast, the free-living planktonic stages can be identified only within zooplankton samples. This operation is quite challenging, given their low abundance, as compared to other planktonts (up to 0.70 ind. per cubic meter, according to [*16*]). On the other end, it would be challenging as well to attempt fighting practices directed towards planktonic stages, due to their wide dispersal in the water column over large areas. Instead, it must be considered that aquaculture practices represent the primary sources of parasites because copepod larvae continuously escape from infested facilities located in northern Europe [*17*] and other world areas of intense production. In Europe, wild Atlantic salmon smolts migrate from rivers towards the Norwegian waters and they are often infected by *L. salmonis*. In fact, the increase of salmon aquaculture practices has been associated with the rise in *L. salmonis* infestations in nature and declines of wild salmonid populations [*18*]. The welfare concerns related to these observations prompted regulatory actions [*19*]. This suggests that disinfestation practices performed within aquaculture farms could not only improve the production efficiency and reduce the mortality of cultured fish but also reduce the lice loads in nature, so reducing their environmental impacts [20].

14.2 Possible treatments in aquaculture and their environmental impacts

Salmon farming is commonly organized into two-phase production systems consisting of a land-based freshwater hatchery, followed by a marine cage site. Young salmons (from 1 to 2 yr after egg hatching) are transferred from the hatchery (as smolts) to sea cages, to increase their weight over a period variable from 18 to 24 months. As expected, sea lice represent a problem only during the second stage of the production cycle; recognising that transfers from land-based stages to the marine stage occur generally in spring or fall, these seasons represent the period of maximum risk of infestation. Various treatments were attempted for this purpose, with variable efficacy. To date, none of them appears to provide complete recovery. For example, Emamectin benzoate (EB, SLICE®) is an Avermectin chemotherapeutant [*21*] and it is often administered to fish as an in-feed treatment lasting 7 days. Typical doses of EB in aquaculture feeds [*22*] correspond to 50 µg/kg fish biomass/day [*23*]. It has been used to treat infestations of sea lice *Lepeophtheirus salmonis* on farmed Atlantic salmon *Salmo salar* in the Bay of Fundy, New Brunswick, Canada, since 1999 [*24*].

On the whole, EB was demonstrated to be effective [*25*] against *Lepeophtheirus salmonis* when applied to North American farmed Atlantic salmon [*26*]. The time necessary to obtain the maximum effect of the treatments may be delayed in winter. In fact, Stone et al. [*27*] observed that treatments applied during colder months reached the maximum effect later and the duration of the effect lasted for approximately 9 to 10 weeks, after the initiation of treatments [*28*]. In contrast, reduced efficacy of EB has been reported in Scotland [*29*]. Similarly, a reduced sensitivity of *Caligus rogercresseyi* to EB has been documented in Chile [*30*], based on laboratory bioassays. The underlying causes of ineffective treatments are still undetermined [*24*]. However, resistance to EB could be the primary cause for treatment failure, but also other reasons for reduced treatment efficacy (e.g., lower feed ingestion by fish in colder months, improper feed application or inappropriate concentration) may contribute to sub-therapeutic dosing and lead to failures. An additional issue is that EB is detected at considerable levels in sediments under coastal fish farm facilities [*31*]. It was estimated that only 5–15% of treated food remains uneaten, and carries EB in the environment [*3*]. However, this must be summed to fish faces, that still contain EB for a period of 3–4 months post-treatment. Consequently, EB may enter the marine environments through various sources, even several months after the completion of treatments. Recent studies [2] show that EB residues are more widely distributed in the benthic environment than previously thought, and have a strong impact on benthic communities. The degradation half-life of EB is longer than 120 days in marine sediments (according to the Authority and European Food Safety Authority, 2012) and its hydrophobic nature (National Center for Environmental Assessment, 2009) produces a long persistence in marine sediments around fish farm cages. This prompts a high risk of exposure for several benthic organisms. In addition, the bioactivity of EB is non-targeted: several crustaceans are subject to the same mode of action [32]. This represents a tremendous impact on several crustacean species [33], given the large use of this compound for aquacultural activities all over the world.

For example, Park (2013) demonstrated a reduction in the abundance and the size at the harvest of the economically important prawn *Pandalus platyceros* immediately following treatments with EB, as compared to two months later [34]. In a further study, Bloodworth et al. (2019) demonstrated an inverse exponential gradient of impacts in the response of benthic communities according to the distance from the cages, with the greatest impact observed under the cages, and widespread detection of EB (in 97% of samples) in the sampled sediments [2]. Remarkably, EB exhibits a single huge negative effect on crustacean abundance and diversity. The species richness of crustaceans is the environmental parameter most closely correlated to the impacts of EB. Small crustaceans, like copepods and amphipods, respond quite negatively to EB gradients [35]. The only crustacean apparently insensitive to EB accumulation in the sediments is the epifaunal hermit crab *Pagurus cuanensis* [36]. Differently, tube-building amphipods have their optima at moderate EB concentrations, probably due to their ability to control the microenvironment in the tubes [37]. Among amphipods, *Tanaopsis graciloides* and *Ampelisca tenuicornis* are able to switch between suspensions-feeding and deposit-feeding [38] which potentially alters the vulnerability of these species to EB concentrations. An additional demonstration of the effects of EB is given by the caprellid amphipod *Pariambus typicus*. In normal conditions, this species appears to respond positively to a moderate organic enrichment accumulated around fish farms [39], probably due to its detritivore feeding habits. However, in presence of EB its abundance rapidly decreases, confirming the toxigenic effect of this drug. Finally, it is important to observe that bioturbation may accelerate the release of EB stored in marine sediments towards the water column, so increasing the impacts on planktonic crustaceans. However, low mobility taxa of crustaceans with a burrowing or a detritivore lifestyle are particularly vulnerable to EB and these findings demonstrate clear effects on all crustaceans, quite below the concentration of EB set by the current Environmental Quality Standards (EQS) at 0.763 µg/kg wet weight of sediment, at a distance of 100 m from the cages.

However, EB is not the only drug impacting natural populations of crustaceans. Another pesticide largely used in medicated feed for salmon for the control of sea lices is Diflubenzuron (DFB), a chitin synthesis inhibitor that interferes with the molting stages during the development of *Lepeophtheirus salmonis*. DFB spreading from salmon feeds may impact the natural populations of non-target crustaceans, such as the northern shrimp *Pandalus borealis*, an economically and ecologically important species in colder parts of the Atlantic [*40*]. These shrimps accumulate DFB through the consumption of residues of medicated feeds for fish but also feeding on salmon faeces, as well as by ingesting other contaminated invertebrates [*3*]. The recommended frequency of treatment with such Flubenzurones as DFB is every six months [*40*] and its application is commonly adopted in spring and autumn, when blooms of the parasitic copepods are expected. Overall, the harmful effect of DFB on the survival of shrimps (both adults and larvae) and on other benthic and planktonic crustaceans has been demonstrated by laboratory experiments [*41*]. However, intensification of salmon aquaculture and of productive farms in general, led to increased use of several drugs and chemicals, including heavy metals, antibiotics and generic pesticides, most of which are detrimental to aquatic ecosystems and in particular to crustacean communities, due to demonstrated acute and chronic toxicity [*42*]. For this reason, it is worth developing a new generation of anti-parasitic agents, based on physical methods (non-polluting) and natural substances (easily biodegradable and not accumulated in sediments and organisms). Physical methods as Catalytic Ozonation may be applied to the tanks for the culture of larvae and younger specimens, to reduce the presence of swimming larvae of parasites. Natural substances having a strong effect on adult sea lice are under investigation in various laboratories of the world, given the high economic impact of these parasites.

14.3 Drugs from nature: cyanobacteria vs. copepods

Cyanobacteria are an ancient group of organisms living in a wide range of marine and freshwater habitats, besides humid environments, and even in extreme conditions, as hot thermal waters and cold Antarctic environments. They are outstanding sources of secondary metabolites, often showing remarkable bioactivity. Besides oligopeptides (predominantly, cyclic peptides), other compound classes have been isolated from cyanobacteria, including terpenes and alkaloids. Although cyanobacteria have been more often studied as threatens for cultured fish, molluscs and other seafood [*43*], due to their ability to intoxicate and kill, or to accumulate in tissues of fish and invertebrates, some of them might have a definite role in the eco-sustainable control of the sea lice and other fish parasites. In fact, besides the production of toxic metabolites [*44*], they are also enriched with several pharmacologically active compounds that have antibacterial [*45*] and antifungal activities. Even anticancerous effects have been demonstrated [*46*]. Interestingly some cyanobacteria exhibit antiplasmodial [*47*] and antiviral [*48*] activities. Other strains exhibit immunosuppressive activities [*49*]. Thus, due to a promising pharmaceutical value, a new perspective of using cyanobacteria and algae in the field of human and animal medicine has grown.

Here we will briefly discuss how the active compounds present in the spent medium of a cyanobacteria culture demonstrated very interesting bioactivity against salmon parasites, so appearing as a promising solution for an eco-sustainable treatment of sea lices. In particular, we cultured a strain of *Halomicronema metazoicum*, isolated from leaves of the seagrass *Posidonia oceanica*, and we cultured it until active compounds were excreted in the culture medium. Further, we tested these compounds on sea lice to detect the possible effects at various concentrations.

14.4 Materials and Methods

14.4.1 Cyanobacteria cultures

A strain of *Halomicronema metazoicum* has been cultivated in continuous conditions at Stazione Zoologica Anton Dohrn. The strain has been isolated from *Posidonia oceanica* leaves and identified through a polyphasic approach using both morphological and molecular tools [50]. The purity of the culture has been ascertained by means of SEM and molecular investigations and the absence of contaminating organisms has been demonstrated [51]. Five 2L Erlenmeyer flasks were used to cultivate for 30 days 5 mats of *H. metazoicum*, each fresh pre-weighed and dosed at 5 (± 0.3) grams. The culture medium was prepared using filtered and sterilized natural seawater added with *f/2* salts (Sigma-Aldrich, Merck KGaA, Darmstadt, Germany). The cyanobacteria were cultured in axenic conditions at a temperature of 22°C with irradiance ranging from 130 to 160 µE and a photoperiod of 12/12 hours, at a salinity of 38 PSU. These culture conditions have been demonstrated to stimulate the production of bioactive compounds [51]. At the end of the culturing period, spent media was collected and filtered using Stericup Filter Units (Sigma-Aldrich, Merck KGaA, Darmstadt, Germany). The collected spent medium was further diluted using sterile seawater to obtain the desired concentrations for bioassays on sea lice and, in particular, it was diluted 1:5, 1:10, 1:100, 1:1000 and 1:10,000.

14.4.2 Bioassays

Toxicity tests were conducted on the sea lice *Lepeophtheirus salmonis* when they were in the phase of copepodite I-II. Copepodites were obtained from a local aquaculture centre in Sweden and they were transferred to the testing laboratories in refrigerated containers. Before each experimental run, copepodites were checked under optical microscopy to check their vitality and then counted before being placed in the multiwell chambers. Only individuals that showed a healthy status were used for the activity bioassays. Copepodites represent the free-swimming infective stage and as above mentioned they do not need any feeding in this phase [11]. Two types of bioassays were performed, the first one using spent medium at various dilutions, the second one using 12 fractions at the same nominal concentrations of the raw spent medium. In the first experiment, five different dilutions of the cyanobacterial spent medium were tested (1:5, 1:10, 1:100, 1:1000, and 1:10,000 dilution of spent medium in water). Negative controls were prepared by adding fresh *f/2* medium to sterile seawater at concentrations corresponding to those of the above-mentioned treatments. Copepods (n = 2) were incubated in a 24-well plate, using 8 replicated wells for each dose and treatment. Mortality and movements of sea lice were checked at time lang of 5 minutes, 60 minutes, 300 minutes, 24 hours, and 48 hours. The second experiment was carried out using the same protocol as the first, by exposing the sea lice to 12 fractions of the spent medium obtained by HPLC separations, at the same nominal concentrations of the medium above reported. The control replicates were obtained by adding diluted HPLC fractions of fresh medium.

14.4.3 Chemical fractionation

The GF/F filtered cyanobacterial culture supernatant and *f/2* medium (as a control treatment) were adjusted to 1% MeOH and subsequently passed through a preconditioned (100% MeOH followed by 1% MeOH) Varian BondElut C18 SPE column. One litre of the cyanobacterial medium was extracted and the final extract was collected up to a volume of 500 μl. Columns were washed with 10 ml 1% MeOH and this aqueous eluate was collected in round-bottomed flasks (assumed 5x concentration *vs.* the original supernatant). Subsequently, the columns were eluted with 10 ml of 100% MeOH into two additional round-bottomed flasks. Each HPLC run was performed with 10 uL of the cyanobacterial extract which resulted in 12 fractions. All fractions were tested on *Lepeophtheirus salmonis salmonis* at nominal concentrations corresponding to the ratios of the spent medium in seawater, as above reported for the bioassays on crude medium, i.e., corresponding to ratios of 1:5, 1:10, 1:100, 1:1000 and 1:10,000 of spent medium:water.

14.4.4 Statistical analysis

Data deriving from various replicates were organized as means with standard deviations for each set of measurements. The differences among various concentrations and controls were analyzed using two-way ANOVA. Dunnett's multiple comparisons posthoc tests were applied to check differences against negative controls of each treatment at each time. Results from spent medium and fractions tests were analyzed by two-way ANOVA with Tukey's multiple comparisons posthoc tests to evaluate the significance of differences between replicates at the same dilution and at the same experimental time. Data were tested for normality by the Shapiro-Wilk normality test. Graphs and statistical analyses were computed using GraphPad Prism version 7.00 for Windows (GraphPad Software, La Jolla California USA).

14.5 Results and conclusive remarks

The replicated bioassays performed using crude spent medium produced a linear pattern of responses in sea lice, according to dilutions (Figure 2). In treatments added with the highest concentration (1:5, 1:10, 1:100) of the spent medium, the survival was null, yet at the first time-lag (5 minutes), indicating the highest toxicity of the cyanobacterial compounds (Figures 2A, B, C). At a dilution of 1:1000, the survival was 100% at the first-time lag (5 minutes), then started to decrease, indicating time-related toxicity. In fact, at the second time lag (60 minutes) the survival was 87.5 (± 13.4) and at the third time lag (300 minutes) it was 75.0 (± 18.9). After 24 hours, however, sea lice were all dead (Figure 2D). In the treatment at a concentration of 1:10,000, no mortality was exhibited until the fifth time lag (48 hours) were survival decreased to 18.75 (± 25.88). Statistical analyses confirmed the significance of differences above reported for a dilution of 1:1000, at time lags 2, 3, 4, 5 (two-way ANOVA, p < 0.0001), and for a dilution of 1:10,000, at the time lag 5 (p < 0.0001).

These results indicate a specific effect dependent on time and doses, of some active compounds present in the spent medium of *H. metazoicum*. It also demonstrated

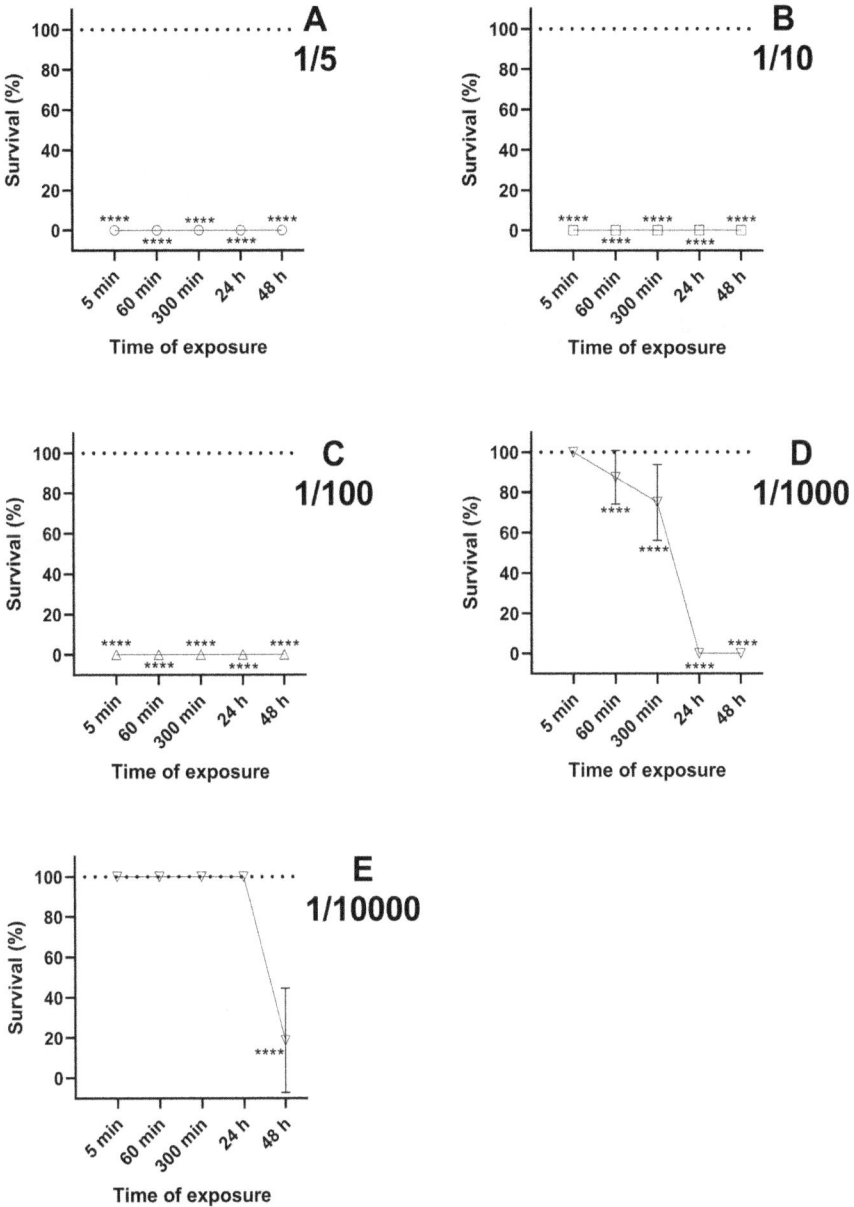

Figure 2. Survival of *Lepeophtheirus salmonis salmonis* during the bioassays conducted at various concentrations of *H. metazoicum* spent medium (A–E), from 1/5 to 1/10,000.

the concentration of 1:1000 as the lowest threshold of activity, able to kill all sea lice during the first 24 hours of treatments. During this time, these concentrations may be toxic also to young fish. However, it must be considered that the compound is easily stored in the body of sea lice. For this reason, shorter treatments (less than 1 hour) may be sufficient to produce effects within 24 hours, when sea lice are transferred in clean water along with their hosts. This study demonstrates that the treatment with cyanobacterial active compounds may be a very promising and eco-sustainable solution to treat the infested fish, by means of short baths in water added with a spent medium of *H. metazoicum*.

To further characterize the nature of the active compound/s present in the spent medium, bioassays were performed using HPLC fractions of the spent medium, at the same nominal concentrations of the previous bioassays (i.e., at concentrations corresponding to 1:5, 1:10, 1:100, 1:1000 and 1:10,000 of the spent medium from which they were extracted). These tests confirmed the total absence of toxicity at nominal concentrations of up to 1:100. The fractions tested at a nominal concentration of 1:1000 consistently demonstrated activity on the survival of sea lice, in agreement with the test performed on the whole spent medium (Figure 3).

In particular, three fractions had a toxic effect on sea lice: fraction n° 8 demonstrated the maximum toxicity, leading to the death of all the tested individuals

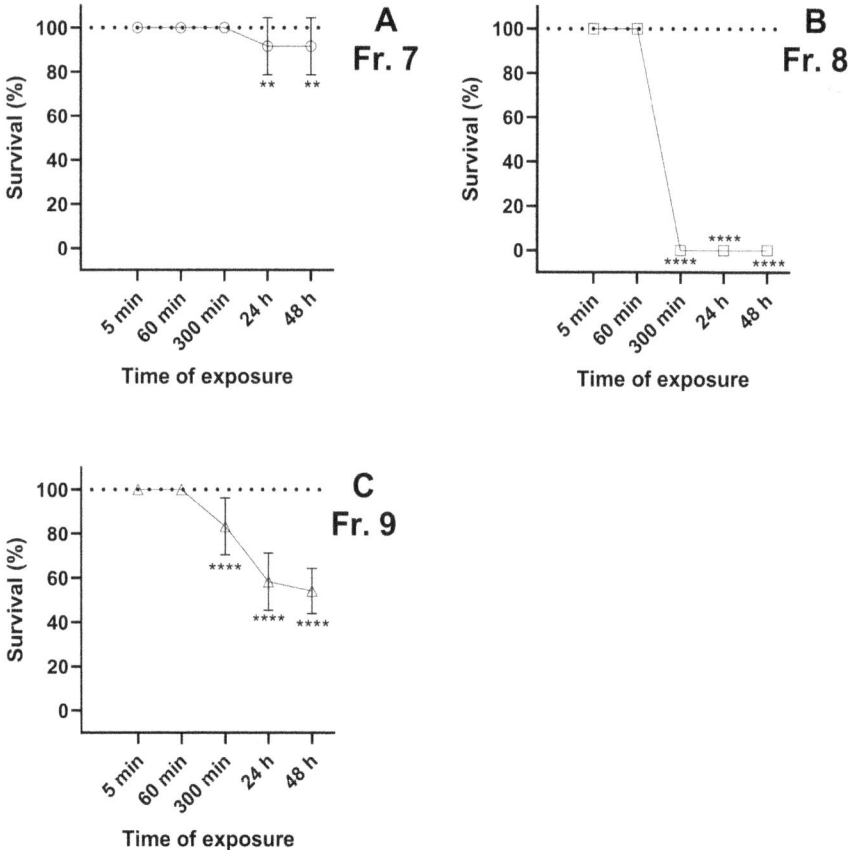

Figure 3. Survival tests performed at a concentration of fractions corresponding to a dilution of 1:1000 of the original spent medium. Only the active fractions have been reported, for brevity, since all the other fractions exhibited no effect on the survival of *Lepeophtheirus salmonis salmonis,* even at maximum doses.

at the third time lag (300 minutes). Fraction 9 demonstrated intermediate toxicity, by exhibiting survival of 83.3 (± 12.9) at the third time lag (300 minutes), and correspondingly, 58.3 (± 12.9) at the fourth time lag (24 hours), and 54.2 (± 10.2) at the fifth time lag. The differences *vs.* negative controls were statistically significant (two-way ANOVA, $p < 0.005$ at the time lag 3; $p < 0.0001$ at the time lags 4 and 5). In contrast, fraction 7 was the least toxic (two-way ANOVA, $p < 0.0001$), exhibiting significant differences among controls only at the fourth time lag (24 hours) and the fifth (48 hours).

On the whole, these results indicate that the compound was maximally collected in fraction 8 and it was a partially polar compound of small molecular weight. The active concentration of 1:1000 may be toxic both for sea lice and young fish, after 24 hours of exposition. However, sea lice quickly accumulate the compound in their body, while fish are able to detoxify and excrete the toxic compound. For this reason, shorter treatments (baths of 45 minutes, followed by a total change of water) were demonstrated to be sufficient to obtain complete detachment of sea lice from the body of their hosts and dead of sea lice sinking on the bottom after some hours. The compound appears to be enough stable in natural conditions, even at higher temperatures, but it is quickly degraded in the natural environment. Since it is active even after short baths, treatment tanks could be set to assure total recovery of cultured fish in a few hours, prior to being returned to the cages.

Such an environmental-free and eco-sustainable treatment is very promising, permitting to reduce the impacts of parasites on cultured fish and, in addition, reducing the possibilities for the parasites to invade the surrounding environments, impacting the natural populations of salmons.

Acknowledgements

We acknowledge Michael Steinke, Senior Lecturer at the University of Essex, for providing copepods for the tests reported in this chapter.

References

1. S. Ohtsuka et al. (2009). *J. Nat. Hist.* **43:** 1779–1804.
2. J.W. Bloodworth, M.C. Baptie, K.F. Preedy, and J. Best. (2019). *Sci. Total Environ.* **669:** 91–102.
3. K.H. Langford, S. Øxnevad, M. Schøyen, and K.V. Thomas. (2014). *Environ. Sci. Technol.* **48:** 7774–7780.
4. A. Macken, A. Lillicrap, and K. Langford. (2015). *Environ. Toxicol. Chem.* **34:** 1533–1542.
5. R. Skern-Mauritzen et al. (2021). *Genomics* **113:** 3666–3680.
6. J. Abolofia, J.E. Wilen, F. Asche. (2017). *Mar. Resour. Econ.* **32:** 329–349.
7. B. Robertsen. (2011). *Aquac. Res.* **42:** 125–131.
8. K. Boxaspen. (2006). *ICES J. Mar. Sci.* **63:** 1304–1316.
9. T. Bjørndal, and A. Tusvik. (2019). *Aquac. Econ. Manag.* **23:** 449–475.
10. I. Madinabeitia, and K. Nagasawa. (2011). *J. Parasitol.* **97:** 221–236.
11. L.A. Hamre et al. (2013). *PLoS One* **8**.
12. J. Ho, and C.-L. Lin. (2004). *The Sueichan Press, Taiwan.*
13. P.A. Heuch, J.R. Nordhagen, and T.A. Schram. (2000). *Aquac. Res.* **31:** 805–814.
14. T.A. Schram. (1993). In *Pathogens of Wild and Farmed Fish: Sea Lice* **1:** 30–47.
15. E.B. Thorstad et al. (2015). *Aquac. Environ. Interact.* **7:** 91–113.
16. J. Skarhamar, M.N. Fagerli, M. Reigstad, A.D. Sandvik, and P.A. Bjørn. (2019). *Aq. Env. Int.* **11:** 701–715.

17. H.B. Fjørtoft et al. (2019). *Aquac. Environ. Interact.* **11:** 459–468.
18. O. Torrissen et al. (2013). *J. Fish Dis.* **36:** 171–194.
19. K.W. Vollset et al. (2018). *ICES J. Mar. Sci.* **75:** 50–60.
20. G.L. Taranger et al. (2015). *ICES J. Mar. Sci.* **72:** 997–1021.
21. T.E. Horsberg. (2012). *Curr. Pharm. Biotechnol.* **13:** 1095–1102.
22. J.N. Kuo, C. Buday, G. Van Aggelen,M.G. Ikonomou, and J. Pasternak. (2010). *Env. Tox. Chem.* **29:** 1816–1820.
23. V. Benson, E. Aldous, and A. Clementson. (2017). *Scottish Environment Protection Agency.* Report Reference: UC12191.03. 62 pp.
24. P.G. Jones, K.L. Hammell, I.R. Dohoo, and C.W. Revie. (2012). *Dis. Aquat. Organ.* **102:** 53–64.
25. S. Saksida, J. Constantine, G.A. Karreman, and A. Donald. (2007). *Aquac. Res.* **38:** 219–231.
26. R. Armstrong, D. MacPhee, T. Katz, and R. Endris. (2000). *Can. Vet. J.* **41:** 607–612.
27. J. Stone, I.H. Sutherland, C. Sommerville, R.H. Richards, and K.J. Varma. (2000). *Aquaculture* **186:** 205–219.
28. F. Lees, M. Baillie, G. Gettinby, and C.W. Revie. (2008). *J. Fish Dis.* **31(12):** 947–951.
29. F. Lees, G. Gettinby, and C.W. Revie. (2008). *J. Fish Dis.* **31:** 259–268.
30. S. Bravo, S. Sevatdal, and T.E. Horsberg. (2008). *Aquaculture* **282:** 7–12.
31. B. Cheng et al. (2020). *Ecotoxicol. Environ. Saf.* **194:** 110452.
32. K.J. Willis, and N. Ling. (2003). *Aquaculture* **221:** 289–297.
33. N. Veldhoen et al. (2012). *Aquat. Toxicol.* **108:** 94–105.
34. A. Park. (2013). *ProQuest Diss. Theses*, 128 pp.
35. J. Hall-Spencer, and R. Bamber. (2007). *Ciencias Mar.* **33(4):** 353–366.
36. T.C. Telfer et al. (2006). *Aquaculture* **260:** 163–180.
37. J.A. de-la-Ossa-Carretero et al. (2016). *Estuar. Coast. Shelf Sci.* **172:** 13–23.
38. M.G. Shojaei, L. Gutow, J. Dannheim, H. Pehlke, and T. Brey. (2015). In *Towards an Interdisciplinary Approach in Earth System Science: Advances of a Helmholtz Graduate Research School*, 183–195.
39. V. Fernandez-Gonzalez, F. Aguado-Giménez, J.I. Gairin, and P. Sanchez-Jerez. (2013). *Aquac. Environ. Interact.* **3:** 93–105.
40. S.J. Moe, D. Hjermann, E. Ravagnan, and R.K. Bechmann. (2019). *Ecol. Modell.* **413:** 108833.
41. R.K. Bechmann et al. (2018). *Aquat. Toxicol.* **198:** 82–91.
42. D. Douet, H. Le Bris, and E. Giraud. (2009). *Opt. Méditerranéennes. Séries A. Mediterr. Semin.* **86:** 105–126.
43. J. Cazenave et al. (2005). *Aquat. Toxicol.* **75:** 178–190.
44. M. Mutalipassi et al. (2021). *Mar. Drugs* **19(4):** 227.
45. T. Malathi, M. Ramesh Babu, T. Mounika, D. Snehalatha, and B.D. Rao. (2014). *Phykos.* **44:** 6–11.
46. N.A. El Semary, and M. Fouda. (2015). *Asian Pac. J. Trop. Biomed.* **5(12):** 992–995.
47. O. Papendorf, G.M. König, and A.D. Wright. (1998). *Phytochemistry* **49:** 2383–2386.
48. G. Riccio et al. (2020). *Biomolecules* **10:** 1–36.
49. S. Vijayakumar, and M. Menakha. (2015). *J. Acute Med.* **5:** 15–23.
50. N. Ruocco et al. (2018). *PLoS One* **13(10):** e0204954.
51. M. Mutalipassi et al. (2019). *Biol. Open* **8(10):** bio.043604.

CHAPTER 15

Automatic Culture of Crustaceans as Models for Science

Francesca Glaviano[1,2,#] *and Mirko Mutalipassi*[3,*,#]

15.1 Introduction: culture of Crustaceans for ornamental market, research, and aquaculture purposes

Crustaceans are emerging model organisms in various fields ranging from marine biology to ecotoxicology and physiology. They are part of a wide variety of aquatic models used to investigate the function of live organisms and ecosystems. Crustaceans have some key characteristics that give them many advantages as models: they are small in size, cheap to culture and maintain if compared to vertebrates, for example mice and rats [1]. Small research centres need facilities able to culture various aquatic organisms that can answer researcher's needs. Non-automatized procedures, although capable of good efficiency, need plenty of space and trained operators [2].

Modern research and aquaculture are increasingly projected towards the integration of new technologies, through the use of automated and intelligent control systems. These implementations can lead to a more efficient control of the culture's conditions, such as water quality, stocking density and feeding rate allowing a constant and automatic regulation of the conditions of cultivated organisms (Figure 1).

The chance to make automatic measurements of values such as temperature, oxygen, pH, ORP, etc., it allows to optimize their efficiency by reducing labour and utility costs.

---

[1] Stazione Zoologica Anton Dohrn, Department of Ecosustainable Marine Biotechnology, Villa Comunale, 80121 Naples, Italy.

[2] Department of Biology, University of Naples Federico II, Complesso Universitario di Monte Sant'Angelo, 80126 Naples, Italy.
Email: Francesca.glaviano@szn.it

[3] Stazione Zoologica Anton Dohrn, Department of Marine Animal Conservation and Public Engagement, Calabria Research Centre, C. da Torre Spaccata, 87071 Amendolara (CS) – Italy.

* Corresponding author: mirko.mutalipassi@szn.it

# Both authors contributed equally to this work

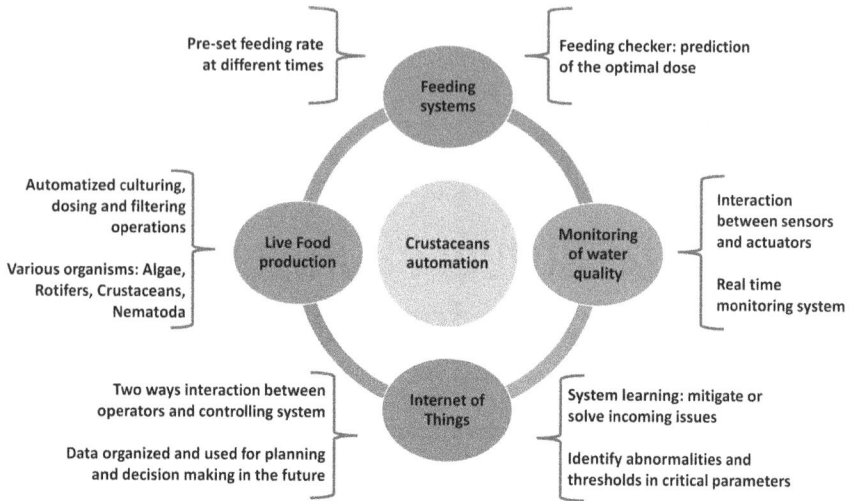

Figure 1. Summary of the potential application related to crustacean's automation.

Automatic devices can not only improve a large-scale system as for aquaculture but they can also largely simplify the culture of small laboratory organisms and it offers interesting applications for the purposes of scientific or biotechnological research [*1*]. If automated culture systems of established aquatic model organisms, such as *Danio rerio*, are widely present on the market, few researchers and industries focused on the development of small, flexible, programmable, and modular culture systems able to culture various aquatic species, such as crustaceans. The culture of Crustaceans, not only for aquaculture but for ornamental and research purposes too, can be considered a multimillionaire market [*3*]. In fact, species belonging to *Palaemon, Lysmata, Stenopus, Periclimenes, Hymenocera, Clibanarius* e *Mithraculus* genera were reared and studied in the frame of the optimization and, in some cases, automation of culture techniques [*2*]. Among this interesting species, many studies focused on *Lysmata* ones due to their interesting physiology, i.e., peculiar sex reversal and hermaphroditism, but also the high selling price in the ornamental market [*3*]. Rearing systems, such as *planktokreisel*, has been extensively used to achieve the reproduction of species with demanding larval phases, such as *Jasus edwardsii*, and it is one of the first devices on the long road to automation of farming systems. The idea of semi-automatic devices, able to remove degrading organic matter or change water to achieve optimal chemical parameters or assure an optimal water flux, has led scientist to accomplish the reproduction and rearing of various marine species of crustaceans, assuring low mortality. Variation of the original productive plant designed in 70', 80' and 90' were made to optimize recirculation, water flow and to simplify maintenance and operation activities [*2*]. Such modification assured a different movement of water inside the cone-truncated cylinders as well as a reduction in potential damage to larval feeding and swimming appendices. Although these rearing apparatuses were not able to culture all the considered species, as in the case of *Mithraculus forceps* larvae, they can in general assure not only a high output in term of biomass and number of individuals, but also the healthy state of reared organisms, making them useful for scientific purposes, where physiological responses of model organisms can be altered by stress.

Small culture systems, able to administer food, to check and maintain chemical and physical water parameters, may dramatically reduce production costs and the need for personnel. In addition, the ability of such devices to be monitored continuously may reduce potential losses related to power failure or caused by breakage of water pumps or other instruments [4]. Mutalipassi developed a compact culture system that can be modulated on the necessity of various cultured aquatic organisms. Various software, in Ladder language, can be uploaded into the central control unit (PLC) to meet the needs of a range of species. In addition, this system can be used to culture different life stages, not only adult individuals but also larvae of delicate and demanding species. This automatic culture system, set to work without the intervention of trained biologists and intensive daily maintenance, has been devised to meet the needs of cultured species and reduce research and laboratory resources while optimizing spaces and the manpower. The efficiency of automatic culture system depends, as demonstrated in many automatic or semi-automatic culture systems [2], on various chemical, biological and mechanical parameters can influence the efficiency of automatic culture systems, as demonstrated in many automatic or semi-automatic culture systems. The efficiency, in terms of health as well as production and costs, can be improved including in the system a robust mechanical and chemical filtration, frequent water changes, a correct feeding, the use of ozone and a careful setting of optimal densities of cultured individuals. Culture process was also improved by a fine tuning of the operational software, providing efficient water changes.

In addition to decapods, various copepod species have been used as model organisms in various research fields. For example, *Tigriopus fulvus* has been used as a model for ecotoxicological studies. Other species, such as *Temora stylifera* and *Centropages typicus*, were used as live feed for a variety of aquaculture species, as substitution of rotifers and other species. Copepods production had to face many issues linked both to feeding unicity and needs and to management difficulties due to the small size of nauplii and larval phases [3].

15.2 How automation and digitalization support aquaculture

Aquaculture researchers produced a great amount of knowledge that contributed to the industrialization of this productive field. Aquaculture operations require sophisticated tools and specially designed facilities, which "have evolved through intensive research and a great deal of innovation". As seen in other fields, advancements in technology have supported the development of aquaculture despite many products of technology not developed specifically for application in farming systems have found applications in this area. The improvements in aquaculture technologies have provided a supporting basis to the evolution of aquaculture systems where artificial intelligence, automation and integrated management are key points of the productive system.

The use of computer monitoring and automation in aquaculture has grown during the last years. Benefits aims at: reduction of energy and water waste; greater process efficiency; reduction of labor costs; optimization of animal health conditions. Robotics has made substantial advances in a wide range of applicative fields including aquaculture [5]. In this field, robotics acquired an important role in many applications, ranging from feeding to monitoring water quality, to the cleaning of fouling on the cages and the monitoring, through ROVs, of cage integrities, finishing with security surveillance in open sea. Automation projects play a fundamental role in improving economic profitability. For example, optimization and appropriate management of feeding improves the growth rate of the marine animal and the real-time monitoring of physicochemical parameters of water provides information used to carry out corrective actions for optimal conditions of cultivation [5]. The main factors to consider in the design of an intelligent control software are ease of use and intuitiveness, otherwise the non-expert user in the informatic sector will not be able to interface with it; flexibility and adaptability, so that the chosen software is interchangeable with other products. Aquaculture is becoming increasingly important for global food production. On one side oceans are dramatically overfished, limiting the future development of fishing activities; on the other side, only 2–3% of the global food production comes from the oceans despite the oceans cover more than 70% of the earth surface. In addition, aquaculture for non-food purposes is a growing industry that can encompass great revenues with its direct and indirect association with human activities. In fact, beyond human consumption, marine organisms provide various services that can be used, such as biotechnology [6]. Non-food aquaculture can provide live feeds, food production, natural restocking and can provide bioactive molecules for pharmaceutical, cosmeceutical and nutraceutical industries, among others. To ensure sustainable and efficient aquaculture production, modern technology should meet aquaculture production needs with the aims to improve energy efficiency, plant availability, product quality, and overall productivity. In an integrated shrimp aquaculture, biological entities (the cultured organisms but also co-cultured species or biological filters) and non-biological components, such as water parameters or instruments, should be monitored together [5]. All the sampled data is channelled to a central command system, which responds by sending signals (motor pathways) to regulators (for example, aerators, water flow control devices) which in turn act according to information fed into the system in the form of algorithms. The continuous monitoring of all the aspects of shrimp aquaculture, from environmental conditions to physiological parameters and metabolic outputs, allow commercial aquaculture facilities to optimize their efficiency by reducing labour and utility costs, and decreasing the environmental impacts.

15.3 Automatic feeding system for aquaculture

Shrimps are popular seafood and among all the cultured species, *Litopenaeus vannamei* is the preferred one due to its culture characteristics and consumer acceptance. The shrimp farming, not differently from the rest of the aquaculture field, relies on intensification, improved feed management and reductions in labour costs to expand and produce wealth. Into the frame of the limitation of production costs, the amount of used food for kilogram of produced shrimp, the source of nutrients and the biological wastes produced are important variables that should be taken into consideration. The cost and quality of shrimp food is generally adequate, and the development of the new techniques is focused on the reduction of the waste, of the personnel work-time and on the environmental improvements in aquaculture farms [7]. In aquaculture, personnel have to face many problems in relation with the feeding routines. These involve the depth of the pond or the shape and culture nets, the differences among solid, liquid feed and, in case, medicines, the risk of working on operational boats in open sea, sometimes miles away from the coast. Automatic feeding systems, expressly conceived for aquaculture, were designed to face these issues in both indoor (plastic or concrete tanks) and outdoor (ponds, cages, river estuaries) aquaculture. First experimental automatic feeders were food dispensers, able to pump a volume of water containing live food or dry food, triggered by an electronic timer. The following step was to develop a system that could be activated, providing food, through photoelectric and motion sensors that detect certain animal behaviour. At the same time, commercial feeding robots were developed to feed different foods, in various tanks, differentially. Usually, these projects are based on a robot that move on a rail and that is managed by a PLC or via Wi-Fi by a computer (Storvik feeding robot) and Arvo-Tec robot feeder system (Arvo-Tec 2010). In some other cases, automation of the aquaculture processes led to the development of an automatic boat, designed to operate in ponds, by untrained personnel [8]. The aim is to sell on the market a system that can be used by farmers, with low or no instruction, but involved in aquaculture activities. The boat can be easily controlled by remote, due to a RF transmitter and receiver that can operate up to a range of 3–5 kilometres. In addition, the inbuilt video transmitter and receiver enable the farmer to have a "first person" point of view of the boat movements. The boat is built with an electric engine, powered by LiPo battery (or by solar plate). This technic solution has two advantages: on one side, it allows to reduce noise that can disturb cultured organisms, on the other side it helps to contain building and maintenance costs. It is provided with two different food-pumping mechanisms, the first mechanism was designed for dosing solid food, the second for dosing liquid food and medicines. In addition, the system can be easily modified, for example changing the RF wireless communication with Wi-Fi module. With Wi-Fi, the operators can manage all the operations from anywhere in the world.

This cheap, low-tech system cannot be easily adapted to large, highly productive farms. Digitalized feed monitoring and distributor is the answer in terms of automation as well as digitalization of the feeding process. The development of integrated feeders is one-step through the optimization of the aquaculture performances, especially if referred to shrimp aquaculture. In fact, generally, benthic shrimps are grazers that externally masticate their feed. Cultured species are usually benthic and they have a limited capacity to store food inside their digestive tract which result in the need to eat reduced amount of food throughout the day. For this reason, as the number of feedings is increased, shrimp performance in aquaculture is improved [9]. This also means that food that has not been eaten immediately is exposed to leaching and, as consequence, to a decrease in nutritional value. In fact, many of the essential nutrients are water-soluble and evidences demonstrate significant reductions in growth for shrimp offered feeds leached for more than one hour [9]. Consequently, industry is moving towards the development of automated feeders that would allow to dose multiply small quickly consumed meals throughout the day minimizing the effect of leaching and addressing the shrimp physiological needs to have access to nourish over a prolonged period [10]. Integrated feeding systems have been developed coupled with automatic aerator and monitoring system. Automatic feeder, working with pre-set feeding rate at different times of the day and night, comprises a control panel and a real-time alarm system able to send SMS messages to operators in case of problems thanks to the embedded microcontroller. The activities of the feeder are subjected to feedback given by temperature, salinity and oxygen probe connected to the monitoring system that, at the same time, can power the aerator in case of needs. Offering multiple meals can be very labour intensive and economically impracticable and the only way is to use automatic and, if possible, digitized devices. Several investigations demonstrated an enhancement of growth performance in shrimp cultured with multiple feeding throughout the day [9]. The enhancement is not only due to an increasing in the availability of feeds but also to a reduced degradation and degeneration of the food that is in contact with water for a reduced amount of time.

In addition, in the frame of the new concept of aquaculture farm where digitalization and automation should be fully integrated in the management of the production process, the automated feeders are developing to be more "accurate" not only in the set dose, but also in the prediction of the "optimized dose" that should be administered, and some methods were developed to detect left over feed to stop feeding. Initially, estimated food waste was measured by suspending a sheet below the sea cage during the feeding period, retrieving it after feeding, and counting the left over feed pellets and then by using an underwater camera and image analysis tool to detect and count leftover pellets. To automatically detect the optimal dose of food that should be administered, acoustic technologies have been applied in this field in fish aquaculture and then, in Tasmania, Australia, in a shrimp aquaculture plant [9]. This sensor-based feed control system uses sonic technology to obtain indirect measure of feeding intensity, thanks to complex filtering algorithms that can analyse and recognise the unique sound of shrimp feeding and identify its intensity. This acoustic system, coupled with temperature and dissolved oxygen probes, allows a complete management of the feeding procedure as well as real time observation of water parameters, such as dissolved oxygen, that can give indirect indications on how the water parameters change over time as consequence of the food dosage. This acoustic system can be applied in many aquaculture farms, with various productive methods ranging from extensive to super intensive conditions, giving the possibility to provide real time adjustment of feed input based on the real needs of the cultured shrimps. Experiment performed in pond culture of Pacific whit shrimp *Litopenaeus vannamei*, comparing acoustic feeding system with standard feeding strategies demonstrated the advantages of integrated, automatic, and digitized system in terms of growth performance, production, water quality and economic returns [9]. In addition, on-demand acoustic feedback feeding system has been developed and has proved to be a reliable tool in shrimp farming. This acoustic on-demand system responds to the signature clicking noise produced by shrimp feeding and produced a higher production and value of *L. vannamei* produced in semi-intensive pond culture [10].

Similar systems, used in fish aquaculture but that can be easily converted for shrimp farm purposes, are commercially available. These systems are based on Doppler pellet sensor, CAS pellet sensor and camera sensor (Akvasmart, Norway). Visual systems were developed to obtain the status of growth and health of the shrimp culture, data that can be integrated with other water parameters. Underwater image visibility technology, developed for aquaculture in ocean environments, were used in association with image defogging technology to work in culture ponds too. The visual system was improved and enhanced with detection technology that can give feeds to automatic feeding devices about the remaining feed at the bottom of the pond [*11*]. The increase of the number of daily feedings in conjunction with a system able to administrate the right amount of food, allow an increase in feed inputs with the consequence of an increased intake and growth as well as an increase value of each production, without the negative consequence related to an overfeeding (increase in productive costs, pollution etc.).

15.4 Automatic live food production and administration

Aquaculture needs live feeds for its productions. Live food production is a very intensive and costly manner. Automatized and/or integrated live food production systems have been developed by many authors to simplify, standardize, and kept constant production of the most common or interesting species. Dehasque et al. (1997) developed two culture systems designed to culture *Artemia salina* and Rotifers, respectively [*12*]. In the first system, the one dedicated to *A. salina*, repeated decapsulation procedure, automatic rinsing, concentration, hatching, enrichment, and feeding are completely automatized and take place in one recipient, reducing of approximately four times the manpower needed if compared with the manual method. Rotifers automated culture system was designated with a reversed filter system, in which, during the harvest, the rotifers first pass through a bigger filter (about 300 µm) with the function of debris removal and provided an aeration collar to avoid clogging of the filter screen. Rotifers are then concentrated outside the central cylindrical filter (about 60 µm). After harvest, the settling of the rotifers occurs in the same recipient. Papandroulakis developed at the Aquaculture Department of the University of Crete, an automatized system able to culture phytoplankton, rotifers, and crustaceans, such as *Artemia salina*, with the aim to produce feeds for aquaculture [*13*]. Another intensive system to culture crustaceans, but also shellfishes and fishes, as live food, was developed by the University of Trømso in collaboration with the Aquaculture Department Polarmiljosenteret of the Norwegian Polar Institute as part of the ALFA project (Development of an automated innovative system for continuous live feed production in aquaculture hatchery units), founded by EU (https://cordis.europa.eu/project/id/512789/reporting). The system is coupled with photobioreactors and rotifers culture system (CROPS), to cover all the aspects of the larval development. A novel optical algal density monitoring system based on colour image analysis techniques was also developed for the continuous assessment of algal density and the control of quality and growth rate of the culture. The project was integrated by system led to the footprint reduction, such as solar panels as electricity source, as well as with devices to control carbon dioxide concentration, pH, nutrient contents, and other parameters. Two full-scale complete systems were built and tested in Greece and Norway and adaptations were made to optimise output according to local conditions.

Alternative live foods have been developed to replace traditionally used living organisms such as *Artemia* nauplii and rotifers. Nematodes appear to be a promising food source for commercial penaeid larval culture and various studies have been performed to find the most promising species. Among various nematode species, *Panagrellus redivivus* demonstrated to be a highly promising species on both mass production and nutritional value point of view. Schlechtriem et al. (2004) investigated various techniques to mass culture nematodes with an enhancement of their nutritional values [*14*]. These techniques, based on oat media, enable shrimp hatchery operators to rely on an inexpensive, standardized, semi-automatized and permanently available live food for first-feeding fish larvae.

Copepods have been demonstrated to be interesting organisms for scientific (as model organisms) and aquaculture (as live food, especially for larvae) purposes. Various systems have been developed for intensive copepod production, taking into consideration that different species need different environments, i.e., harpacticoid copepods need larger surfaces instead of deeper tanks needed by calanoids [*15*]. Whatever the species considered, tanks for copepod culture are usually of a volume of 100–1000 litres, with gentle aeration and they are renewed every 3 weeks, and re-circulating systems are equipped with specific filtration systems [*3*]. Payne and Rippingale developed a semiautomatic system designed for the brackish water copepod *Gladioferens imparipes* [*16*]. This semiautomatic system is designed to separate the various larval stages thanks to a complex system of decreasing mesh nets. A first net, with a pore diameter of 150 μm, was used in the cylinder dedicated to the culture of males and ovigerous females. A second net, with a pore diameter of 53 μm was used to collect nauplii and it works, in addition, as an overflow system. The system was automatized by a PLC produced by Toshiba that allows the routine activities of valves, pumps, lights etc. Buttino and colleagues developed in 2012 at the Stazione Zoologica Anton Dohrn of Naples a complex system of intensive aquaculture designed for the copepods *Temora stylifera* and *Centropages typicus* [*17*]. This system used and improved the ideas of Payne and Rippingale allowing the concentration of larval stages using positive phototaxis and the selective division of different larval stages using nets with various pore diameters. The system was designed with an efficient filtration system made by a protein skimmer, UV sterilizer lamp and a bio-mechanical filter and it was managed by a PLC.

Finally, automatic culture and administration of marine micro-algae should be taken into consideration. In fact, marine microalgae are increasingly used as both feed for marine organism, constituting both a source of energy as well as the essential vitamins and PUFAs (as in the case of Penaeid larviculture), and as source of high value fine chemicals, therapeutics, and health foods. Large scale microalgae culture systems provide the possibility of delivering a continuous supply of high-quality microalgae although the newest culture systems appear quite expensive and with high operational costs. Control over the growing culture is necessary to maximize the production and an automated control system should take important decisions with respect to fertilising, harvesting, lighting and temperature to prevent the reduction of the efficiency and economical losses. Several control systems have been developed for this purpose, especially for the in-situ growth monitoring of the unicellular cultures. Most promising technologies are flow injection analysis, based on turbidimetric measurements, technique based on the monitoring of oxygen production, based on the measurement of the pressure inside a closed reactor and the use of optical density as a turbidimetric measure of biomass through spectrophotometry [*18*]. Sananurak designed a small scale (260 litres), highly sophisticated, although expensive, integrated continuous production system for *Tetraselmis suecica* and the zooplanktonic rotifer *Brachionus plicatilis* [*28*]. Considering the huge needs of microalgae in Penaeidae aquaculture plant, the increase in efficiency scaling up the production system and the need for an automatic control of culture conditions, Erbland and colleagues designed a large scale photobioreactor (170 litres of volume) built using cone-bottom, polyethylene tank, equipped with fluorescent lamps, monitoring and control system that measured temperature, pH and optical density of the microalgal culture [*20*].

15.5 Automatic monitoring and control of shrimp aquaculture

Real time monitoring of environmental parameters are very important for both shrimp aquaculture and paddy farming. Constant control of the water quality to keep the concentration of the water environment parameters in the ideal range can enhance the growth rate of cultured organisms, affect dietary utilization, and reduce the probability of diseases. Gathering information of water physic-chemical parameters is the key activity to perform the appropriate technical intervention to prevent harm to aquaculture production and is crucial to keep up sufficient conditions and avoid unfortunate circumstances that cause the failure of aquaculture [*21*]. Automatic monitoring and control systems can be used to face some serious issues like wastage of water. An integrated aquaculture can be controlled using aquaponics plants, which requires consistent water quality checking procedures that depend on intense information securing, communication, and handling. The connection between the hydroponic and aquaculture sections, that are the two main components of aquaponics, relies on ideal water quality conditions and on the monitoring of them.

Real time monitoring take advantage of various probes and sensors, which are the source of data for the automatic system, and actuators. Sensors and actuators are usually connected to microcontrollers built using various boards, for example Arduino or Raspberry [*22*]. The latter is preferred by some authors due to the intrinsic Wi-Fi module. Microcontroller monitors the output of sensors and it logs all the data using a data logging system. Actuators are activated or stopped, or in some cases their activity can be modulated, on the base of the software loaded on the microcontrollers and according with sensors output and pre-set thresholds. Several authors focused their attention on few kinds of sensors, such as turbidity, dissolved oxygen, and pH, although modern aquaculture relies on the real time monitoring of a large variety of physical and chemical parameters, such as Ammonia, Carbonates, Nitrate, etc.

The environmental monitoring of the water parameters, in both aquaculture plant or ponds, can be improved including Internet of Things (IoT) ZigBee-based wireless sensor network based on low power microcontrollers that are capable of collecting, analysing and presenting the whole data using an easy-to-access Graphical User Interface. Preetham proposed an IOT based aquaculture monitoring and control system, able to continuously observe the water quality factor and to take preventive steps early to harm for water animals [21]. The proposed architecture is composed by power module, sensor module, controller module and output module. In addition, a large variety of innovative sensors, that use new concepts and techniques, are replacing the traditional methods of water quality measurement, based mainly on UV measurement, Mass Spectrometry and amperometric sensors. Optical sensors, Microelectronic Mechanical System and Biosensors are used to measure different water quality parameters. These sensors that use emerging techniques can be merged in a single system as demonstrated by some authors [23]. They are, if compared with the previous generation of probes and sensors, more selective, sensitive, cheap and user friendly. The combined use of ZigBee wireless network and new probes allow aquaculture plants to apply wireless and highly efficient systems for real time monitoring to aquatic animal production. To make the communication network more efficient, sensors and microcontrollers should be connected using a MESH topology, where all nodes cooperate to distribute data amongst each other, with routers linked to end-devices and a coordinator node. These technologies have been applied successfully on aquaculture shrimp. Internet of Things in many cases was applied in rural context, such as tiger shrimp aquaculture in south east of Asia thanks to the low required power, the redundancy of nodes and the user-friendly graphical user interface. The automatic monitoring is strictly linked to the real-time alarms provided by the control system. On one side, automatic monitoring and control systems for aquaculture should monitor data coming from probes in continuous to identify abnormalities and identify thresholds in critical parameters, acting consequently on the actuators to mitigate or solve incoming issues. On the other side the automatic monitoring and control system, using the Internet of Things, should interact with the operator(s), that can be based all over the world, communicating issues and action taken to solve them. The interaction between operator(s) and controlling system via open-source app, such as instant messenger service, allows a two-ways communication, where operators are not only passive spectators of the controller routines, but they can issue orders based on the analysis of data available online [22]. For these reasons, monitoring and control system in aquaculture plant should meet four fundamental requirements:

- The operating routing should be systematic, performing the programmed activities at regular intervals with minimum or no deviation.

- Information should be always available and easy to access also for low trained personnel.

- Basing on the data collected by the system and available online, personnel should be able to interact with the system, giving new commands and reacting promptly to the encountered issues, via open-source software(s).

- Data should be organised and used for planning and decision making in the future, with the prospect of software implementation of new commands and routines to face common issues.

Considering the attention that has been given in recent years in the implementation of automation and intelligent systems applied to aquaculture and industrial systems, several models have been designed and tested [*24*]. There are automated control systems used in aquaculture whose main purpose is to acquire and record data; programmable logic controller (PLC) systems [*4*] and systems that integrate artificial intelligence for a more complex control, advanced interpretation and problem solving.

Understanding the basic architecture of these systems, making it intuitive and integrable, can allow not only to use new and completely innovative systems, but also to update and renew systems already in use without excessive economic expenditure.

As a guideline, a typical process control software application is composed of various modules in which the different tasks are sorted. It includes: data acquisition, communication between the different hardware components, information acquisition and management, creation of an easily accessible database, graphic interface for interaction with the user who can control the system through specific functions. The acquisition of information, through probes and smart devices, uses analogue and digital inputs and outputs, each with its own reference protocols [*5*]. The man-machine interface is made up of blocks (that is, a set of functions that encoded the instructions that perform specific activities) in turn connected in such a way that they can cooperate in creating an effector monitoring and/or control circuit. The process control software must be able to make available both data spreadsheets and graphical diagrams that are simpler and more immediate to interpret, which at the same time allow simple consultation even to the interaction of processes [*21*].

Through these automated control systems, it is possible to access both real-time data and historical databases and any alarm signals. Furthermore, if, in addition to automation, one can integrate an artificial intelligence, it may be able to process information collected, immediately and based on historical data, to develop useful statistics and future forecasts. To maximize the efficiency of these systems, a remote controller software is required. It allows you to interact with the system remotely, through dedicated control nodes. In this way, in addition to receiving alerts and messages in real time changes can be made to database blocks.

15.6 The internet of things revolution

Industrialization processes are associated with technology improvements. No other period has been as rich in scientific breakthroughs and technological innovations as the 20th century. Science and technology have been protagonists of these changes: the technology of the twentieth century has improved the lives of billions of people, while science has changed the very conception of man has of himself and of his role in the universe. At the same time and in sync with the latest technological developments that have revolutionized the last few decades, radically changing our lives, a new means of communication has also developed: the Internet, which in just over twenty years has grown exponentially, passing from a few thousands of connections, estimates of the International Telecommunications Union, which counted at the end of the 1980s and a few billion people connected all over the world through the use of different terminals such as computers, smartphones and tablets. The evolution of the

Internet has experienced two distinct phases that have revolutionized the lifestyles, habits, and behaviours of everyone, from citizens to institutions to companies: the World Wide Web phase in the 1990s and the Mobile Internet phase in the 2000s. Today, the Internet is entering its third phase of development: The Internet of Things or "Internet of Things" which, as Kevin Ashton, co-founder of the Auto-ID Centre at the Massachusetts Institute of Technologies (MIT) argues, "the Internet of Things have the potential to change the world, just like the internet did. Maybe even more".

The potential of this new evolution lies in the fact that with the Internet of Things the devices connected to the network, which are therefore able to exchange information, will grow exponentially since they will be able to connect to the network not only computers and mobile phones, but also everyday objects from means of transport to household appliances. This will generate a mole of information many orders of magnitude higher than the one the Internet has generated which, if used correctly, will allow the creation of new products and services for users.

The Internet of Things (IoT) is the extension of the internet in the sense known today, from people to objects. It is the construction of a network that places objects and their interaction in the centre [25]. More properly, it is the passage of a connection network for end-user-devices to an interconnection network made of physical objects able to communicate and cooperate (Figure 2) with each other through the internet connection, becoming smart devices (intelligent devices). To apply the IoT concepts it is necessary that various conditions are met. The most important is to make sure that all the devices used, which need to make communications on the network, have the ability to access the Internet, directly or indirectly. Moreover, they must have a unique digital identity to be recognized and distinguished. In the presence of simpler

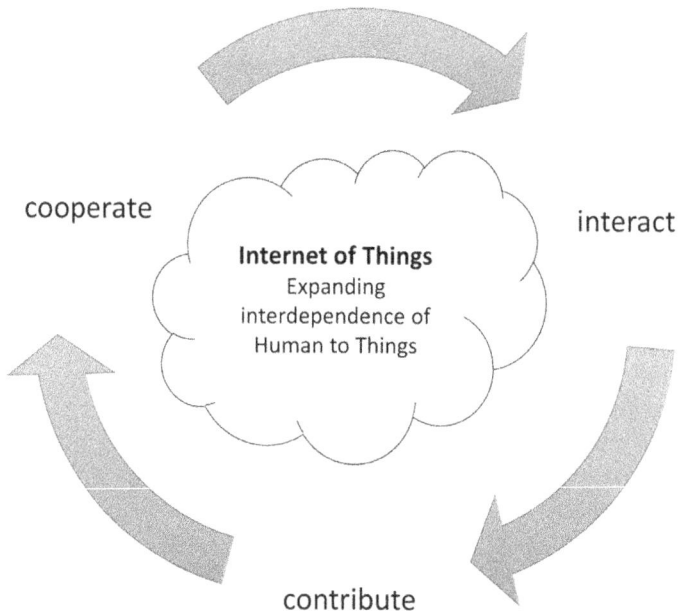

Figure 2. Interdependence of Internet of Things.

devices (such as simple sensors) which do not natively have network cards, it will be possible to connect them to a device that has gateway functions, or that is responsible for forwarding to the central server all the information received.

The integration of sensors in the network gives the possibility to obtain and elaborate data. When we refer to the elaboration of the collected data, a fundamental part concerns the management of Big Data. Big Data is a field that deals with the ways to analyse and then extract information from a large amount of data that would otherwise be too large or complex to be managed. Furthermore, although a survey with many cases offers greater statistical power, at the same time a very extensive data collection in terms of volume, speed and variety is very difficult to consult and can potentially lead to a higher rate of false interpretations.

The life sciences have a long history of dealing with large quantities of data, and recent advances in experimental capabilities have vastly increased the amount of data that needs to be stored and analysed. Biology is notoriously fragmented in its methods, goals, instruments, and conceptual frameworks. Often, different groups— even within the same subfield—disagree over preferred terminology, research organisms, and experimental methods and protocols [26]. Using interoperable databases and file formats to integrate data from different sources so that they can be used and re-used across a variety of research contexts. Databases need to be accessed through a common 'query' system, and this raises the question of which terminologies should be used to classify the data and integrate them with other data, and what are the implications of such choices. We should acknowledge that no data are 'raw' in the sense of being independent from human interpretation, assessing which data are reliable and which are not. But it does not take into account that data are often extensively processed artefacts resulting from highly planned interactions with the world; nor does it do justice to the observation that biologists have different views of what counts as reliable data, or what counts as data in the first place [27].

In details, the data are collected through the sensors, stored generally in a cloud where they are analysed, and the information obtained is available for the users according to the designer's purposes. The sensory network that the IoT allows to implement is able to produce data that, compared to the current ones, are more reliable and derive from statistically more relevant samplings [28]. In fact, many of the barriers due to the various passages that the information must go through from the moment of its acquisition to use/interpretation are removed. This is understood as the possibility of constant monitoring of all events and processes of a given environment/ business. In general, the IoT allows devices to be monitored and controlled remotely through special network infrastructures, creating a direct interaction between the real world and computer-based systems. The application of these concepts to the world of industry makes it possible to improve the efficiency and accuracy of the processes, as well as obtaining economic benefits, thanks also to the fact that the cases in which human intervention is required are reduced. The Internet of Things is in fact revolutionizing the market and numerous opportunities for growth and development are being created. The use of a system implemented by IoT applied to the culture of marine organisms such as crustaceans, both as model organisms for scientific research purposes and for production for commercial purposes, could be extremely useful and revolutionary. The idea is to have an intelligent system with special sensors for measuring the fundamental parameters for breeding and special effectors to intervene autonomously in such a way that all measured values always remain within the optimal limits (Figure 3).

Therefore, the primary purpose would focus on optimizing the production processes, making them on the one hand more efficient in terms of quantity and quality of the product and at the same time lowering production costs (intended both from an economic and temporal point of view since it would reduce the use of human resources). Considering a crustacean culture, with the use of IoT it could be possible to optimize the use of water and primary resources, in consequence of an adequate real-time analysis of the values in the tanks and an intelligent management of food additions and supplements. The changes of water, if necessary, will be regulated by the intelligent system itself which will consider every variable, even the weather forecast (which, in some cases, can affect the quality of the incoming water where it is possible to directly draw on sources of external natural water), consulting both local sensors and open-data environmental/meteorological services.

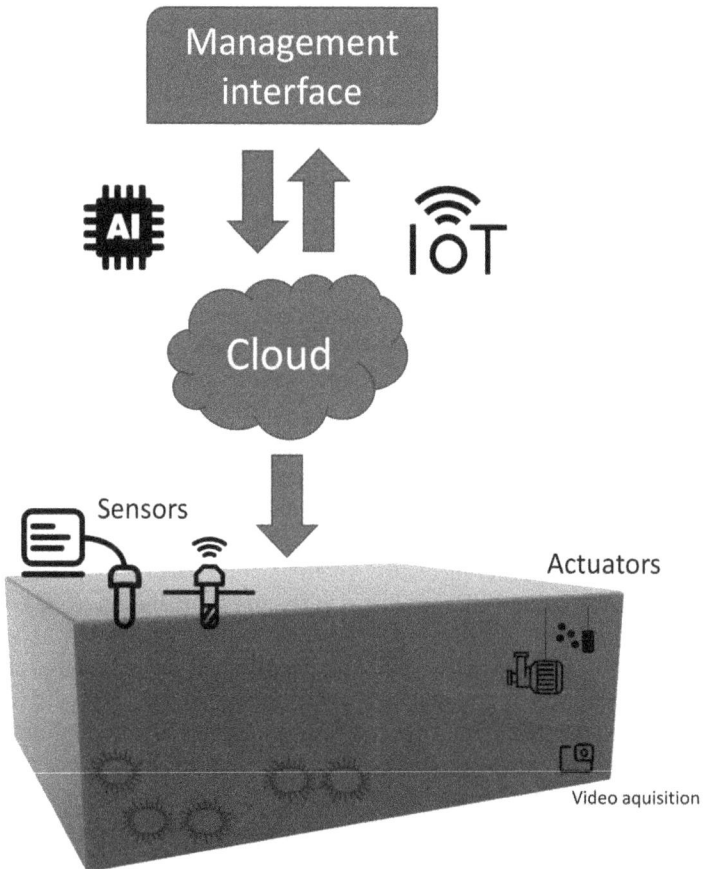

Figure 3. "Interactive operators" capable of reading environmental signals, carry out assessments and respond with programmed actions.

These sensors must be connected to the internet and the measurements made will be sent to an online platform, which will have the purpose of collecting the data. This platform will give the possibility to clearly view the data collected, for example by generating intuitive and easily understandable graphs. The system itself will also be able to analyse these data and understand where it will be necessary to intervene (however, leaving manual intervention possible) to correct them, for example through specific actuators for each value. Furthermore, the platform will be able to autonomously interrogate open-access platforms or private databases to consult data collected in previous comparable farms, to be able to reason on the actual need to intervene based on current conditions. In details, the data are collected through the sensors, stored generally in a cloud where they are analysed, and the information obtained is available for the users according to the designer's purposes. The sensory network that the IoT allows to implement is able to produce data that, compared to the current ones, are more reliable and derive from statistically more relevant samplings [28]. In fact, many of the barriers due to the various passages that the information must go through from the moment of its acquisition to use/interpretation are removed. This is understood as the possibility of constant monitoring of all events and processes of a given environment/business.

In general, the IoT allows devices to be monitored and controlled remotely through special network infrastructures, creating a direct interaction between the real world and computer-based systems. The application of these concepts to the world of industry makes it possible to improve the efficiency and accuracy of the processes, as well as obtaining economic benefits, thanks also to the fact that the cases in which human intervention is required are reduced. The Internet of Things is in fact revolutionizing the market and numerous opportunities for growth and development are being created. The use of a system implemented by IoT applied to the culture of marine organisms such as crustaceans, both as model organisms for scientific research purposes and for production for commercial purposes, could be extremely useful and revolutionary. The idea is to have an intelligent system with special sensors for measuring the fundamental parameters for breeding and special effectors to intervene autonomously in such a way that all measured values always remain within the optimal limits. Therefore, the primary purpose would focus on optimizing the production processes, making them on the one hand more efficient in terms of quantity and quality of the product and at the same time lowering production costs (intended both from an economic and temporal point of view since it would reduce the use of human resources). Considering a crustacean culture, with the use of IoT it could be possible to optimize the use of water and primary resources, in consequence of an adequate real-time analysis of the values in the tanks and an intelligent management of food additions and supplements. The changes of water, if necessary, will be regulated by the intelligent system itself which will consider every variable, even the weather forecast (which can affect the quality of the incoming water where it is possible to directly draw on sources of external natural water), consulting both local sensors and open-data environmental/meteorological services. These sensors must be connected to the internet and the measurements made will be sent to an online platform, which will have the purpose of collecting the data. This platform will give the possibility to clearly view the data collected, for example by generating intuitive and easily understandable graphs. The system itself will also be able to analyse these data and understand where it will be necessary to intervene (however, leaving manual intervention possible) to correct them, for example through specific actuators for each value. Furthermore, the platform will be able to autonomously interrogate open-access platforms or private databases to consult data collected in previous comparable farms, to be able to reason on the actual need to intervene based on current conditions.

References

1. M. Mutalipassi. (2019). Ph.D. Thesis, The Open University, 231 pp.
2. R. Calado, T. Pimentel, A. Vitorino, G. Dionísio, and M.T. Dinis. (2008). *Aquaculture* **285:** 1–4, 264–269.
3. R. Calado, I. Olivotto, M.P. Oliver, and G.J. Holt. (2017). *J. Fish Biol.* **91:** 1250–1251.
4. M. Mutalipassi, M. Di Natale, V. Mazzella, and V. Zupo. (2018). *Animal.* **12(1):** 155–163.
5. F. Antonucci, and C. Costa. (2020). *Aquaculture International.* **28:** 41–57.
6. C. Campanati, D. Willer, J. Schubert, and D.C. Aldridge. (2021). *Rev. Fish. Sci. Aquac.* 1–50.
7. T.P.T.H. Van, M.A. Rhodes, Y. Zhou, and D.A. Davis. (2017). *Aquac. Res.* **48:** 5346–5355.
8. K. Krishna Kishore, P. Vamsi Krishna, and D. Srikanth. (2017). *ICSSS* **4:** 26–429.
9. C. Ullman, M. Rhodes, T. Hanson, D. Cline, and D.A. Davis. (2019). *J. World Aquac. Soc.* **50:** 54–64.
10. J. Reis, R. Novriadi, A. Swanepoel, G. Jingping, M. Rhodes, and D.A. Davis. (2020). *Aquaculture* **519:** 734759.
11. I.J. Huang, C.C. Hung, S.R. Kuang, Y.N. Chang, K.Y. Huang, C.R. Tsai, and K.L. Feng. (2018). *ICAM* 177–180.
12. M. Dehasque, B. Ooghe, M. Wille, P. Candreva, Y. Cladas, and P. Lavens. (1997). *Aquacult. Int.* **5:** 179–182.
13. N. Papandroulakis, P. Dimitris, and D. Pascal. (2002). *Aquacul. Eng.* **26(1):** 13–26.
14. C. Schlechtriem, M. Ricci, U. Focken, and K. Becker. (2004). *J. Appl. Ichthyol.* **20:** 161–168.
15. J.G. Stottrup. (2000). *Aquacult. Res.* **31:** 703–711.
16. M.F. Payne, and R.J. Rippingale. (2001). *Aquaculture* **201:** 329–342.
17. I. Buttino, A. Ianora, S. Buono, V. Vitiello, M.G. Malzone et al. (2012). *Aqua. Res.* **43:** 247–259.
18. J.M. Sandnes, T. Ringstad, D. Wenner, P.H. Heyerdahl, T. Källqvist, and H.R. Gislerød. (2006). *J. Biotechnol.* **122:** 209–215.
19. C. Sananurak, T. Lirdwitayaprasit, and P. Menasveta. (2009). *Science Asia.* **35:** 118–124.
20. P. Erbland, S. Caron, M. Peterson, and A. Alyokhin. (2020). *Aquacult. Eng.* **90:** 102103.
21. K. Preetham, B.C. Mallikarjun, K. Umesha, F.M. Mahesh, and S. Neethan. (2019). *Int. J. Adv. Res. Ideas Inn. Tech.* **5:** 4–7.
22. P.S. Sneha, and V.S. Rakesh. (2018). *Proc. Int. Conf. Inven. Comput. Inf.* ICICI 1085–1089.
23. Y. Qin, A.U. Alam, S. Pan, M.M.R. Howlader, R. Ghosh et al. (2018). *Sens. Actuators Chemic.* **255:** 781–790.
24. G. Xu, Y. Shi, X. Sun, and W. Shen. (2019). *Sensors* **19:** 12–22.
25. D. Miorandi, S. Sicari, F. De Pellegrini, and I. Chlamtac. (2012). *Ad. Hoc Net.* **10:** 1497–1516.
26. S. Leonelli. (2019). *eLif.* **8:** 123–145.
27. J. Calvert. (2018). *New Genet. Soc.* **37:** 275–277.
28. L. Atzori, A. Iera, and G. Morabito. (2010). *Comput. Netw.* **54:** 2787–2805.

CHAPTER 16

Current Issues on Freshwater Crayfish Aquaculture with a Focus on Crustacean Welfare

Marina Paolucci,[1,*] *Elena Coccia,*[2] *Gianluca Polese*[3]
and *Anna Di Cosmo*[3,*]

16.1 Introduction

The biology of freshwater crayfish as well as its farming potential has been the subject of numerous publications since the 50s and 60s of the last century. Basic biology has been extensively covered by experts and researchers from around the world and therefore will not be discussed in this chapter. References to the biology of the main critical aspects of farming, namely reproduction and feeding, will be kept at a minimum, and only those aspects, functional to the comprehension of this chapter will be reported.

16.1.1 Classification and distribution of freshwater crayfish

Freshwater crayfish constitute a relatively homogeneous group of Decapod crustaceans, which also include typically marine species of considerable economic interest such as lobsters (genus Homarus), Norway lobster or langoustine (genus Nephrops), spiny lobsters (genus Panulirus and Jasus) and prawns (genus Penaeus). There are approximately 600 freshwater crayfish species that occur naturally on every continent except Africa and Antarctica. Freshwater crayfish also include brackish crayfish from the Northern Hemisphere (Central-Northern America, Europe, Far East) and the Southern Hemisphere (South America, Madagascar, New Guinea, New

[1] Department of Science and Technologies, University of Sannio, Italy.
[2] Dipartimento di Chimica e Biologia "Adolfo Zambelli", University of Salerno, Italy.
[3] Department of Biology, University of Napoli, Federico II, Italy.
* Corresponding authors: paolucci@unisannio.it; dicosmo@unina.it

Zealand, and Australia). Freshwater crayfish have undergone numerous taxonomic reviews in the last years also on the basis of the recent advances in morphological and molecular phylogenetic studies. However, the classification backbone has undergone a few changes, with the exception of some relevant exceptions reported in the excellent recent work of Crandall and De Grave [*1*], in which the authors integrated, updated, and revised crayfish taxonomy with insights gained through systematic and taxonomic studies. For the purposes of this chapter, it will be enough to remember that the systematic classification of freshwater crayfish divides them into two main Superfamilies: Astacoidea, with two families: Astacidae and Cambaridae, and Parastacoidea with one family: Parastacidae. The two superfamilies correspond to Northern and Southern hemisphere divisions.

Astacidae are distributed in a large part of the European continent, in the western areas (Turkey, Siberia) of the Asian continent, and West of the Rocky Mountains from the Northwestern United States into British Columbia, Canada. Cambaridae are distributed in North America, from Southern Canada through Mexico, as well as in the far East of Asia, where they are present in the Chinese and Japanese inland waters. The Superfamily of Parastacoidea occurs in Australia, New Zealand, and South America.

Central-North America is home to the largest number of living freshwater crayfish: 12 genera belonging to the families of Astacidae and Cambaridae and comprising over 382 species. Of the 15 genera of the Parastacidae family, 3 are located in South America (Brazil, Chile, Argentina), 1 in New Zealand, 1 is present exclusively in Madagascar, and 10 are part of the Australian fauna, one of which is also present in New Guinea. Interestingly, only a few crayfish species are currently of commercial importance. These species are *Procambarus clarkii* and *Faxonius limosus*, belonging to the Cambaridae family; *Astacus astacus*, *Astacus leptodactylus*, and *Pacifastacus leniusculus*, belonging to the Astacidae family; *Cherax destructor* and *Cherax albidus*, known by the name of yabbies, *Cherax tenuimanus* and *Cherax cainii*, referred to as marron, and *Cherax. quadricarinatus* referred to as redclaw, belonging to the Parastacidae family.

Freshwater crayfish can be found in nature in a large variety of environments. The robustness, tolerance to adverse conditions, and the adaptability of some species have certainly played a key role in the worldwide distribution. Generally speaking, Astacids can be defined as crayfish of temperate waters, with some species, typical of calm waters (lakes, slow stretch of rivers, ponds, and collections of waters in general), and others, including the indigenous crayfish of Italian waters, more related to the running water. Cambarids and Parastacids include both tropical species and temperate climates, widespread in environments ranging from alluvial marshes to mountain streams. From a practical point of view, it is important to point out that the crayfish of calm waters are better suited to exploitation in aquaculture than those of running water, by virtue of the greater growth, adaptability, and prolificity that generally characterize them. In addition, the same aquatic environmental conditions necessary for their breeding guarantee easier reproducibility in captivity.

The current distribution of freshwater crayfish was undoubtedly influenced to a significant extent by man, who already in prehistoric times contributed to the expansion of the diffusion area of some species, an event that is not at all inexplicable if we consider the remarkable palatability associated with the ease of catching, housing, and transporting these crustaceans. In more recent times there have been spectacular cases of introduction and acclimatization of various species of crayfish in inland waters around the world. The translocation essentially involves the species of commercial interest for food purposes and in more recent times those of aquariological interest. As for the species of commercial interest for food purposes, the most cosmopolitan freshwater crayfish is by far the North American Cambarid *Procambarus clarkii*, or killer crayfish of the Lousiana. As the name goes, its native area is Luouisiana from where it was firstly translocated in the 1920s. It is currently distributed across all continents except Australia and Antarctica. In Europe, it is widespread in Spain, France, Germany, and Italy. This species prefers slow-moving stagnant waters and marshes, even subjected to summer drying; it can be found in rice fields, lakes, and brackish coastal ponds. It is able to survive the drying of its habitat by digging deep burrows in the bed of the canals, shelters where it finds the humidity necessary to keep it alive until the next wet season. It is also able to tolerate rather high temperatures and relatively modest concentrations of dissolved oxygen. Disease resistance, rapid growth rate, and high fertility make this species the most farmed and fished freshwater crayfish in the world, as well as the most economically valuable aquatic species to be farmed. Unfortunately, its ease of breeding also makes it a species with high weed potential.

Faxonius limosus (Rafinesque, 1817), or spinycheek crayfish, previously known as *Orconectes limosus*, is also one of the most successful crayfish invaders of Europe and the most frequently encountered species in Central Europe. Native to Northern and Central America, it was first introduced from North America into a fish farm pond in Poland in 1890 for commercial purposes. It is one of the most common crayfish alien species in European inland waters and by 2014, *Faxonius limosus* was already present in 20 European countries. It is highly prolific, aggressive, and tolerant with respect to the water quality. The ideal habitat is represented by streams with little current and stagnant waters, which present a muddy or sandy substrate in which it is possible to dig holes. *Faxonius limosus* also manages to tolerate long periods outside the water (even a few weeks) and a moderate level of water salinity.

Pacifastacus leniusculus or signal crayfish has been heavily translocated from North America since the beginning of the last century. It was introduced into streams or lakes in Hokkaido (the northernmost island of Japan) and Honshu (the main island of Japan) between 1926 and 1930, from where it rapidly expanded throughout the country. *Pacifastacus leniusculus* was massively imported into Europe between 1960 and 1970 through Sweden, from where it was introduced in lakes and streams of Northern, Central, and Eastern Europe. It is a very tolerant species towards environmental adversities, in particular to brackish waters and thermal variations.

Astacus astacus or noble crayfish is the European autochthonous crayfish species with the highest commercial value. Despite being very sensitive to the disease called crayfish plague, transmitted by the oomycete *Aphanomyces astaci*, *Astacus astacus* is still of commercial interest because of its abundant meat content and high quality. For this reason, it has been introduced from continental Europe in numerous countries, including Greece, Cyprus, the United Kingdom, and even Morocco. After the introduction of the American crayfish species, *Pacifastacus leniusculus* and *Faxonius limosus*, and the concomitant spread of the crayfish plague, to which American species are resistant and act as vectors, vast populations of *Astacus astacus* and the total number of individuals in Europe dramatically decreased at the end of the 19th century. Water pollution and habitat alteration and destruction have further aggravated the survival of the species, which today appears endangered and remains with small and isolated populations. Thus, the limited availability has increased the economic value of this good, considered today as luxury food product and available in small quantities only on local markets.

Astacus leptodactylus or narrow-clawed crayfish comes from rivers that flow into the Caspian and Black Seas. It moved naturally from the Ponto-Caspian basin through rivers and canals to colonize new areas in eastern and northern Europe and has been translocated for stocking purposes in at least 30 countries, although not in the Iberian Peninsula and Nordic countries. Its introduction into Europe dates back to the beginning of the last century, probably with the aim of repopulating some water bodies. It is a highly fertile species (200–400 eggs by a female) and has a fairly fast growth rate. Unlike the European indigenous species, *Astacus leptodactylus* does not need good quality water, lives mostly in ponds and lakes, and tolerates low oxygen levels, high levels of suspended solids, brackish water, temperature changes, and temperatures above 21°C. *Astacus leptodactylus* has a widespread distribution in lakes and ponds in many parts of Turkey and Iran. Turkey for many years was the main exporter of *Astacus leptodactylus* to Europe. In the early years of this century (2003–2004) the production reached a maximum value which, however, began to decline, and today the production is very limited and only achieved through fishing. In the inland waters of Turkey, the existence of the crayfish has been confirmed in more than one hundred localities and although the consumption of crayfish as food is quite low in Turkey, it is in demand in touristic regions. Indeed, the crayfish produced in Turkey are exported to various European nations such as France and Sweden. In Iran, *Astacus leptodactylus* is present with two subspecies: *Astacus leptodactylus eichwaldi* in brackish water and *Astacus leptodactylus leptodactylus* in freshwater sources, but only *Astacus leptodactylus leptodactylus* is commonly distributed in Iranian water bodies. In Iran, the first crayfish introduction was successfully carried out by the Iranian Fisheries Research Organization back in 1988. All production is obtained from the wild.

The *Cherax* genus includes several species of commercial interest in Australia. Known to indigenous peoples for its tasty meats, *Cherax* has been widely translocated around the world due to its use in aquaculture and aquariology. In particular, *Cherax quadricariatus* is highly valued as a food resource and was first introduced to the public in the late 1980s in southeastern Queensland. Since then, populations of *Cherax quadricarinatus* have been introduced into several countries for research on its aquaculture potential, together with other *Cherax* species. Today, *Cherax quadricarinatus* presence and its exploitation for aquaculture purposes have been reported in numerous countries such as Europe, Africa, Indonesian territories, China, Singapore, Taiwan, Japan, Malaysia, Israel, United States, Mexico, Puerto Rico, Jamaica, Ecuador, and Argentina.

16.2 Freshwater crayfish aquaculture

16.2.1 Global overview

Crustaceans are very important for aquaculture. Shellfish aquaculture developed rapidly, and global production reached 7.9 million tonnes in 2016 (FAO report 2018, https://www.fao.org/documents/card/en/c/I9540EN/). The marine decapod *Peneaus vannamei* positioned itself in 2017 as the most farmed crustacean in the world, although no crustacean is among the top ten most farmed species in the world in terms of quantity. Unexpectedly, *Procambarus clarkii*, is the second most-produced species, representing 12% of the total aquaculture production of crustaceans, and ranks among the 10 species with the highest economic value in the world (FAO report 2020, https://www.fao.org/documents/card/en/c/ca9229en/).

Freshwater crayfish aquaculture or "astacicolture" has been practiced successfully, and for a long time in many countries, particularly in the United States, Australia and Europe, and more recently in China. Astacicolture is well-developed in North America, targeting mainly three species: *Procambarus clarkii* in Louisiana, *Faxonius limosus* in the Midwest, and *Pacifastacus leniusculus* in Pacific Northwest. The most cultivated species is undoubtedly the *Procambarus clarkii*, an indigenous species of the Southern United States from where it is exported alive for human consumption in many other countries. The farming of *Procambarus clarkii* is a very flourishing commercial activity with a long history. Commercial sales of *Procambarus clarkii* from natural waters began in Louisiana in the late 1800s. Crayfish are cultivated and consumed for food in several southern states, but Louisiana dominates the crayfish industry and became the most successful producer and seller of crayfish in North America with reference to aquaculture and wild capture fisheries. Total production is rather limited in absolute terms, but still able to satisfy the domestic market. Approximately 48 000 hectares are devoted to the culture of crayfish in Louisiana and the state accounts for about 90–95 percent of the total production in the USA (FAO, http://www.fao.org/fishery/culturedspecies/Procambarus_clarkii/en).

The crayfish farmed in Australia belong to the Cherax genus present with the species *Cherax quadricarinatus*, *Cherax destructor*, *Cherax albidus*, and *Cherax tenuimanus*. They are highly appreciated for the consistency and flavor of their meat, competing favorably with commonly consumed marine crustaceans. In addition, their size and appearance makes them particularly popular with consumers of marine crustaceans. *Cherax quadricarinatus*, popularly indicated by its Australian synonym "redclaw", is a tropical species native to the rivers of northwest Queensland and Northern and eastern parts of Northern Territory in Australia. Although well known to the local inhabitants of this isolated region, it remained almost unknown to the rest of the world until the late 1980s, when it began to be farmed on a large scale. Farming developed rapidly and spread throughout Northern Australia, and soon after abroad. Rigg et al. [2] report that "redclaw are farmed commercially on 22 licensed farms in Queensland (Queensland Government 2020) stretching from the Atherton Tablelands in the far north, down to the Sunshine Coast and State border areas in the extreme south of Queensland". The production is, however, largely fluctuating. The highest production was reached in the years 2005–2006, with 105 tons, but dropped in the following years, reaching a minimum in the years 2013-2014 when the production was less than 40 tons. In more recent years (2015–2018) it stabilized at around 50 tons. The reasons for such variability are mainly related to environmental variables, although Redclaw farmers complain about the lack of seedstock (juvenile redclaw) suppliers, underlying the need for hatcheries to produce juveniles as occurs for other aquaculture sectors. In any case, the numbers are too small to support an export market. The redclaw crayfish has biological characteristics that make it an excellent candidate for farming. It is physically robust with a short life cycle, fast growth, reaching 100 to 200 g within 12 months of growth, and simple production technology, it requires a low protein diet and is inexpensive to produce. Such adaptive traits make the redclaw one of the most farmed crayfish all around the world. However, redclaw crayfish represents a minor aquaculture industry, with limited production. Mexico, one of the largest redclaw crayfish-producing countries outside Australia, produces around 50 tons per year (FAO report 2016, https://www.fao.org/3/i5555e/i5555e.pdf). Mismanagement caused the escape of *Cherax quadricarinatus* in the environment, where well-established populations are present and represent a significant proportion of the freshwater decapod catch with a high commercial value. Apart from Mexico, this species is cultivated in New Caledonia, Africa, China, Taiwan, Japan, Malaysia, Israel, Italy, United States, Mexico, the Caribbean, Puerto Rico, Ecuador, and Argentina.

Another Australian crayfish that also has received some attention from farmers around the world is the *Cherax destructor*, known by the vernacular name of Yabby or Yabbie. The term Yabby is used to indicate both the *Cherax destructor* and the *Cherax albidus*, although their taxonomic status is still under debate. The Yabby is widely distributed throughout Australia, being present in most of Victoria and New South Wales, as well as Southern Queensland, South Australia, and parts of the Northern Territory. *Cherax destructor* is an aggressive digger and can become an ecological problem (invasive species).

The third species of the *Cherax* genus is *Cherax tenuimanus*, also known as Marron. Its geographical distribution is rather narrow, being limited to a range within the Southwestern region of South-West of Western Australia. However, *Cherax tenuimanus* has been introduced by people into water bodies throughout the Australian continent. Marron is relatively slow-growing, taking 2 or more years to achieve a minimum marketable size. It is advantaged, however, by reaching 500 g or more (over several years), making them comparable to marine lobsters in the marketplace. The Cherax genus has a great commercial aquaculture potential, but its expansion has still not occurred and its farming industry remains in a developmental stage. In Europe, the aquaculture of freshwater crayfish is based on a limited number of species. Only two are autochthonous: *Astacus astacus* and *Astacus leptodactylus*, while the others two are allochthonous: *Pacifastacus leniusculus* and *Procambarus clarkii*. In addition, among the introduced species *Cherax destructor*, *Cherax quadricarinatus* and *Faxonius limosus* are cultured in small quantities. Freshwater crayfish production in Europe is very limited and there are no official data on recent production. Spain is an exception and can boast a good production of *Procambarus clarkii.*

Although Asia is one of the main producers of marine crustaceans, the cultivation of freshwater crayfish is still in its infancy. In particular, China is the leading country in the production of *Procambarus clarkii.* Chinese production has increased from 6700 tonnes in the early 1990s up to more than one million tonnes in 2017, with a current commercial value of $ 42 billion (China's Ministry of Agriculture and Rural Affairs 2018). *Procambarus clarkii* has long been considered an invasive and therefore harmful species. This is not surprising as crayfish have long been the scourge of rice farmers, due to their burrowing behavior undermining rice paddies and making them unsustainable. However, China is currently experiencing the "crayfish craze" as small red lobsters have become a favorite of Chinese cuisine. Chinese production of *Procambarus clarkii* increased from 0.26 million tons in 2007 to 0.85 million tons in 2016 ($ 8.14 billion, Ministry of Agriculture Fisheries Department, 2017), and exceeded 1 million tons with 37 billion dollars of production in 2017, equal to 80% of global production according to FAO statistics (FAOSTAT, https://scihub.wikicn.top/http://www.fao.org/faostat/it/) and Crayfish Industry Report 2018 (https://scihub.wikicn.top/http://www.moa.gov.cn/), becoming an important food producing sector in China. It ha been estimated that more than 55 percent of crayfish consumed globally is cultured in Hubei Province, China.

16.2.2 *Farming techniques*

Despite being widely practiced all over the world, freshwater crayfish farming is still in its infancy, especially from a technological perspective. The different farming techniques are based on the diet rather than on the density of individuals which depends on the species bred and the quantity of food supplied. The most used farming techniques are extensive, and semi-intensive, which require minimal human intervention, but whose productivity is difficult to control. On the contrary, intensive farming requires greater human commitment, technologically and for the need of trained dedicated staff. It represents, however, especially in countries where there are no large areas of territories suitable for other forms of farming, the only possible solution alongside perhaps semi-intensive farming, practiced in countries with limited availability of large and flat portions of territory. The role played by the ever-increasing environmental degradation which constitutes a strong limitation to the applicability of extensive farming must also be assessed, and therefore even where the conditions of the territory allow these types of farming, the semi-intensive, and intensive farming should be considered, because all phases are under human control, especially the water quality. Aquaculture techniques are discussed in this contest only in relation to their implication for astacicolture. For their in-depth treatment, we suggest the reader to refer the numerous technical texts available on the market.

16.2.3 *Extensive farming*

Extensive farming is a long-established method consisting of vast ponds (natural or artificial) where low-numbered adult animals are introduced in order to make them reproduce. It is generally polycultural. When possible, ponds are drained and fertilized to boost the aquatic food pyramid by stimulating the aquatic vegetation and intensifying the presence of microorganisms, small mollusks, and crustaceans, larvae, and worms, which form the base of the aquatic food pyramid. For crayfish, the production cycle has an average duration of three years, with a very high degree of juvenile mortality and low production, also determined by the fact that no supplement is provided to improve the environmental conditions or to feed the animals. Profits are variable according to the quality of the water or the surrounding environment. The production undergoes fluctuations due to the presence of ichthyophagous birds, water quality, difficulties in finding juveniles, and management costs. 70–80% of the costs are represented by the work and sowing, due to the lack of artificial reproduction systems. This type of farming does not provide for any type of human intervention, except, depending on the case, a limited supply of food to supplement the natural diet. In this case, we speak of semi-intensive farming.

16.3 Semi-intensive farming

The semi-intensive method instead involves the use of ponds of limited size, around the hectare, with monoculture farming and minimum water exchange. In a semi-intensive system, the production of the pond is increased beyond the level of extensive aquaculture by adding a supplementary feed, usually in the form of dry pellets, to integrate the feed naturally available in the pond, allowing for higher stocking density and production per hectare.

16.4 Intensive farming

The intensive farming method is generally composed of open-air concrete tanks, raceways, or earth ponds of different sizes and depths suited to the different stages of growth of the crayfish. It is usually a monocultivation carried out in ponds of size around 0.1 hectares, 1.5–2 meters deep, or in tanks of varying sizes and materials. In both cases, a certain change of water must be ensured. This method requires the administration of either natural or pelleted food. In this type of farming, the presence of many hiding places on the bottom (for example pieces of plastic pipe) is fundamental in order to avoid cannibalism by providing shelters for the individuals. It is important, in order to avoid excess toxic nitrogen catabolites and low oxygen levels at the bottom of the tanks or ponds, to ensure limited water movement and a certain supply of oxygen, especially during the night, when the lack of photosynthetic activity of phytoplankton leads to a rapid decline of this precious dissolved gas. A problem of the intensive breeding system is given by overcrowding as several couplings often occur which cause the presence of a very high number of individuals per tank, individuals of much smaller size and weight than the potential of the species. To deal with this situation, one option is to rely on single-sex populations which, given the larger size reached by the specimens, is proved to be much more profitable.

A variant of the intensive system is defined as recirculation o RAS (recirculation aquaculture system). It is a system relying on more complex technology. Into the recirculation system, the water is pumped into a tube system and is accumulated in tanks. It is an advantageous system in which water, isolated from the external environment, can be controlled in all its parameters: temperature, hardness, pH, microbial load, etc. There are, however, disadvantages, above all related to the cost of the investment and the consumption of energy.

A super-intensive system is also under experimentation, it accommodates each specimen in a separate tank with its own water supply and water exchange. This method is able to guarantee a production of over 280 large specimens for a cubic meter of water, against the ~ 5 of the semi-intensive system.

16.5 Integrated multi-trophic aquaculture systems (IMTA)

The term Integrated Multi-Trophic Aquaculture systems (IMTA) was born in 2004 and, as the name implies, it indicates the incorporation of species from different trophic positions or nutritional levels into the same system. With the integrated multi-trophic aquaculture systems (IMTA), the by-products of the feed used by one species, including nutrient-rich feces, become the food source for other species. Fish waste naturally produces very high concentrations of nitrogen. In the recirculating system of IMTA, the nitrogen-rich waste is also used to fertilize the plants. As a result, this system reduces waste accumulation and helps improve water quality by providing higher productivity and reduced environmental impact.

There are different ways of declining the concept of IMTA, depending on the degree of interrelation between the variables that make up the integrated system. IMTA is a more complex system than combining hydroponics (growing plants in water without soil) with the process of aquaculture (raising aquatic species) a type of practice referred to as aquaponics. "Multi-Trophic" refers to the incorporation of species from different trophic or nutritional levels in the same system. This is one crucial distinction from the well-known and widely practiced since ancient times, aquatic polyculture, which is the co-culture of different fish species from the same trophic level.

Sometimes the term "Integrated Aquaculture" is used as a synonym for IMTA to describe the integration of monocultures through water transfer. Both terms define hydroponics and aquaculture systems which are variously integrated. Other terms such as Fractionated Aquaculture, IAAS (integrated agriculture-aquaculture systems), IPUAS (integrated peri-urban-aquaculture systems), and IFAS (integrated fisheries-aquaculture systems) are all variations of the IMTA concept.

Multi-trophic aquaculture has been widely used in China for centuries, employing various species, among which crayfishes. Crayfish species are omnivores and detritivores and thus are low trophic feeders and therefore can be considered promising candidates for IMTA systems. A working IMTA system can result in greater total production based on mutual benefits to the co-cultured species and improved ecosystem health. In the integrated crayfish-rice cultivation model, wastes (e.g., weeds, insects, leaves) are a good food sources for crayfish, while crayfish digging burrows increase soil permeability and material/energy circulation and its feces are used as high-quality fertilizer and thereby, advance the rice growth.

16.6 Biofloc system

The term biofloc defines aggregates (flakes) of organic material (feces, uneaten feed), algae, bacteria, and protozoa, but can also include zooplankton and nematodes. These characteristics allow the flakes to be visible even to the naked eye, even if in most cases the dimensions are microscopic, ranging from about 50 to 200 microns. The aggregates are held together thanks to a mucous matrix secreted by bacteria, filamentous microorganisms or electrostatic attraction. The management of the biofloc systems is not simple, and it is necessary to have a certain degree of technical sophistication for the system to be fully functional and productive. In fact, there must be sufficient mixing and aeration to keep flakes in suspension, and the quality of the water must be constantly monitored. The biofloc systems are useful in environments where water is scarce and intensive forms of aquaculture are required. The biofloc system has also been developed to prevent the introduction of pathogens carried by the incoming water. On farms, it is generally necessary to achieve a minimum water exchange of around 10% per day, and the reduction of water exchange could be a useful strategy to improve biosecurity.

The biofloc system is particularly suitable for the breeding of species that are able to derive some nutritional benefit from the direct consumption of bioflocs. Since the concentration of bioflocs in the water is high, this system is also useful for species that can tolerate a high concentration of solids in the water and tolerate low water quality. Almost all biofloc systems are in fact used to grow shrimp, tilapia or carp which have biological characteristics that allow them to consume bioflocs and digest microbial proteins, while channel catfish and hybrid striped sea bass are fish that do not adapt to biofloc systems because do not tolerate water with very high solids concentrations.

Biofloc technology has been applied to the farming of different species of marine and freshwater crustaceans such as *Litopenaeus vanname*, *Penaeus monodon*, *Macrobrachium rosenbergii*, and *Artemia franciscana*. However, experiments with crayfish are still very rare, although crayfish species are able to consume various types of feed including algae and detritus and therefore constitute excellent candidates for the biofloc technology that, moreover, may contain major nutrients such as protein and lipids that could contribute to the growth of the cultured crayfish. *Procambarus clarkii* is able to adapt to the biofloc environment and to feed on biofloc. Juveniles were raised for 60 days in zero exchange water, showing growth performance; digestive enzyme activity, immune response and antioxidant activity higher than the traditional farming system. *Cherax quadricarinatus* successfully bred in a co-culture system based on biofloc technology with Nile Tilapia (*Oreochromis niloticus*) demonstrated positive results on the production performance.

In general, crayfish are raised with extensive and semi-intensive techniques in monoculture in an open pond system, where crayfish obtain the bulk of the food from decaying matter and associated microbes contained in the pond bottom mud. In particular, in the case of *Procambarus clarkii* it is farmed in rotation with vegetable crops, most of the time rice, but also with wheat, rapeseed, and soy. That, according to an integrated model of crayfish and rice cultivation, in which the waste (for example, weeds, insects, leaves and plankton) is used as a food source that can be eaten by the crayfish. While *Procambarus clarkii* digs burrows that increase soil permeability and facilitate material/energy circulation, its feces are used as high-quality fertilizers and, therefore, improve rice growth. This model is mainly practiced in China from ancient times and in the southern states of the United States. The integrated rice system has also become popular in some countries such as Spain, where the invasion of the allochthonous species has been exploited for production purposes thanks to the favorable topography of the territory, although the intense burrowing activity causes severe damage to the soil and deterioration of draining systems.

The other techniques described so far, namely RAS, super-intensive and biofloc still remain largely experimental, despite the numerous efforts made to try to find solutions for the cultivation of freshwater shrimp in tanks and under an intensive regime. Australia, but also other countries, have applied this approach for years, yet the only commercially viable operations appear to be based on the open pond system. There are no confirmed or documented cases of commercial success for crayfish production in an intensive regime tank system.

16.7 Favorable aquaculture attributes

The biological characteristics of the crayfish and their wide diffusion have attracted the attention of farmers over time. Several species have been subject to farming, with not always satisfactory results. At first, the native species are the ones that receive the most attention, especially for the ease with which they can be found in the environment. However, indigenous species do not always adapt to breeding, not presenting that constellation of suitable morphological and functional characteristics. For example, the species *Austropotamobius pallipes*, widespread in most of Europe, and traditionally consumed by local populations, is ill-suited to being farmed for commercial purposes due to the slow growth, limited size of adult individuals, prolonged reproductive cycle, high sensitivity to pathogens, and demand for stringent environmental parameters. The next choice is therefore to resort to allochthonous species, with biological characteristics suitable for farming, such as fast growth rate, easy adaptation to the diet, disease resistance, short reproductive cycle, and tolerance to environmental parameters. Consequently, the allochthonous species that correspond to this description have been transferred to countries other than their regions of origin, often generating ecological emergencies which are difficult if not impossible to solve. The farming of allochthonous species is a phenomenon of difficult control and seems inevitable in an era of globalization, however the ecological consequences must be seriously evaluated, especially when the topography of the territory does not allow the isolation of the area intended for farming.

Below, we propose the detailed description of two of the most relevant biological characteristics for crayfish aquaculture: reproduction, and feeding and nutrition.

16.8 Reproduction

The biological characteristics of crayfish are well suited for aquaculture, since they show an uncomplicated reproductive biology with no free-living larval stages. Sexual maturity is reached significantly earlier in warm water species than in temperate or cold ones. Cold-adapted species can generally complete only one reproductive cycle per year, characterized by a long period (3–6 months) of incubation of the eggs; tropical crayfish, on the other hand, can also reproduce several times a year, with much shorter incubation periods.

The reproductive cycle of the two families (Astacidae and Cambaridae) of the greatest interest to the astaciculture, will be described below. In this chapter, the description of the reproductive cycle will be limited to those practical aspects beneficial for astacicolture.

In Astacids—and in crayfish of temperate waters—seasonal thermal variations significantly influence the maturation of the gametes: an early spring and a hot summer favor a faster maturation and anticipate the mating as well as the hatching of the eggs, delayed by a late and cold spring. Astacids mate in the autumn months (September–December), in response to the thermal drop that follows the end of summer. The male lays its own spermatophores between the 3rd and 5th pair of thoracic limbs where, solidifying on contact with water, form a milky white mass, clearly visible in all fertilized females. The female, after an interval depending on the temperature and earliness of the mating (from 2 days to about a month), turns on the back by folding the abdomen on itself and practically forming a cavity closed on all sides. In the meantime, the integumentary glands produce an abundant mucous secretion which invades the abdominal concavity and incorporates the eggs that gradually emerge from the oviducts; the mucus also mixes with the secretion produced by the sexual glands. This secretion, thanks also to the incessant movement of the pleopods, comes into contact with the spermatophores, dissolving them and freeing the spermatozoa which thus fertilize the eggs inside the mucous mass. The mucus, in contact with water, gradually solidifies, forming thin elastic filaments with which the eggs attach to the pleopods and the carapace of the abdomen. The whole process lasts on average 2–3 hours, at the end the female hides up until the time of hatching, oxygenating and continuously cleaning her eggs.

In *Astacus astacus*—and probably in all Astacids—the females do not lay all the years; the number of sterile females seems to increase in unfavorable environmental conditions. Normally, the percentage of sexually mature females who do not participate in reproduction is low (1–4%). In the crayfish of temperate waters the incubation of the eggs generally coincides with the colder season and, as a rule, is very long: on average about 4–5 months, however, it can continue in certain species—or in particularly unfavorable environmental conditions—up to 6–7 months. Immediately after birth, the crayfish (1st stage larva) remains 2–3 days passively hung for the telson on the egg peduncle through a thin filament formed by the solidified secretions of the telson glands. Then the filament breaks and the larvae actively hang on the mother's pleopods with their hooked-tip claws. For 1–4 days (maximum 10 days) the larva does not feed, consuming the reserves contained in the yolk sack. The carapace of the 1st stage larva is still relatively soft and elastic, allowing the crayfish to grow in length and weight. The 1st phase larva is quite different from the adult, although it resembles it in broad lines (overall the development can be defined as direct, unlike most other crustaceans): cephalothorax disproportionately large and rounded, curved rostrum, hooked claws, telson free of uropods; it weighs around 20 mg and is 8–9 mm long. After about a week the first molt takes place and the crayfish (2nd phase larva) is now quite similar to the adult.

The yolk sack is reabsorbed, and active feeding begins, even if the crayfish never go too far from the mother. Even the female, in this period, starts to feed normally, the feeding being very reduced during the winter months. The 2nd stage larvae are about 12 mm long and weigh 35–40 mg. About 3 weeks after hatching, the second molt takes place, from which the 3rd phase larva comes out more and more similar to the adult. The 3rd phase larvae, now completely autonomous and independent of the mother, are 12–13 mm long and weigh about 50 mg. In European crayfish during the first summer of life the juveniles change 5–6 times, reaching at the beginning of autumn about 16–20 mm in length for a weight ranging between 130 and 230 mg; at this stage they are called 0+ and are ready for sowing. The first summer of life is undoubtedly the most critical period in the life of crayfish, the one in which the high mortality is recorded. In astacicolture, a survival (from the fertilized egg at stage 0+) of 30% is to be considered acceptable with traditional methods, while adopting the modern techniques of artificial incubation and following with particular attention the development of the larvae is certainly possible to exceed 50% of survival. On the other hand, from stage 0+ to sexual maturity (or reaching market size), livestock survival is significantly higher, hypothesized in the order of 60–80% against the 20–40% probably found in nature. In *Astacus astacus* (but the data can roughly be considered valid for all Astacids), during the second summer, after a period of relative winter stasis, the juveniles change 5 more times, usually once a month from May to September. In August of the second summer, there is a growing difference between the males and the females, albeit not very markedly; at this time the crayfish measure between 3 and 4 cm. During the third summer 3 molts are observed, at the beginning of autumn the prawns are 5–6 cm long and weigh 4–6 g. At the fourth summer, only 2 molts take place; the males (already sexually mature) are, on average, 8 cm long and weigh 14 g, the females 7.5 cm on average for 12 g of weight. Finally, from the fifth summer onwards, the males molt 2 times and the breeding females 1 time; older subjects (from 6–7 years of age) generally molt only once a year and not every year.

In Cambarids, once sexual maturity is reached, the alternation of two different forms is observed, defined as form I and form II. The form I correspond to the sexually active crayfish and is maintained throughout the reproductive period; it involves some important morphological changes, particularly evident in the male: the claws lengthen and become stronger, the gonopods harden and, an exclusive peculiarity of this family, small hooked teeth appear at the base of the 3rd and 4th pair of pereiopods, whose function is to help the male retain the female during mating. In the female, the changes are less spectacular and are limited to an enlargement of the claws. In normal conditions, a few weeks after mating (in females about 3 weeks after hatching of the eggs) the crayfish undergo a molting passing to form II, sexually inactive: shorter and thinner claws, less accentuated coloring, absence of hooks and slightly sclerified gonopods in males. The form I will take over again, with a molt, to the next reproductive period.

In species such as *Procambarus clarkii* only form I is frequently observed—once sexual maturity is reached—in particular, if the crayfish are farmed in constantly warm waters. Mating and egg-laying occur in a way not unlike what already described for Astacids. In the annulum ventralis of Cambarid females, the sperm can remain active for more than 8 months, even if the egg-laying generally follows the actual mating for a few days. In *Procambarus clarkii*, after mating, the female takes refuge in holes dug at the bottom, near whose entrance a male usually stays. The maturation cycle of the eggs lasts from 6 weeks to 8 months, the incubation spans from 2–3 weeks to 2–3 months, depending on the temperature. During the incubation the female does not keep the eggs continuously submerged since, in the characteristic habitat of this species, the water (in the reproductive period, that is in the warmer months) is frequently too low in oxygen, so it is preferable that the eggs are often exposed to the air where, thanks to the incessant movement of the pleopods, they find sufficient oxygen for the development. The first larval period takes place as already seen for Astacids. The larvae can stay in the hole together with the mother for 6–12 weeks; the female leaves the den with the rains, when the water level begins to rise. After 2 molts, reached a length of about 1 cm, the juveniles detaches themselves more and more often from the mother, although they continue for a few days to take refuge under its abdomen in case of danger. The growth is very rapid, at temperatures between 20 and 30°C the juveniles can molt every 5–10 days: in practice, from abandoning the mother to sexual maturity, at least 9 molts are needed, for a total of 11 molts of which 2 while still attached to the mother. In nature sexual maturity is reached, in *Procambarus clarkii*, in 3–5 months, with sizes oscillating—according to environmental conditions—between 5.5 and 12.5 cm.

Freshwater crayfish farms are experiencing serious difficulties related to reproductive processes and the poor survival of youthful forms. Increasing reproductive efficiency is an essential aspect in any animal farming program, in light of the relationship between the reproductive cycle and the production cycle. For this reason, the development of new techniques such as artificial incubation of eggs and their maintenance and transport has received a lot of attention in recent year and is regarded in a following section of this chapter. In fact, the particular reproductive biology of the freshwater crayfish does not allow the eggs to be removed from the mother after they are laid. They remain attached to the female pleopods for long periods, varying from species to species. The larvae detach themselves definitively from the mother after having undergone some molts. For this reason, it is necessary to consider an area dedicated to the housing of berried females, which must be isolated from other individuals, especially males. This aspect of reproductive biology is in some ways an advantage for the farmer, since it relieves him of the heavy task of feeding the larvae just after hatching. This has always been a problem in aquaculture since juvenile forms have a different diet from adult forms which are often difficult to identify.

16.9 Feeding and nutrition

An adequate knowledge of the feeding strategy and nutritional requirements of freshwater crayfish is fundamental for the development of a correct farming. Undoubtedly, the increase in crayfish production can be achieved through the proper diet formulation, and this requires the sound knowledge in two related areas: feeding and nutrition. In fact, the formulation of effective diets depends on our level of knowledge of the feeding strategy and of the biochemistry and nutritional physiology of the cultivated species. Freshwater crayfish play an ecological role of particular relevance in freshwater ecosystems, thanks to their detritivorous and omnivorous feeding strategy. Their feeding strategy allows for the incorporation of a broad range of animal and plant-based ingredients into formulations of practical diets. In many cases, plants are considered the main form of feeding, while food of animal origin would be limited to a negligible quota [*3*]. More recently, however, studies on the feeding strategy of crayfish questioned this historic assumption and suggest that *Cherax quadricarinatus* prefers animal proteins and when they are not available it may become an optional herbivore.

Our knowledge of nutrient requirement is not as advanced as it is for fish, and currently employed diets are based on formulations of other aquatic species, mainly marine shrimps and freshwater fishes. However, it is clear that their nutritional requirements differ in several important ways from those of farmed fish. The numerous studies conducted on different species indicate that, in general, the crayfish diet should consist of 25–30% of proteins of which 15–20% of animal origin. Fats should be around 6%. The remaining percentage should be made up of carbohydrates. Differences are possible according to the species and the life stage. Here we propose an update on the nutritional requirement in terms of proteins, lipids and carbohydrates of the main farmed crayfish species (Table 1).

The assimilation of nutrients depends on their complex digestive machinery. Their plasticity to obtain nutrients from a broad range of food sources derives largely from physiological processes occurring in the digestive tract, including the profile and activity of the digestive enzymes that are present.

Since crayfish have the ability to modify their digestive enzyme secretion in response to different ingredients in the diet over time, we can therefore predict a positive relationship between the levels (including quantity and structure) of carbohydrates, proteins and lipids in the natural diet and the presence or quantity of intestinal enzymes required for digestion. In crustaceans, the intestinal gland or hepatopancreas synthesizes digestive enzymes, which are subsequently released into the gastric chamber forming the gastric juice. Furthermore, it is well known that the digestive enzyme profile of decapod crustaceans varies according to their particular feeding strategies. The properties of the associated enzymes will determine the digestive abilities of organisms. High levels of proteolytic enzymes such as trypsin and chymotrypsin are produced in the presence of a diet mainly composed of proteins. When the diet is mixed, i.e., based on proteins of animal and vegetable origin, high levels of proteinase and carbohydrase are produced, while in the presence of an entirely vegetable diet, there are high quantities of cellulases and hemicellulases, necessary to digest the carbohydrates from plant cell walls.

Feeding trials are conducted to determine digestive enzyme activity in response to diet composition, an important strategy to verify the impact of the formulated diet on the ability to digest nutrients. Knowledge of digestive enzymes is essential to determine digestive capacity, and information regarding the type of enzymes and their activity can guide us in selecting the ingredients to include in a diet. In fact, changes in digestive enzyme activity indicate physiological responses to different diet composition. A wide range of digestive enzymes including protease, lipase, carbohydrates, cellulase is found in the hepatopancreas, gastric fluid and intestine of crayfish [20]. Equally important is the evaluation of enzymatic activity during

Table 1. Suggested amounts of proteins, lipids an carbohydrate in the diet based on best growth performance.

Species	Life stage	Duration (days)	Manufacturing	Feeding rate (%body weight)	Optimal percentages			Gross energy	Rearing system	References
					protein	lipid	Carbohydrates			
Procambarus clarkii	Juvenile	56	RT extrusion	Fed to satiation	30	-	-	2.5–3.5 (k cal/g)	RAS, individually caged	[4]
		165	Cooked-extrusion	4	22–26	6	36–41	12.5–12.6 (kJ/g)	RAS, individually caged	[5]
		56	RT extrusion	5	27	4–7	-		Floating cages in the pond	[6]
Astacus astacus	Juvenile	259	RT extrusion	Fed to satiation	31–40	5.5–10	9–26	251–321 (kcal/100 g)	RAS, individually caged	[7]
		56		4	> 30	< 13	-		RAS, group caged	[8]
Astacus leptodactylus	Adult	56	RT extrusion	3–4	30	-	-	370 (kcal/100 gr food)	RAS, group caged	[9]
	Juvenile	56	RT extrusion	Fed to satiation	-	13	-	380 (kcal/100 gr food)	RAS, group caged	[10]
Pacifastacus leniusculus	Juvenile	180	Commercial pellets	4	55.5–40.3	8.6–14	14.5–33.1	20.5–18.8 (kJ g^{-1})	RAS, intensive rearing	[11]
		100	Cooked-extrusion	3	45–50	11.5–11.9	-	19.3–19.6 (kJ g^{-1})	RAS, individually and group caged	[12]

Cherax destructor	Juvenile	140	RT extrusion	-	30	4.7	-	19 (kJ g⁻¹)	RAS, group caged	[13]
Cherax quadricarinatus	Juvenile	60	RT extrusion	Fed to satiation	31	-	-	18.9 (kJ g⁻¹)	RAS, group caged	[14]
	Juvenile	84	RT extrusion	10 at the beginning and 2.5 the end	30	4.2	-	17.6 (kJ g⁻¹)	Flow-through, semi-intensive	[15]
	Pre-adult	70	RT extrusion	Fed to satiation	27	-	-	21.01 ± 0.05 (kJ g⁻¹)	RAS, group caged	[16]
	Juvenile	117	RT extrusion	10 at the beginning and 4 the end	22	-	-		Pond	[17]
	Juvenile	60	RT extrusion	Fed to satiation	31	8	-	18.2 (kJ g⁻¹)	RAS, group caged	[18]
	Sub-adult	84	RT extrusion	3	25	6.5	-	11.6 (kJ g⁻¹)	RAS, group caged	[19]

ontogenesis, capable of providing useful information on whether different diet formulations are needed for juveniles and adults. Changes in the activity of digestive enzymes during ontogenesis and in response to changes in the diet of the main farmed crayfish species are shown in Table 2.

16.10   Management practices in freshwater crayfish aquaculture

Crayfish are primarily cultured extensively in farm dams, or in semi-intensive, purpose-built, earthen ponds. Crayfish hatcheries are virtually non-existent unless for repopulation or reintroduction purposes. However, attempts have been made by researcher to facilitate and improve the reproduction and increase the number of hatching eggs/female. One possible strategy is the artificial incubation of eggs, that has reported interesting results over the years, but is not employed in a production model of modern aquaculture system for crayfish.

16.10.1 Artificial incubation

Crayfish farming is advantaged by its uncomplicated reproductive biology, with no free-living larval stages. Even in countries where farming is successfully practiced, there is no hatchery production. Juvenile generation is based on natural reproduction and individuals are reared directly in the juvenile ponds. However, this simple strategy to generate juveniles for growth purposes is inefficient and perceived by breeders as an obstacle to production. Indeed, the development of optimal artificial reproduction techniques is crucial for improving reproductive performances and juvenile production. The possibility of adopting artificial reproduction techniques in crayfish farms would lead to numerous advantages such as the possibility of manipulating the speed of embryonic development through temperature, reduction of food and space costs, reduced dependence on females and loss of eggs. Good results were obtained using the *Austropotamobius pallipes*. Eggs can be detached from females in the early stages of embryonic development, and after 34 days of laying the survival rate of stage 2 juveniles was over 50%. A critical aspect of the artificial incubation techniques is the transmission of pathogens. Fungal infections of incubated crayfish eggs due to *Saprolegnia* sp. and other oomycetes can cause 100% mortality. In this case, the addition of biocides and the immediate removal of dead eggs, during the incubation of the eggs, becomes a mandatory step to stop the growth of fungi. The use of experimental treatments such as UV lighting and exposure to formaldehyde of Astacus astacus eggs did not lead to satisfactory results. In fact, an inability of the juveniles to moult was recorded, and limb deformities occurred in the groups treated with UV, as well as a high juvenile mortality. On the other hand, the treatment of eggs by repeated saline solution bathing seems more effective.

16.10.2 Diets

There are still many gaps regarding the knowledge of the nutritional needs of freshwater crayfish, especially in conditions of intensive farming, in which the feeding depends entirely on man. One of the reasons why the intensive farming of freshwater crayfish is not yet practiced on a large scale, lies precisely in the lack of

Table 2. Ontogenetic changes of the digestive enzymes.

Species	Life stage	Anatomical site	Enzyme	Initial activity	Activity behavior	References
Procambarus clarkii	Juvenile (0–65 days after hatching)	Whole thorax	Trypsin U g^{-1}	I = 0 F = approx 2500*	Increase from day 0 to day 40, then a decrease up to day 65	[21]
			Lipase U g^{-1}	I = 0 F = approx 4.2	Increase from day 0 to day 40, then a decrease up to day 65	
			Amylase U g^{-1}	I = 0 F = approx 3500	Increase from day 0 to day 40, then a decrease up to day 65	
	Eggs and embryos (Stage I to stage VI)	Whole egg and embryo	Pepsin U mg protein^{-1}	I = 1.85 ± 0.16 F = 0.52 ± 0.07	Decrease from stage I to VI	[22]
			Trypsin U mg protein^{-1}	I = 0.22 ± 0.02 F = 0.59 ± 0.05	Increase at stage IV and VI	
			Lipase U mg protein^{-1}	I = approx 0.016 F = 0.002 ± 0.000	Decrease from stage II to VI	
			Amylase U mg protein^{-1}	I = approx 0.02 F = approx 0.01	Very low, limited increase at stage III	
	Juveniles (1–31 days after hatching)	Whole body	Pepsin U mg protein^{-1}	I = approx 20 F = approx 40	Increase from day 1 to day 31	[23]
			Trypsin U mg protein^{-1}	I = 0 F = approx 1.000	Increase from day 5 to day 31	
			Lipase U mg protein^{-1}	I = approx 1 F = approx 11	Increase from day 5 to day 31	
			Amylase U mg protein^{-1}	I = approx 0.75 F = approx 0.5	Decrease from day 5 to day 31	
Astacus Leptodactylus	Phase III (blastula), Phase X (embryo), Phase XIV (embryo with hepatopancreas), Stage I (post-embryonic no feeding), Stage II (prior to the onset of feeding).	Whole embryo	Protease U mg protein^{-1}	I = 1.0376 ± 0.1083 F = 2.3214 ± 0.0556	Increase from phase III to stage I (stage III not detected)	[24]
			Amylase U mg protein^{-1}	I = approx. 0.0001 F = 0.0529 ± 0.015	Increase from phase XIV to stage III	
			Lipase U mg protein^{-1}	I = approx. 0.0005 F = 0.0059 ± 0.0019	Increase from phase XIV to stage III	

Table 2 contd. ...

...Table 2 contd.

Species	Life stage	Anatomical site	Enzyme	Initial activity	Activity behavior	References
Cherax quadricarinatus	Juveniles (From 2 to 14 cm total length)	Mid intestine	Protease U mg protein	I = approx 0.4 F = 0	Decrease from cm 2 to cm 14	[25]
			Trypsin U mg protein	I = approx 0.75 F = approx 0.15	Decrease from cm 2 to cm 4, then constant up to cm 14	
			Carbossipeptidase A U mg protein	I = approx 3.9 F = 0	Decrease from cm 2 to cm 14	
			Carbossipeptidase B U mg protein	I = approx. 1.2 F = approx 0.8	Decrease from cm 2 to cm 14	
			Leucine aminopeptidase U mg protein	I = approx 3.9 F = approx 1.1	Decrease from cm 2 to cm 14	
			Amilase U mg protein	I = approx 1.4 F = approx. 2.5	Decrease from cm 2 to cm 10, then increase	
			Cellulose U/mg protein	I = approx 2.0 F = approx 1.9	Decrease from cm 2 to cm 10, then increase	
			Maltase U/mg protein	I = approx 0.25 F = approx 1.0	Constant from cm 2 to cm 10, then increase	
			Laminarase U/mg protein	I = approx 0.004 F = approx 0.014	Constant from cm 2 to cm 10, then increase	
			Invertase U/mg protein	Not detected		
	1, fertilized egg stage; 2, cleavage and blastula stage; 3, gastrula stage; 4, egg nauplius stage; 5, embryo with well-formed pigments stage; 6, prepare-hatching stage		Trypsin	I = 0 F = approx. 12.5	Increase from stage 1 to stage 5, then decrease	[26]
			Chimotripsin U/mg protein	I = approx. 0.5 F = (2.52 ± 1.82)	Increase from stage 1 to stage 4, then decrease	

* the exact value was not reported. Approximate value based on the extrapolation on the figure.

availability of an artificial food capable of covering all the nutritional requirements of freshwater crayfish in the various phases of the life cycle.

In the lack of a standardized artificial diet for farmed crayfish, a wide variety of ingredients, including vegetables, potatoes, carrots, fish, cereals, meat, zooplankton, and pelleted food for aquatic species are used. The administration of the artificial food, in the early stages of development of the crayfish, is difficult, especially for the ephemeral weight of the larvae, combined with the fact that they are lazy eaters, and need some time to feed. To overcome this drawback, the feed in the form of pellets must be dosed and placed on trays present on the bottom of the tank. This useful and indispensable action, in addition to optimizing the quantity of artificial feed, will help avoid unnecessary harmful waste and undesired eutrophication and facilitation of the onset of any pathologies of bacterial origin. Moreover, the pellets must resist in the water for some time without crumbling down and releasing the nutrients, with consequent impoverishment of the diet. A possible solution is to add binders to the pellets. This subject will be treated in a following section. Equally interesting is the use of live nauplii of *Artemia salina* and/or defrosted adults, which have shown a high palatability especially in the first phase of larval weaning.

16.10.3 Feed ingredient substitution

The use of particular nutrients capable of accelerating the growth and achievement of the commercial size has always been desirable for any animal production activity. Feed is the main item of expenditure in aquaculture activities, therefore most of the effort made in the aquaculture sector have as their object the search for the optimal composition of the diet, using inexpensive proteins, lipids and carbohydrates, possibly deriving from sustainable sources.

The issue of the last decades on which the efforts of the scientific world and feed companies are focusing, namely the replacement of fish meal and fish oil with plant meal and plant oil in fish farming, especially the carnivores that require high quantities of fish derived nutrients, is only marginal in the case of astaciculture, both for the quantity of crayfish produced and for their adaptability to diet based on vegetable ingredients, with good growth performance. When considering alternative ingredients, however, it must be borne in mind that the formulation of well-balanced diets is very important for successful aquaculture and that the nutritional value of the formulated feed depends on the digestibility of the individual components. In Table 3 are reported the most employed feed ingredients in freshwater crayfish artificial diet.

16.10.4 Functional feeds

As it has been pointed out, astaciculture is an important but not predominant production activity in the aquaculture sector. However, the development of this activity is destined to meet the same challenges of the aquaculture sector, which are invariably linked to environmental sustainability, animal welfare and profitability.

Diseases are often identified as the main threat to the aquaculture activity. Pathogenic bacteria, viruses and parasites can cause huge losses causing slow growth, product downgrading or massive death. The use of antibiotic, vaccines and other chemical drugs to manage those pathogens is not always desirable nor

Table 3. Most employed feed ingredients in the freshwater crayfish artificial diet.

Species	Life stage	Dietary ingredients	Rearing system	Manufacturing methods	References
Astacus leptodactylus	Juvenile 10.13 ± 0.38 g	Casein, Gelatin, Dextrin, Starch, Sucrose, Soybean oil, Kilka fish oil, Cholesterol, Lecithin, Cellulose	Tanks (individually held)	Single extrusion. Pellets dried at 70°C	[10]
	Adult (15–20 gr)	Casein, Gelatin, Dextrin, Wheat meal, Fish meal, Corn meal, Soybean meal, Cod oil, Cellulose, Crayfish Meal	Tanks	Meat grinder. Pellets dried at 70°C	[9]
	Sub-adult (13.35 ± 1.47 g)	Anchovy fish meal, Sardine fish meal, Blood meal-poultry, Feather meal, Poultry by-product meal, Soy protein concentrate, Corn gluten meal, Wheat gluten meal, Full fat canola meal, Full fat soybean meal, Canola meal, Fermented canola meal, Soybean meal, Fermented soybean meal, Spirulina meal , Corn flour, Wheat flour, Wheat, whole	Tanks with a semi-recirculating system	Feedstuffs ground to a particle size of <250 μm. Hand pelletizer. Pellets dried at 30°C	[27]
Cherax quadricarinatus	Juvenile (about 1 g)	Sardine meal, Sorghum meal, Soy bean meal, Red crabmeal, Squidmeal, Wheatmeal, Grenetine 40 (binder), Fish oil, Soya lecithin, Mineral premix, Vitamin premix, Ascorbic acid, Choline chloride, Calcium carbonate	Static water experimental system	Feedstuffs ground and sieved through a 500 μm mesh. Pelleted by a meat grinder. Pellets dried at 40°C	[14]
	Juvenile (0.5 ± 0.1 g)	Soya bean meal, Peruvian fishmeal, Rice bran, Wheat flour, Vitamin mix, Fish oil, Salt, Binder (alginic acid)	Indoor aquaria, closed system	Feedstuffs were grounded and sieved through a 100 μm sieve, pelleted by a meat grinder. Pellets dried at 60°C	[28]
	Juvenile (4.08 ± 0.2 g)	Fish meal, Soybean meal, Fish oil, Corn oil, Wheat starch, Whole wheat meal, Mineral pre-mix, Vitamin mix, Rovimix Stay C-35, di-calcium phosphate, Soy-lecithin, Cholesterolj, Cellufil	outdoor flow-through culture system consisting of circular fiberglass tanks	Feedstuffs pelleted by a grinder. Pellets dried at 40°C	[15]
	Juvenile (4.6 ± 2.2 g)	Menhaden fish meal, Soybean meal, BGY, Wheat flour, Menhaden oil, Corn oil, Vitamin mix, Mineral mix, Wheat gluten	pond	Pellets manufactured by a commercial feed mill	[17]
	Juveniles (3.62 ± 1.35 g)	Sardine meal, Sardine meal, Red crab meal 4, Squid meal, Wheat meal, Sorghum meal, Soy paste meal, Grenetine (binder), Fish oil, Soy lecithin	Aquaria	Feedstuffs ground to a particle size <250 μm. Pelleted by a meat mill at temperatures under 90°C	[29]
	Adult (29.6 ± 1.44 g)	Fishmeal, Wheat starch, cellulose, Fuller's earth, Gelatin, Cod liver oil, CaHPO4, Minerals, Vitamins.	Aquaria with ricirculating water system	Feedstuffs pelleted by a mechanical mincer. Pellets dried at 60°C	[30]

Species	Stage	Diet ingredients	System	Processing	Ref.
	Juveniles (1.54 ± 0.02 g)	Fish meal, Soybean meal, Dextrin, Fish oil, Rapeseed meal, Cottonseed meal, Wheat bran, Wheat middlings	Tanks with a recirculating water stem	Feedstuffs pelleted by a meat grinder. Pellets dried at 50°C	[31]
Pacifastacus leniusculus	Juvenile (30.1 ± 0.3 mg)	Fish meal, Corn meal, Cod liver oil, Soy lecithin, Cholesterol, L-ascorbyl-2-monophosphate–Na, Choline chloride, Dicalcium phosphate, Decapsulated Artemia cysts, Carboxymethyl cellulose, Astaxanthin, Mineral premix, Vitamin premix	Tanks with a flow through system	Extruded between 75°C and 90°C	[12]
	Stage 2 juvenile (30.3 ± 0.2 mg)	Fish meal, Corn meal, Poultry by-product meal, Cod liver oil, Soy lecithine, Cholesterol, L-ascorbyl-2-monophosphate–Na, Choline chloride, Dicalcium phosphate, Decapsulated Artemia cysts, Carboxymethylcellulose, Astaxanthin, Mineral and vitamin premix	Tanks with a flow through system	Extruded between 95°C and 115°C	[32]
	Stage 2 juvenile (28.1 ± 0.4 mg)	Fish meal, Corn meal, Pea protein concentrate, Cod liver oil, Soy lecithine, Cholesterol, L-Ascorbyl-2-monophosphate–Na, Choline chloride, Dicalcium phosphate, Decapsulated Artemia cysts, Carboxymethylcellulose, Astaxanthina, Mineral premix, Vitamin premix	Tanks with a flow through system	Extruded between 75°C and 100°C	[33]
	Stage 2 juvenile (27.2 ± 0.2 mg)	Fish meal, Corn meal, Feather meal, Cod liver oil, Soy lecithin, Cholesterol, L-ascorbyl-2-monophosphate–Na, Choline chloride, Dicalcium phosphate, Decapsulated Artemia cysts, Carboxymethylcellulose, Astaxanthin, Mineral premix, Vitamin premix	Tanks with a flow through system	Extruded between 75°C and 90°C	[34]

Table 3 contd. ....

...Table 3 contd.

Species	Life stage	Dietary ingredients	Rearing system	Manufacturing methods	References
Procambarus clarkii	Juveniles (2.52 ± 0.04 g)	Fish meal, Soybean meal, Rapeseed meal, Cottonseed meal, Wheat middlings, Soybean oil, Monocalcium phosphate, Salt, Premix, Starch, Shrimp shell meal, Ecdysone premix, CMC	Floating cages anchored in an outdoor concrete pond with a flow-through system.	Feedstuffs ground into a fine powder. Pelleted by a laboratory pellet machine . Dried at room temperature	[35]
	Adult	Fish meal, Sodium Microcrystalline Cellulose, Squid extract, Betaine, Squid meal, Stone powder, Biofeed	Indoor tanks		[36]
	Juvenile (5.43 ± 0.21 g)	Fish meal, Soybean meal, Rapeseed meal, Cottonseed meal, Flour, Soybean oil, Fish oil, Attapulgite, Salt, Ecdysone premix, Shrimp shell meal, Calcium biphosphate, Carboxyl-methy cellulos, Premix	Tanks	Feedstuffs ground and sieved through 60 µm mesh. Pelleted by a small feed machine.	[37]
	Juvenile (0.39 ± 0.00 g)	White fishmeal, Fish oil, Corn starch, Cellulose, Mineral & vitamin premix, Choline chloride, CaH2PO4, Cholesterol, Lecithin, Swine liver power, Sodium alginate	Tanks	Extrusion machine	[38]
	Juvenile (5.39 ± 0.12 g)	Fish meal, Soybean meal, Phytase (IU/kg) 1000, Extruded soybean meal, Fermented soybean meal, Rapeseed meal, Cottonseed meal, Soybean oil, Fish oil, Attapulgite, Salt, Ecdysone premix, Shrimp shell meal, Calcium biphosphate, Carboxymethyl cellulose, Premix	Indoor tanks with a flow-through system.	Feedstuffs ground and sieved through 60 µm mesh. Pelleted by a small feed machine. Dried at 40°C	[39]
	Juvenile (about 5 g)	Fish meal, soybean meal, rapeseed meal, Corn starch, Soya oil, Calcium biphosphate, Choline chloride, Cholesterol, Astaxanthin Premix, Premix for crayfish, L-Lysine hydrochloride, DL-Methionine, Vitamin C, Sodium chloride, Carboxymethyl Cellulose, Microcrystalline cellulose	Indoor culture system	Feedstuffs ground and sieved through 60 µm mesh. Pelleted by a small feed machine.	[40]

possible. Aquaculture feed additives are useful tools to help aquatic animal face those pathogens. Several studies have monitored the effects of orally administered dietary supplements of various nature (prebiotics, probiotics and synbiotics) on crayfish species, and they found that treated crayfish showed improved growth, as well as antioxidant and immunological parameters (Table 4).

16.10.5 Feed manufacturing

The production of well-balanced feed at low cost is an essential prerequisite for profitable aquaculture production. The small-scale production sector, such as that of crayfish, is penalized by the lack of formulated feed suitable for the farmed species. The feed is often produced locally, with variable ingredients at a low technological level. The development of public-private partnerships with groups or associations of farmers could provide access to ad hoc financing and improve production capacity. In extensive and semi-intensive production systems, it is necessary to establish the relationship between the productivity of the natural pond and the impact of supplementary feed. This is crucial for the optimization of formulated feeds and reduces feed costs. The diet manufacturing of aquatic animals must take into account the feeding behavior of the species. In fact, in accordance with the biological characteristics of the species, the food must float, sink with variable speed, absorb water, do not disperse too many nutrients and at the same time allow the release of attractants. In the case of crayfish, the feed requires a certain level of treatment to ensure good stability in the water, for a period long enough to allow the animals to consume it. In order to lengthen the duration of the feed in water, the addition of the so-called binders is considered. Binders are generally biopolymers of different nature, the most used are proteins and carbohydrates. Carbohydrates are particularly useful as binders for the manufacturing of crayfish pellets since they derive from sustainable sources, are biodegradable, create three-dimensional networks or hydrogels that trap nutrients, are part of their diet and can also be used and a source of energy [58,59]. Alginate, agar, chitosan and pectin caused good growth performance in *Cherax albidus* and *Cherax destructor* with respect to control animals fed a natural diet [59–64]. The gelatinization of starch is also often used to improve the stability of the pellets in water, but this is achieved through a heating process at high temperatures that alter the quality of the nutrients present in the pellets.

Often, the production of feed for small-scale crayfish farms takes place on site and through the use of simple equipment. The ingredients are ground and mixed by hand or by extrusion at room temperature (cold extrusion). The grinding of the ingredients to be used in feed is a crucial operation, since the pellets made with finely ground ingredients will be more durable and water-resistant. A simple and widespread method is cold extrusion, which has the advantage of preserving the nutritional properties of the diet compared to high temperature extrusion. In fact, high temperature extrusion can modify the nutritional properties of nutrients, depleting them, even if it is often used to gelatinize starch as mentioned above, in order to give further strength to the pellets. In cold extrusion, binders normally used with a final percentage between 1 and 10% are used.

Table 4. *In vivo* studies of several dietary supplements on growth and immune response of crayfish. The feed additives were orally administrated.

Species	Life stage	Feed additives	Length of treatment (days)	Growth promoter	Immune stimulant	Antibacterial	References
Cherax destructor	Adult (35.14 ± 0.48 g)	Mannanoligosaccharide (MOS)	56	+	+		[41]
Cherax tenuimamus	Juvenile (0.47 ± 0.02 g)	β-1,3-glucan	84	No effect	+		[42]
Cherax tenuimamus	Juvenile (4.44 ± 0.20 g)	MOS	112	+	+		[43]
Astacus leptodactylus	Juvenile (10.56 ± 0.32 g)	MOS and Fructooligosaccharide (FOS) (alone and in combination)	126	+	+	*Aeromonas hydrophila*	[20]
Astacus leptodactylus	Juvenile (8.25 ± 0.39 g)	Nucleotides (Adenosine-5-monophosphate (AMP) Cytidine-5-monophosphate (CMP) Guanidine-5-monophosphate (GMP) Inosine-5-monophosphate (IMP) Uridine-5-monophosphate (UMP)).	126	+	+		[44]
Astacus leptodactylus	Juvenile (7.58 ± 0.39 g)	L-carnitine	126	+	+		[45]
Astacus leptodactylus	Adult (26.85 ± 5.0 g)	Polyphenols from olive mill waste water	168	+	Not evaluated		[46]
Astacus leptodactylus	Juvenile (5.62 ± 0.39 g)	Onion powder	126	+	+		[47]
Astacus leptodactylus	Sub-adult (11.45 ± 1.87 g)	Prebiotics (MOS and Xylooligosaccharide (XOS)), probiotics (*Enterococcus faecalis* and *Pediococcus acidilactici*), and synbiotics (MOS+*Enterococcus faecalis*, XOS+*Enterococcus faecalis*, MOS+*Pediococcus acidilactici*, XOS *Pediococcus acidilactici*)	126	+	+	*Aeromonas hydrophila*	[48]

Species	Stage (weight)	Treatment				Pathogen	Ref.
Astacus leptodactylus	Juvenile (6.18 ± 0.31 g)	Prebiotics (galactooligosaccharide) (GOS), probiotics (*Enterococcus faecalis* and *Pediococcus acidilactici*), and synbiotics (GOS +*Enterococcus faecalis* and GOS +*Pediococcus acidilactici*)	126	+	+	*Aeromonas hydrophila*	[49]
Astacus leptodactylus	Juvenile (4.13 ± 0.12 g)	MOS+ *Enterococcus faecalis* GOS+ *Enterococcus faecalis*	126	+	+	*Aeromonas hydrophila*	[50]
Astacus leptodactylus	Adult (27.88 ± 0.27 g)	GOS	97	No effect	+		[51]
Astacus leptodactylus	Adult (27.88 ± 0.27 g)	Probiotics (*Lactobacillus plantarum*)	97	No effect	+		[52]
Procambarus clarkii	Juveniles (7.15 ± 0.21 g)	Astaxanthin	56	+	+	*Aeromonas hydrophilia*	[53]
Procambarus clarkii	Sub-adult (10.31 ± 1.9 g)	Polysaccharides from *Rhodiola rosea* roots	56	+	+		[54]
Procambarus clarkii	Juvenile (5.80 ± 0.1 g)	*Codonopsis pilosula* polysaccharide	56	+	+		[55]
Procambarus clarkii	Juvenile (5.80 ± 0.10 g)	Hesperidin	56	+	+	White Spot Syndrome Virus (WSSV)	[56]
Astacus leptodactylus	Juvenile (4.38 ± 0.08 g)	Organic salts (Na-acetate, Na-butyrate, Na-lactate and Na-propionate)	63	+	Not evaluated		[57]

16.11 Welfare in crustaceans

16.11.1 Pain in crustaceans

When it comes to defining pain for crustaceans and other invertebrates there are usually two thinking streams: one is based on the presence of neurophysiological basis that scientists define as a necessary substrate to feel the pain, the other is based instead on the avoidance behavior that animals show when facing nociceptive stimulus and its memory.

Up today, there is no proof of pain described in appropriate ways in crustaceans. The idea of pain in these animals was traditionally rejected because they were thought to respond to noxious stimuli purely by reflex [65]. Clarify this issue become then a priority. A reflex is usually defined as a short-term reaction to a stimulus without integrating information about motivational requirements. What we can define as no simple reflexes are behavioral responses that result influenced by different feelings coming from central processing that swift avoidance learning, and long-term behavioral changes essential to enhance future avoidance of tissue damage. Pain in crustaceans should be defined in ways that do not depend on human pain experience. So far, collected data regarding crustaceans are broadly consistent with criteria for pain as described above.

16.11.2 Welfare legislation

Even though all farmed crustaceans belong to the order of Decapoda, the current European legislation does not include any minimal animal care requirement for them. Nevertheless, it has been recently reported [66] that from 2005 the Animal Health and Animal Welfare Panel of the European Food Safety Authority (EFSA, 2005) recommend their inclusion in the legislation that provides protection for animals: "all decapods should receive protection". However, despite the recommendations, the decapods were excluded from the subsequent 2010 EU Directive. Nevertheless, there are considerable evidence of a change in views regarding sentience in decapods.

In Table 5 we report the documents produced by several countries and international organizations regarding decapods' welfare.

More recently Birch et al. [67] tried to develop a framework for analyzing the evidence for sentience, the capacity to experience pain, distress and/or harm, in cephalopod mollusks, and decapod crustaceans including crabs, crayfish, lobsters, prawns, and shrimps. They developed 8 criteria to assess the evidence from 300 scientific publications in order to investigate the welfare and current commercial implications.

The eight criteria are as follows:

(1) possession of nociceptors;
(2) possession of integrative brain regions;
(3) connections between nociceptors and integrative brain regions;

Table 5. List of legislation and codes of practice relevant to Crustaceans used or proposed by Organizations in different countries inside and outside Europe.

Country	Legislation	Statement	Year
Norway	Norwegian Animal Welfare Act	The Act applies to conditions which affect welfare of or respect for mammals, birds, reptiles, amphibians, fish, decapods, squid, octopi and honey bees. The Act applies equally to the development stages of the animals referred to in cases where the sensory apparatus is equivalent to the developmental level in living animals.	2010
Swiss	Swiss Animal Welfare Act	"Live crustaceans, including the lobster, may no longer be transported on ice or in ice water. Aquatic species must always be kept in their natural environment. Crustaceans must now be stunned before they are killed".	2018
New Zealand	New Zealand Animal Welfare Act	- You must not kill any farmed or commercially caught crab, rock lobster, crayfish, or kōura (freshwater crayfish) for commercial purposes unless it is insensible first (for example stunned or chilled). - This doesn't apply if you catch it in the wild and kill it immediately (for example, by splitting crayfish). - If you don't comply, you could face a criminal conviction and fine of up to $5,000 for an individual, or $25,000 for the business."	2018

Organizations	Document	Statement	Year
UK – British Veterinary Association	Welfare of Animals at Slaughter. Available online	"Evidence indicates that decapods (e.g., lobsters, crabs) and cephalopods (e.g., octopus, squid) are sentient, and experience pain and distress. We therefore support the principle that commercially caught decapods and cephalopods should be stunned before slaughter".	2020
UK – Royal Society for the Protection from Cruelty to Animals	Animal Sentience	"There is currently debate about whether species like decapod crustaceans (crabs, lobsters etc.) and cephalopods (octopus, squid etc.) are sentient. The RSPCA and many others believe that there is sufficient scientific evidence to indicate that these animals should be considered as sentient, and therefore protected appropriately by legislation. This would help ensure they are no longer subjected to some of the current practices, like boiling crabs and lobster alive, that cause serious pain and distress".	2020

(4) responses affected by potential local anaesthetics or analgesics:

    (a) the animal possesses an endogenous neurotransmitter system that modulates (in a way consistent with the experience of pain, distress or harm) its responses to threatened or actual noxious stimuli.

    (b) putative local anaesthetics, analgesic (such as opioids), anxiolytics or anti-depressants modify an animal's responses to threatened or actual noxious stimuli in a way consistent with the hypothesis that these compounds attenuate the experience of pain, distress or harm.

(5) motivational trade-offs that show a balancing of threat against opportunity for reward; [SEP]

(6) flexible self-protective behavior in response to injury and threat; [SEP]

(7) associative learning that goes beyond habituation and sensitization; [SEP]

(8) behavior that shows the animal values local anaesthetics or analgesics when injured [SEP] [67].

The approach is to assess the evidence in terms of a confidence level per criterion for each species, ranging from no confidence to very high confidence. They suggest that very strong evidence of sentience should be assumed if the animal in question satisfies at least seven of the eight criteria, whereas a high confidence level for five or more criteria would be classified as strong evidence, and a high confidence level for three or more criteria.

For the decapods, the authors found strong evidence in Brachyura (true crabs), with high or very high confidence that the crabs satisfy criteria 1, 2, 4, 6 and 7. They also found evidence in Anomura (crabs), Astacida, lobsters and crayfish, and in Caridea (shrimps). These findings have been interpreted on how much scientific research has been conducted on the various species and taxa and that absence of evidence is not evidence of absence. On the basis of this evaluation, the authors recommended that all decapod crustaceans should be regarded as sentient animals for the purposes of UK animal welfare law.

They provide recommendations regarding commercial practices. In particular, they recommend avoiding declawing, nicking, eyestalk ablation and the sale of live decapod crustaceans to untrained, non-expert handlers, and they include suggestions for best practices for transport, stunning and slaughter.

This is certainly a useful guide for decision makers, however it does not collect all the references produced in recent years aimed at demonstrating that cephalopod mollusks and crustaceans are sentient animals. In particular, not all information relating to the use of true anesthetics is considered. These guidelines do not exhaust the topic, other and more targeted experiments are needed that take into account all the research already conducted and that effectively identify the areas of the nervous system responsible for the response to pain.

16.12 Future perspectives

It is clear that we need care and maintenance guidelines on decapod for an ethical farming as well as for scientific research, similar to those developed for vertebrates and cephalopods.

These desirable guidelines should contain species-specific requirements that can be summarized as follows:

- supply, capture and transportation;
- environmental characteristics and the design of facilities (e.g., water quality control (salinity, temperature, O2, CO2, pH), lighting requirements (wavelengths and intensity of lighting));
- housing, feeding, environmental enrichment and care;
- assessment of health and welfare (i.e., monitoring physical and behavioral signs);
- approaches to severity assessment;
- disease (causes, prevention and treatment);
- scientific procedures, general anesthesia and analgesia, methods of humane killing and confirmation of death.

Furthermore, the guidelines should also include a sections covering risk assessment for operators and education and training requirements.

References

1. K.A. Crandall, and S. De Grave. (2017). *J. Crustacean Biol.* **37**: 615–653.
2. D.P. Rigg, J.E. Seymour, R.L. Courtney, and C.M. Jones. (2020). *Freshw. Crayfish* **25(1)**: 13–30.
3. K. Johnston, B.J. Robson, and P.G. Fairweather. (2010). *Austral Ecology* **36(3)**: 269–279.
4. D.M. Hubbard, E.H. Robinson, P.B. Brown, and W.H. Daniels. (1986). *Progress. Fish Cult.* **48(4)**: 233–237.
5. M.J. Jover, M. Fernández-Carmona, M. Del Río, and M. Soler. (1999). *Aquaculture* **178(1-2)**: 127–137.
6. W.-N. Xu, W.-B. Liu, M. Shen, G.-F. Li, Y. Wang, and W. Zhang. (2012). *Aquac. Int.* **21(3)**: 687–697.
7. H. Ackefors, J.D. Castell, L.D. Boston, P. Räty, and M. Svensson. (1992). *Aquaculture* **104(3-4)**: 341–356.
8. U.B. Seemann, K. Lorkowski, M.J. Slater, F. Buchholz, and B.H. Buck. (2014). *Aquac. Int.* **23(4)**: 997–1012.
9. Z. Ghiasvand, A. Matinfar, A. Valipour, M. Soltani, and A. Kamali. (2012). *J. Fish. Sci.* **11(1)**: 63–77.
10. A. Valipour, R.O.A. Ozorio, F. Shariatmadari, A. Abedian, J. Seyfabadi, and A. Zahmatkesh. (2012). *J. App. Aquac.* **24(4)**: 316–325.
11. A. González, J.D. Celada, M. Sáez-Royuela, R. González, J.M. Carral, and V. García. (2011). *Aquac. Res.* **43(1)**: 99–105.
12. J.B. Fuertes, J.D. Celada, J.M. Carral, M. Sáez-Royuela, and A. González-Rodríguez. (2012). *Aquaculture* **364-365**: 338–344.
13. P.L. Jones, S.S. Silva, and B.D. Mitchell. (1996). *Aquac. Nutr.* **2(3)**: 141–150.
14. E. Cortes-Jacinto, H. Villarreal-Colmenares, R. Civera-Cerecedo, and R. Martinez-Cordova. (2003). *Aquac. Nutr.* **9(4)**: 207–213.
15. M.P. Hernández-Vergara, D.B. Rouse, M.A. Olvera-Novoa, and D.A. Davis. (2003). *Aquaculture* **223(1-4)**: 107–115.
16. E. Cortes-Jacinto, H. Villarreal-Colmenares, R. Civera-cerecedo, and J. Naranjo-Paramo. (2004). *Aquac. Res.* **35(1)**: 71–79.
17. K.R. Thompson, L.A. Muzinic, L.S. Engler, S.-R. Morton, and C.D. Webster. (2004). *Aquac. Res.* **35(7)**: 659–668.

18. E. Cortes-Jacinto, H. Villarreal-Colmenares, L.E. Cruz-Suarez, R. Civera-Cerecedo, H. Nolasco-Soria, and A. Hernandez-Llamas. (2005). *Aquac. Nutr.* **11(4):** 283–291.
19. A. Pavasovic, A.J. Anderson, P.B. Mather, and N.A. Richardson. (2007). *Aquac. Res.* **38(6):** 644–652.
20. O. Safari, D. Shahsavani, M. Paolucci, and M.M.S. Atash. (2014). *Aquaculture* **432:** 192–203.
21. H.S. Hammer, C.D. Bishop, and S.A. Watts. (2000). *J. Crustacean Biol.* **20(4):** 614–620.
22. Y. Dai, T.-T. Wang, Y.-F. Wang, X.-J. Gong, and C.-F. Yue. (2009). *Aquac. Res.* **40(12):** 1394–1399.
23. J. Chen, C. Chen, and Q. Tan. (2017). *Aquac. Res.* **49(2):** 676–683.
24. S.B. Koca, N. Mehmet, and A. Esra. (2018). *Indian J. Fish.* **65(2):** 129–132.
25. M.S.R.B Figueiredo, and A.J. Anderson. (2003). *Aquac. Res.* **34(13):** 1235–1239.
26. W. Luo, Y. Zhao, Z. Zhongliang, A.N. Chuanguang, and M.A. Qiang. (2008). *Chin. J. Oceanol. Limn.* **26(1):** 62–68.
27. O. Safari, D. Shahsavani, M. Paolucci, and M.M.S. Atash. (2014). *Aquaculture* **420-421:** 211–218.
28. G.M. Garcia-Ulloa, H.M. Lopez-Chavarin, H. Rodriguez-Gonzalez, and H. Villarreal-Colmenares. (2003). *Aquac. Nutr.* **9(1):** 25–31.
29. A. Campana-Torres, L.R. Martinez-Cordova, H. Villareal-Colmenares, and R. Civera-Cerecedo. (2006). *Aquac. Nutr.* **12:** 103–109.
30. A. Pavasovic, N.A. Richardson, P.B. Mather, and A.J. Anderson. (2006). *Aquac. Res.* **37(1):** 25–32.
31. H. Zhu, Q. Jiang, Q. Wang, J. Yang, S. Dong, and J. Yang. (2013). *J. World Aquacult. Soc.* **44(2):** 173–186.
32. J.B. Fuertes, J.D. Celada, J.M. Carral, M. Sáez-Royuela, and A. González-Rodríguez. (2013). *Aquac. Nutr.* **20(1):** 36–43.
33. J.B. Fuertes, J.D. Celada, J.M. Carral, M. Sáez-Royuela, and A. González-Rodríguez. (2013). *Aquaculture* **404-405:** 22–27.
34. J.B. Fuertes, J.D. Celada, J.M. Carral, M. Sáez-Royuela, and A. González-Rodríguez. (2013). *Aquaculture* **388-391:** 159–164.
35. J.-Y. Xu, T.-T. Wang, Y.-F. Wang, and Y. Peng. (2010). *Aquac. Res.* **41(9):** e252–e259.
36. X.M. Hua, C. Shui, Y.D. He, S.H. Xing, N. Yu, Z.Y. Zhu et al. (2014). *Aquac. Nutr.* **21(1):** 113–120.
37. J.J. Wan, X. Ai-jun, S. Mei-fang, D. Zheng-feng, X. Hui, H. Hong-bing et al. (2014). *J. Aquac. Bamidgeh, IJA* **67.2015.1141:** 1–9.
38. X. Xiao, D. Han, X. Zhu, Y. Yang, S. Xie, and Y. Huang. (2014). *Aquaculture* **426-427:** 112–119.
39. J. Wan, M. Shen, J. Tang, H. Lin, W. Yan, J. Li et al. (2016). *Aquac. Int.* **25(2):** 543–554.
40. Q. Tan, D. Song, X. Chen, S. Xie, and X. Shu. (2017). *Aquac. Nutr.* **24(2):** 858–864.
41. H.M. Sang, and R. Fotedar. (2010). *Fish Shellfish Immun.* **28(5-6):** 957–960.
42. H.M. Sang, R. Fotedar, and K. Filer. (2010). *Aquac Nutr.* **17(2):** e629–e635.
43. H.M. Sang, R. Fotedar, and K. Filer. (2011). *J. World Aquacult Soc.* **42(2):** 230–241.
44. O. Safari, D. Shahsavani, M. Paolucci, and M.M.S. Atash. (2014). *Aquac. Res.* **46(11):** 2685–2697.
45. O. Safari, M.M.S. Atash, and M. Paolucci. (2015). *Aquaculture* **439:** 20–28.
46. L. Parrillo, E. Coccia, M.G. Volpe, F. Siano, C. Pagliarulo et al. (2017). *Aquaculture* **473:** 161–168.
47. O. Safari, and M. Paolucci. (2017). *Aquac. Nutr.* **23(6):** 1418–1428.
48. O. Safari, M. Paolucci, and H.A. Motlagh. (2017). *Fish Shellfish Immun.* **64:** 392–400.
49. O. Safari, and M. Paolucci. (2017). *Aquaculture* **479:** 333–341.
50. O. Safari, and M. Paolucci. (2018). *Aquac Nutr.* **24(1):** 247–259.
51. S. Nedaei, A. Noori, A. Valipour, A.A. Khanipour, and S.H. Hoseinifar. (2018). *Aquaculture* **499:** 80–89.
52. A. Valipour, S. Nedaei, A. Noori, A.A. Khanipour, and S.H. Hoseinifar. (2019). *Aquaculture* **504:** 121–130.
53. Y. Cheng, and S. Wu. (2019). *Aquaculture* 734341.
54. Y. Cheng. (2019). *Fish Shellfish Immunol.* **93:** 796–800.
55. F. Liu, C. Geng, Y.K. Qu, B.X. Cheng, Y. Zhang et al. (2020). *Fish Shellfish Immun.* **103:** 321–331.
56. F. Liu, Y.K. Qu, C. Geng, A.M. Wang, J.H. Zhang et al. (2020). *Fish Shellfish Immun.* **99:** 54–166.
57. O. Safari, M. Paolucci, and H.A. Motlagh. (2020). *Aquac Nutr.* **27(1):** 91–104.
58. M. Paolucci, A. Fabbrocini, M.G. Volpe, E. Varricchio, and E. Coccia. (2012). *Zainal Abidin Muchlisin,* IntechOpen p. 4–34.
59. M. Paolucci, G. Fasulo, and M.G. Volpe. (2015). *Mar Drugs* **13(5):** 2680–2693.

60. M.G. Volpe, M. Monetta, M. Di Stasio, and M. Paolucci. (2008). *Aquaculture* **274:** 339–346.
61. M.G. Volpe, M. Malinconico, E. Varicchio, and M. Paolucci. (2010). *Recent Pat. Food Nutr. Agric.* **2:** 129–139.
62. M.G. Volpe, E. Varricchio, E. Coccia, G. Santagata, M. Di Stasio et al. (2012). *Aquaculture* **324-325:** 104–110.
63. M.G. Volpe, G. Santagata, E. Coccia, M. Di Stasio, M. Malinconico et al. (2014). *Aquac. Nutr.* **21(6):** 814–823.
64. E. Coccia, G. Santagata, M. Malinconico, M.G. Volpe, M. Di Stasio et al. (2010). *Freshw Crayfish.* **17:** 13–18.
65. C.M. Sherwin. (2001). *Anim Welf.* **10:** 103–108.
66. A. Passantino, R.W. Elwood, and P. Coluccio. (2021). *Animals.* **11:** 73.
67. J. Birch, C. Burn, A. Schnell, H. Browning, and A. Crump. (2021). London School of Economics and Political Science. LSE Consulting, London, WC2A 2AE; p. 1–107.

CHAPTER 17

Advanced Molecular Biology Techniques Applied to Crustacean Aquaculture

Maria Costantini,[1,*] *Roberta Esposito,*[1,2] *Serena Federico*[1]
and *Valerio Zupo*[3]

17.1 Introduction on advanced molecular biology techniques applied to aquaculture

Aquaculture is the fastest growing animal food sector worldwide and expected to further increase to feed the growing human population. Although aquaculture management becomes challenging because of overuse, pollution and human activities that reduced resources and genetic variations, current developments in molecular technology have lead an enormous improvement in this field [1]. Even if aquaculture is widely growing in the world, there are still some limiting factors linked to economic loss due to diseases, unavailability of specific feeds and lack of genetically improved varieties. Starting from 2005, genome sequencing through Next Generation Sequencing (NGS) and its applications, including metagenomics, transcriptomics and metabolomics helped scientists to tackle many challenges in aquaculture). These "omic" techniques addressed aquaculture in improving growth rates and cost-effectiveness, increasing resistance to pathogens and stressors, improving quality of broodstock and also the opportunity of making new or different products through altering their genetic make-up [2]. In fact, genomic information represents powerful tools to enhance physiological research, the results of which may be used for optimization of feeding and feed formulations, breeding technologies, or non-genetic selection or screening (e.g., epigenetics, proteomics, and metabolomics).

[1] Stazione Zoologica Anton Dohrn, Department of Ecosustainable Marine Biotechnology, Villa Comunale, 80121 Napoli, Italy.
[2] Department of Biology, University of Naples Federico II, Complesso Universitario di Monte Sant'Angelo, Via Cinthia 21, 80126 Naples, Italy.
[3] Stazione Zoologica Anton Dohrn, Department of Ecosustainable Marine Biotechnology, Villa Dohrn, Punta San Pietro, 80077 Naples, Italy.
* Corresponding author: maria.costantini@szn.it

Whole genome sequences are now available for many aquaculture species, helping research devoted to the identification of genomic variations insertions/deletions, single nucleotide polymorphisms and differentially methylated regions [3]. All this information is useful in predicting phenotypes and genotypic variants, which can have a positive impact on production and/or product quality, as well as in breeding programs. For some aquaculture species genome-based technologies can be useful in enhancing aquaculture traits, aiming to understand functional polymorphisms and the gene regulatory networks involved in growth, reproduction, and disease resistance of commercially important traits.

Technologies and genomic information applied in genetic improvement programs varies across aquaculture species. Private sector investment in research and development for the implementation of new technologies in aquaculture is dependent on unique industry structure and the level of vertical integration. In addition, the approach used for germplasm improvement and status of existing breeding programs dictates whether and which genome enabled technologies are suitable for a given industry. In fact, industries with centralized breeding, as for example rainbow trout and salmon, have greater potential to benefit from new molecular technologies compared to industries where breeding activities are widely distributed. Finally, the current demand for species-specific genomic tools among the diverse aquaculture industry sectors is low, rendering them commercially unaffordable. This prompts some industries interested in genetic improvement to rely on the public sector for financial resources that enable application of state of-the-art genomic technologies.

Taking into consideration all the topics discussed above in evolving of molecular approaches applied to aquaculture, the need to link the genomic information to aquaculture has become extremely evident for a better understanding of breeding improvement [4]. Thus, -omic approaches including genomics, metagenomics, proteomics and metabolomics data, give a good challenge in aquaculture improvement (Figure 1).

In the following paragraphs, we report in general on the application of molecular biology and metabolomics approaches to aquaculture. In more details, we focused on more advanced techniques in the files of genomics, metagenomics, proteomics and metabolomics. In the second part of the chapter, the attention will be focused on significance of crustacean aquaculture and on advanced molecular biology techniques applied to this field, useful in breeding problems linked to inflammation, diseases and sex determination.

17.1.2 Genomics

The genomes of several major aquaculture species are actually sequenced or are being sequenced, including catfish, Atlantic salmon, rainbow trout, tilapia, striped bass, Pacific oyster, eastern oyster, and Pacific white shrimp, as well as yellow perch and bluegill sunfish [3] (Figure 2).

These accomplishments were achieved through support of different agency, such as U.S. Department of Agriculture (USDA), National Oceanic and Atmospheric Administration (NOAA), National Institute of Food and Agriculture (NIFA), and AFRI programs, especially the Animal Genomics, Genetics and Breeding program,

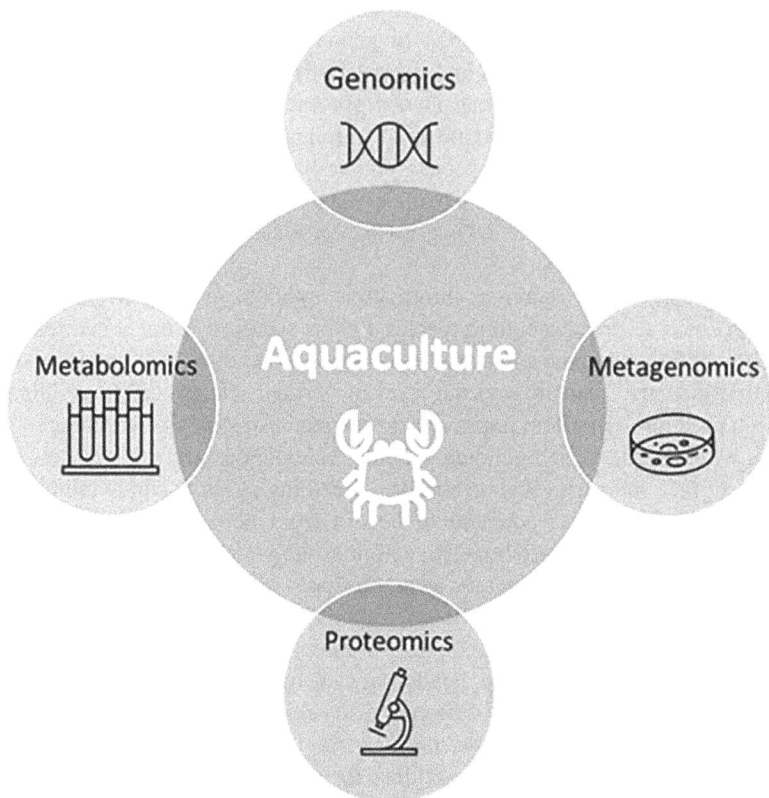

Figure 1. "Omics" approaches applied for aquaculture improvement: genomics, metagenomics, proteomics and metabolomics.

were central to the historical achievements of generating the reference genome sequences.

Different genomic approaches were used for the generation of the whole genome sequence (WGS) of aquaculture species. However, the Illumina and PacBio platforms mostly contributed to the progress of aquaculture genome sequencing. Illumina sequencing is based on a technique known as "bridge amplification" wherein DNA molecules (about 500 bp) with appropriate adapters ligated on each end are used as substrates for repeated amplification synthesis reactions on a solid support (glass slide), which contains oligonucleotide sequences complementary to a ligated adapter. The oligonucleotides on the slide are spaced such that the DNA, being subjected to repeated rounds of amplification, creates clonal "clusters" consisting of about 1000 copies of each oligonucleotide fragment. Each glass slide can support millions of parallel cluster reactions. During the synthesis reactions, proprietary modified nucleotides, corresponding to each of the four bases, each with a different fluorescent label, are incorporated and then detected. The nucleotides also act as terminators of synthesis for each reaction, which are unblocked after detection for the next round of synthesis. The reactions are repeated for 300 or more rounds.

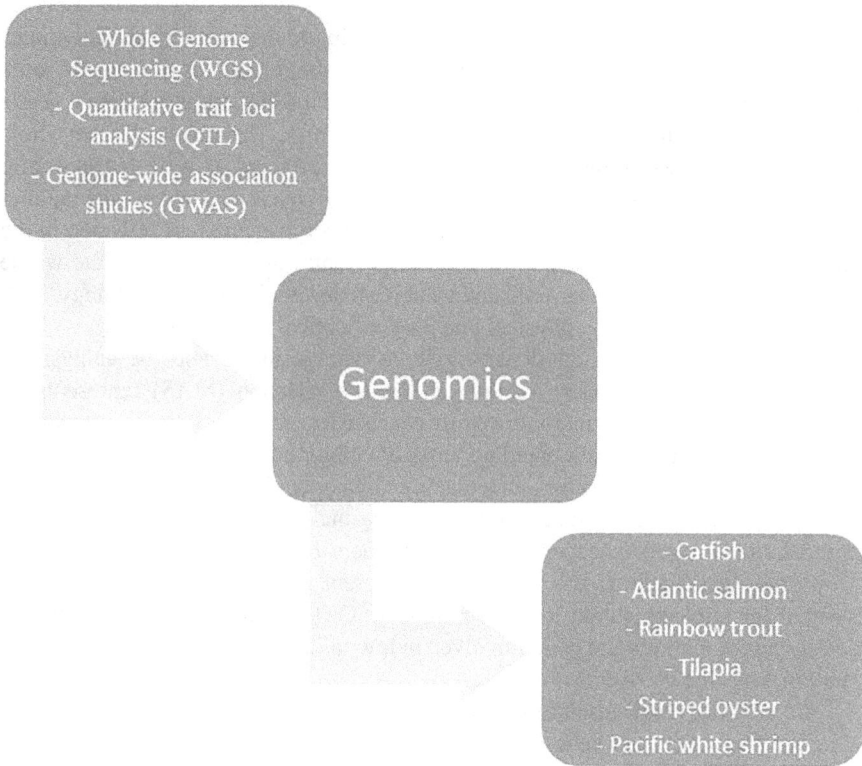

Figure 2. Genomic approaches applied in aquaculture: main tools and species.

The current commercialized technology leader in this area is Pacific Biosciences (PacBio), which has commercialized two sequencing systems, the original RSII model and more recently, the SequelTM. PacBio sequencing, also referred to as Single Molecule Real Time (SMRT) sequencing, enables very long fragments to be sequenced, up to 30 to 50 kb, or longer. The SMRT method involves binding an engineered DNA polymerase, with bound DNA to be sequenced, to the bottom of a well a zero-mode wave-guides (ZMW) is a small chamber that guides light energy into an area, whose dimensions are small relative to the wavelength of the illuminating light. Because of the ZMW design and wavelength of light utilized, imaging occurs only at the bottom of the ZMW where the DNA polymerase, bound to the DNA, incorporates each base in a growing chain. The four nucleotides are labeled with different phospho-linked fluorophores for differential detection. When a nucleotide is incorporated into the growing chain, imaging occurs on the millisecond time scale as the correct fluorescently-labeled nucleotide is bound. After incorporation, the phosphate-linked fluorescent moiety is released, which "floats away" from the bottom of the ZMW and can no longer be detected. The next nucleotide can then be incorporated [5]. Illumina sequencing generates accurate, but short reads at a relatively low cost, while PacBio sequencing generates longer, but less accurate reads at a higher cost. Various sequencing templates were used for the generation of the whole genome sequence assemblies in aquaculture species including mixtures of outbred individuals, single diploid male or female, individuals from inbred lines, and completely homozygous doubled haploids. Often, for some species, multiple individuals must be used because the DNA extracted from a single individual is not sufficient for the sequencing process. The quality of the whole genome sequence assemblies is evaluated by (i) contiguity, as reflected in contig numbers and distribution of contig sizes; (ii) connectivity, as reflected in the number of scaffolds and distribution of scaffold sizes; (iii) completeness, as reflected in the total size of the genome assemblies and the percentage of coverage of the whole genome; and (iv) accuracy, as validated by at least one additional methodology, such as genetic linkage mapping, physical mapping, or optical mapping [*3*].

The construction of high-density genetic linkage maps, and the analysis of quantitative trait loci (QTLs) for marker-assisted selection (MAS) represents an important challenge in aquaculture, mainly because it accelerates breeding processes, reduces population sizes and breeding costs of bringing progeny to maturity of the species of interest [*6*]. QTLs for molecular breeding operations require that molecular markers are closely linked to target genes, being the quantitative trait practical and reproducible, as well detectable in large numbers of individuals. However, only a few study validated across families and different genetic backgrounds, a factor essential for implementation in MAS programs. QTL mapping was also applied in identifying the markers and genes involved in low salinity tolerance of the swimming crab *Portunus trituberculatus*.

An accurate high-resolution genetic linkage map is an essential tool for addressing genetics and genomics questions [*7*]. The development of such maps represents an useful tool for the genetic breeding of a species, and is indispensable for MAS [*8,9*]. With advances in sequencing technologies over the past 5 years, high-resolution linkage maps of several aquacultured crustaceans, as the case of *Penaeus monodon* and *Litopenaeus vannamei* (most widely cultured shrimps in the world) were constructed with thousands of markers identified [*10,11*]. Recently, a high-density *P. trituberculatus* linkage map with 10,963 markers was mapped to 53 sex-averaged linkage groups, and had an average marker distance of 0.51 cM. Many important economic traits, including growth, disease resistance, and sex determination were mapped based on high-resolution linkage maps.

Ammonia is one of the most common toxicological environment factors affecting shrimp health. Although ammonia tolerance in shrimp is closely related to successful industrial production, few genetic studies of this trait are available. A high-density genetic map of *L. vannamei* was constructed in order to identify a QTL for ammonia tolerance [*12*]. By combining QTL and transcriptome analyses, a candidate gene associated with ammonia tolerance was identified, so providing the basis for future genetic studies focused on molecular marker-assisted selective breeding.

Genome-wide association study (GWAS) is a tool to dissect the genetic basis of the traits. A GWAS approach was applied to *L. vannamei*, aiming to improve its body weight at harvest through genetic selection for decades. Genes related to body weight were identified through genotyping 94,113 single nucleotide polymorphisms (SNPs) in 200 individuals from a breeding population. Four body weight-related SNPs located in LG19 and LG39 were identified. Through further candidate gene association analysis, the SNPs in two candidate genes, *deoxycytidylate deaminase* and *non-receptor protein tyrosine kinase*, were found to be related with the body weight of the shrimp. Marker-assisted best linear unbiased prediction (MA-BLUP) based on the SNPs in these two genes was used to estimate the breeding values and to provide useful information for the marker-assisted breeding in *L. vannamei* [*13*].

17.1.3 *Metagenomics*

Metagenomics is the study of the entire genetic content of all microbiota members in a natural habitat by utilization of the whole genome sequencing technique, including also metataxonomics based only on 16S rRNA gene analysis. Metagenomics is a relative recent genomics subdiscipline, which is emerging as a promising scientific tool to analyse the complex genomes contained within microbial communities. However, despite the potential of metagenomics, its use is not yet common in some agro-industrial disciplines, such as aquaculture [*14*], because of the high cost of running metagenomics techniques. Its use is justified by the significance applied to human health and the urgency of microbiologists to understand the microbial dynamics in the environment [*15*]. The flow chart of the main steps consists in the DNA extraction directly from a microbiome with a consequent high-throughput DNA sequencing (Figure 3).

Sequences data are used (i) to define the microbiome abundance and to perform phylogenetic analysis, identifying variance within and between microbiome; (ii) microbiome functional analysis to study the functional diversity, in terms of variance of genes/proteins.

Traditionally, microorganisms are collected from a microbial community and observed in the laboratory. With metagenomics the DNA (16S rRNA) is extracted through different techniques [*16*], so providing information on the diversity of microorganisms, which thrive in a certain area and can reveal information relating to

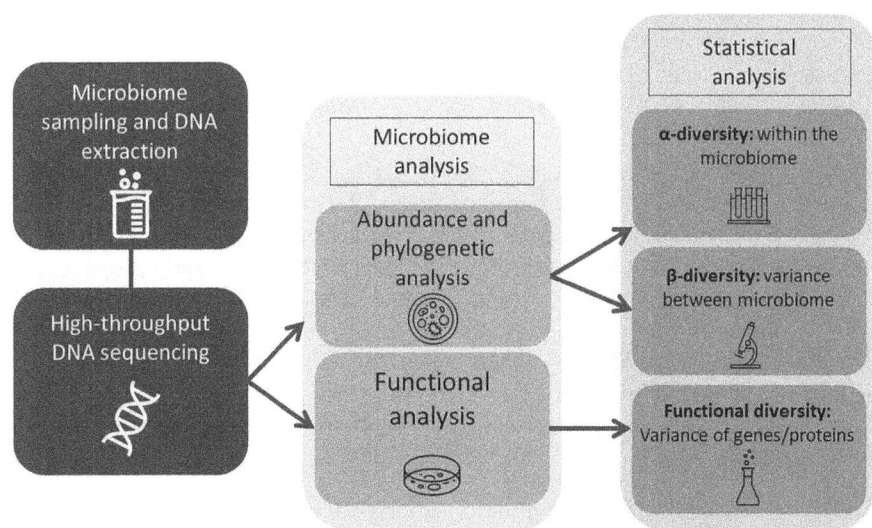

Figure 3. The flow chart of the main steps applied in metagenomics analysis.

their biological functions and roles. In particular, [17] used this method for the first time to perform phylogenetic studies of complete sequences of the 16S ribosomal RNA gene obtained directly from micro-organisms thriving in environmental samples, using recombinant DNA techniques without the need for isolation or previous culture. From this finding and forward, the development of methods for the analysis of 16S rRNA fragments was accelerated by the design of electrophoretic methods in denaturing gels, such as T/DGGE (temperature/denaturing gradient gel electrophoresis; a method useful to separate fragments of similar size but with different nucleotide sequences. Other techniques based on restriction enzymes or RFLP (restriction fragment length polymorphisms) were also used during a considerable period. Moreover, single-strand conformation polymorphism (SSCP), terminal restriction fragment length polymorphisms (T-RFLP), qPCR (quantitative PCR) and non-PCR-based methods, including microarrays [18] and fluorescence *in situ* hybridization (FISH), were used to study the microbial communities, and some of these are still in use. The use of microorganisms in aquaculture as environmental biomarkers or sentinels, effluent-bioremediators, probiotics and a direct food source for the cultured species is expanding in the last decade [19]. In this field metagenomics has allowed researchers to study the diversity and the quantity of particular microbes or genes along spatio-temporal patterns and to make closer associations among given microbial communities and host genotype or phenotype. Recent studies demonstrated that there are a large number of species involved in the recycling of residual organic matter and the recycling of a wide range of nitrogenous compounds, influencing physicochemical characteristics of water and participating in bioremediation processes of aquaculture effluents. Besides, fish and crustacean pathogens causing massive mortalities and economic loss have been detected and identified [20]. The current and probably most important limitation is not obtaining the information, but how to process and analyse the huge data generated using current high-throughput sequencing technologies. In the last few years, different softwares were introduced able in predicting microbial functional content from the presence of detected 16S rRNA genes. With Piphillin. PICRUst, Tax4Fun representative nucleic acid sequences from candidate operational taxonomic units (OTU) are compared directly with 16S rRNA gene sequences from genomes in the database to infer genome content, and thus functional potential. An example is represented by the application of PICRUSt27 to predict the functionality of the shrimp metagenome in cultured animals with acute hepatopancreatic necrosis disease (AHPND). Data reported that (i) microbiota and their predicted metagenomic functions were different between wild type and cultured shrimps; (ii) independent of the shrimp source, the microbiota of the hepatopancreas and intestine was different; (iii) the microbial diversity between the sediment and intestines of cultured shrimp was similar; (iv) changes in the microbiome and the appearance of disease-specific bacteria were associated to an early development of AHPND disease.

17.1.4 *Proteomics*

The proteome can provide relevant information of an organism physiological state, which is eventually missed by the transcriptome. In fact, the transcriptome does not account for the posttranscriptional and post-translational regulation of protein expression. Interpretation of proteomic data requires availability of information on genomic DNA and expressed RNAs, so a major limiting factor in aquaculture proteomics is still the lack of information at the genome level in most of cultured species. The first step is usually protein extraction, since most analytical techniques used in proteomics require prior solubilization of proteins in an appropriate solvent (aqueous buffers, organic solvents; see Figure 4).

The two most important analytical techniques in proteomics are two-dimensional gel electrophoresis (2-DE) and mass spectrometry (MS). Classically, detection and quantification methods for 2-DE are usually based on Coomassie Brilliant Blue (CBB) or silver staining, which enable estimation of protein quantity by scanning 2-DE gels in the visible range. The development of multiplex 2-DE (dubbed "difference gel electrophoresis" or DIGE), which involves tagging the protein samples with different fluorophores prior to 2-DE, not only allows several samples to be run on a single gel, but also significantly improves gel-to-gel variability, by providing a common reference channel across all gels of an experiment [21]. Although immunostaining methods are a valid choice for protein identification after 2-DE, the overwhelming majority of gel-based proteomic studies rely on digestion of detected proteins and analysis of the resulting peptides by MS for their identification and characterization. Instruments currently employed for this purpose include

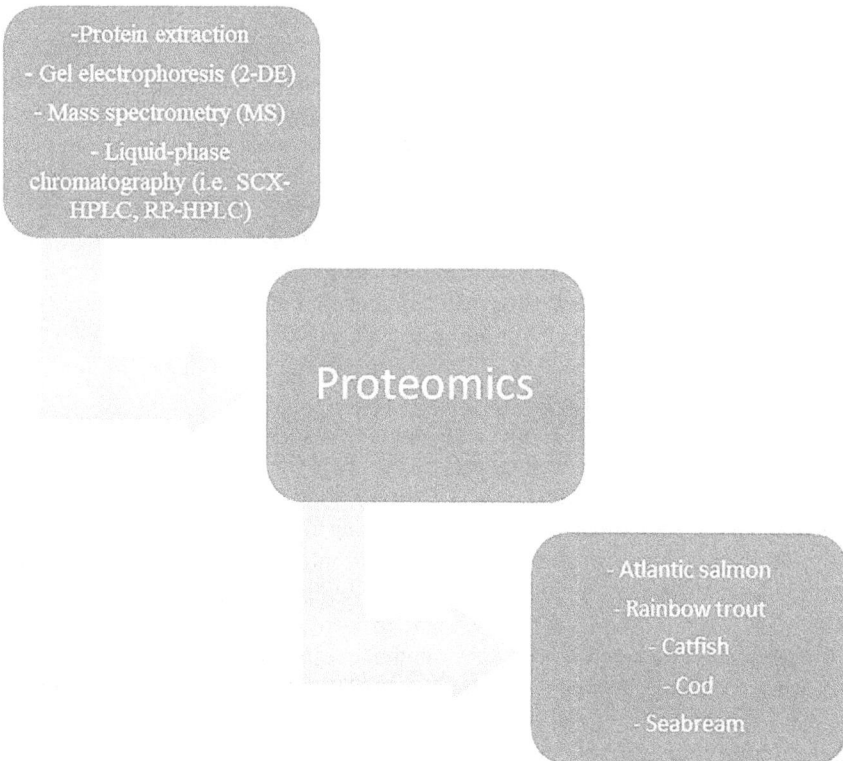

Figure 4. Proteomic approaches applied in aquaculture: main tools and species.

ESI-Ion Trap, MALDI-TOF/TOF and ESI-QTOF mass spectrometers to a lesser extent. Identification of proteins can be assessed either directly through its peptide mass fingerprint (PMF), in the case of organisms with fully sequenced genome, or by analysis of the fragmentation spectra of such peptides (PFF, peptide fragment fingerprinting or even *de novo* sequencing) obtained through tandem MS. In contrast to the usual "top-down" approach of gel-based proteomic studies, where proteins are separated and quantified before being digested and identified, most gel-free studies employ a "bottom-up" approach. Here, proteins are digested from the start and analyses (separation, quantification, characterization) are done at the peptide level. Since digestion of a complex protein mixture dramatically increases its complexity, gel-free methods are usually combined with fractionation methods to reduce the number of different peptides entering the mass spectrometer at any given moment, maximizing the total number of distinct peptides detected over the course of a sample run. Methods used in gel-free proteomics are based on liquid-phase chromatography procedures, which can readily be coupled to ESI-based mass spectrometers. Even multidimensional chromatographic separations are commonly used, as the case of MudPIT, where peptides are separated by charge (SCX-HPLC) and hydrophobicity (RP-HPLC) prior to MS analysis. Despite the fact that label-free methods exist, most gel-free workflows rely on stable isotope labeling for peptide quantification, either by metabolic incorporation of radioactive amino acids in proteins (SILAC) or by post-extraction chemical modification (ICAT, TMT, iTRAQ). It is therefore important to underline that protein identifications depend not only on the quality of spectra, but also on the quality of the sequence database used. Generally, a good approach is to start by searching genomic databases (since they tend to be smaller and therefore fast to query) and then EST databases. The results obtained are often different, since genomic databases reflect genomic DNA sequences while EST databases reflect mRNA sequences (which are closer to the actually translated amino acid sequences). Integrated with other advanced technologies (genomics, transcriptomics, metabolomics, and bioinformatics) and systems biology, proteomics will greatly facilitate the discovery of key proteins, which regulate metabolic pathways and synthesis, degradation and modifications affected by specific nutrients or other dietary factors. This will aid in rapidly enhancing our knowledge of the complex mechanisms responsible for nutrient utilization, identifying new biomarkers for nutritional status and disease progression [22].

So far, 2-DE was used in studies related to protein expression upon *Vibrio harveyi* infection in *Penaeus monodon* hemocytes, White Spot Syndrome Virus (WSSV) *L. vannamei* stomachs, in hemocytes of *Penaeus vannamei* during Taura Syndrome Virus (TSV) infection, and also studies involving functional lumphooid organ of Chinese shrimp *Fenneropenaeus chinensis*. The proteomic approach is a useful method to study prawn immunity, which involves the proteome events associated with infectious hypodermal and hematopoietic necrosis virus (IHHNV) infection in *Macrobrachium rosenbergii*. In a prawn immune response study, the comparative a gel-based proteomics approach was used to identify differentially expressed proteins in the hemocytes of *M. rosenbergii* during IHHNV. Identified proteins improved the understanding of the cellular pathways necessary for resistance against IHHNV infection.

17.1.5 *Metabolomics*

Metabolomics is a fast-evolving field that provides qualitative and quantitative analyses of metabolites within cells, tissues or biofluids. Metabolomic approaches to aquaculture have highlighted the enormous potential to solve aquaculture-related problems such as nutrition, diet, disease and post-harvest quality control (Figure 5).

In general, the most commonly applied solvent extraction methods include the following: (i) extraction of polar and/or nonpolar metabolites with a mixture of methanol, water and chloroform; (ii) extraction of polar metabolites with methanol or in combination with water; and (iii) extraction of polar metabolites with perchloric acid. The analytical measurement phase can take place through different techniques such as nuclear magnetic resonance, mass spectrometry, gas chromatography mass spectrometry and liquid chromatography mass spectrometry, capillary electrophoresis mass spectrometry, vibrational spectroscopy, infrared spectroscopy. The following step consists with the statistical analysis and biological interpretation of the data. In many cases, concentrations of particular metabolites within a tissue, biofluid or organism are correlated very well with current understanding of biochemical networks and the functional relationships among metabolites, enzymes and genes within normal or perturbed systems thanks to numerous software packages available to perform correlation network analysis, such as DPClus, Metscape, COVAIN, 3Omics and MetaMapR [23]. The results of metabolomics are starting to deliver new

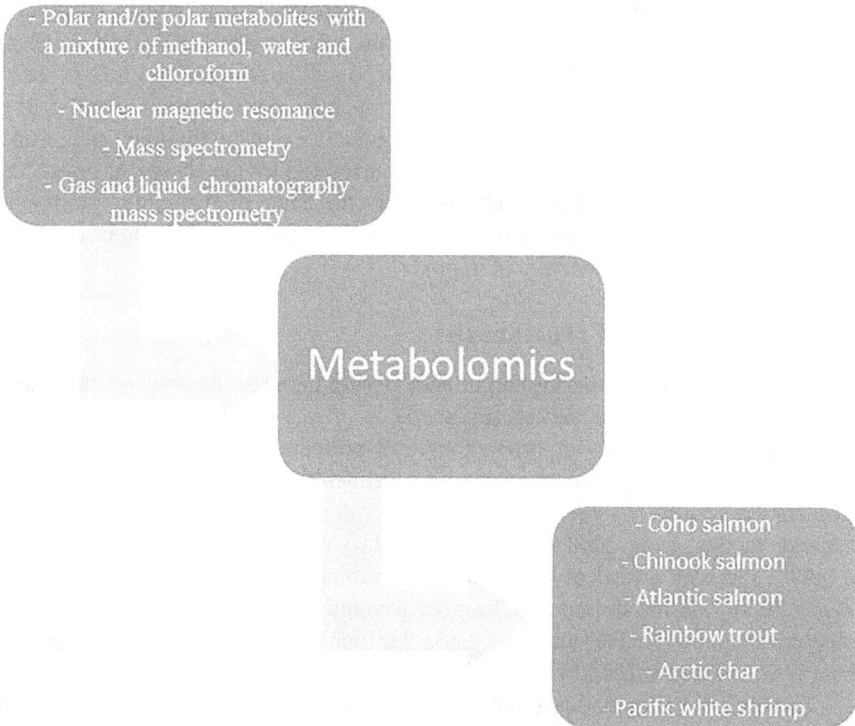

Figure 5. Metabolomics approaches applied in aquaculture: main tools and species.

information that is helping to push forward the understanding of metabolic networks at an astonishing rate and also the highlighting the usefulness and efficiency of omics-based approaches for generating novel data.

Recently, metabolomics studies were widely applied in aquatic animals to elucidate the biological effects of hypoxic stressors on organisms, as well as the case of *Macrobrachium nipponense* (Crustacea; Decapoda; Palaemonidae), also called the oriental river prawn, an important aquaculture species that is distributed widely in freshwater and low-salinity estuarine regions in Asia [24]. Prawns are relatively susceptible to hypoxia compared with most crustaceans, causing large economic losses due to increased mortality and decreased growth rate. In crustaceans, the functions of the hepatopancreas include carbohydrate and lipid metabolism, oxidative stress, energy storage and breakdown. Therefore, the *M. nipponense* hepatopancreas undergoes marked metabolomic changes in response to hypoxia, even if the specific mechanisms are largely unknown and hypoxia-related metabolomics information remains limited. Thus, the effects of hypoxia and subsequent recovery on *M. nipponense* hepatopancreas were investigated using a GC-MS-based metabolomics approach. The activities of metabolic enzymes were compared to electron transport chain-related gene expression variations induced by hypoxia between *M. nipponense* and other species [25,26]. Prawns under hypoxic stress displayed higher glycolysis-related enzyme activities and lower mRNA expression levels of aerobic respiratory enzymes than those in the normoxic control group, and those parameters returned to control levels in the reoxygenated group. Results showed that hypoxia induced significant metabolomic alterations in the prawn hepatopancreas, with the depletion of amino acids and 2-hydroxybutanoic acid and accumulation of lactate. Further, hypoxia affected energy metabolism and induced antioxidant defense regulation in prawns. Surprisingly, recovery from hypoxia, such as reoxygenation, significantly decreased valine, leucine, isoleucine, lysine, glutamate, and methionine, suggesting that increased degradation of amino acids occurred to provide energy in prawns at reoxygenation conditions. This study describes the acute metabolomic alterations that occur in prawns in response to hypoxia and demonstrates the potential of the altered metabolites as biomarkers of hypoxia.

17.2 Aquaculture of crustaceans

Crustaceans acquired increasing importance during the years because of their large demand on the worldwide markets (Figure 6).

These animals are an important source of proteins coming from the sea, also being a fundamental source of economic revenues for developing countries [27]. Efficient aquaculture systems were developed in order to supply the increasing demand. In fact, from 2000 to 2017 the crustacean production has grown yearly of 9.92% reaching a total of about 8.4 million tonnes valued at US $61.06 billion (Tacon 2020). Marine shrimps are the most accounting species in species in the total crustacean culture (65.3% valued at US$34.2 billion) followed by freshwater species that (29.9% valued at US $24.3 billion) [28].

Taking into account this trend, it is evident that the crustacean aquaculture represents an unlimited source of economic revenues for worldwide fish farmers, so

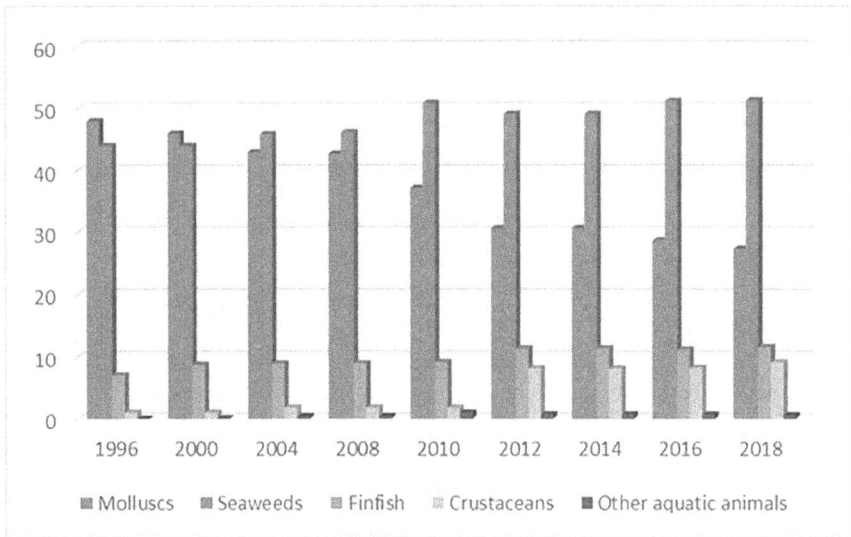

Figure 6. Histogram reporting the distribution of worldwide mariculture production between the main types of cultivated organisms from 1996 to 2018, based on FAO live weight data between 1998 and 2020.

being of extreme significance to improve all the different issues that could emerge during the farming. One of the main issues that is leading to economic losses is the spreading of diseases, which can be responsible of mass mortalities events in the aquaculture facilities. Crustaceans possess several defensive mechanisms that are different depending on the stimuli activated by the pathogen's characteristics [29]. The foreign pathogen is detected by germline encoded proteins named Pattern Recognition Receptors (PRRs). These molecules can recognize both Pathogen Associated Molecular Patterns (PAMPs) and Damage Associated Molecular Patterns (DAMPs) [30].

In the following paragraphs, we report on advanced molecular biology techniques applied in crustacean breeding to solve problems linked to diseases due to bacteria, viruses and fungi, and in sex determination.

17.3 Diseases in crustacean aquaculture

One of the major problems for the development of an efficient aquaculture facility is the possibility that during the massive production of animals some disease may outbreak. These would cause peaks of mortality leading to huge economic losses. For this reason, it is fundamental to understand how to have a precise diagnosis of the pathogen that is causing the disease in order to have a treatment as soon as possible. In aquaculture the diseases can be mainly of bacterial, viral and fungal origin.

17.3.1 Bacterial and viral diseases

Several examples are reported in literature on diseases due to bacterial and viral infections in crustacean aquaculture (see Table 1).

Table 1. The pathogen and the host that is able to infect, the molecular biology techniques used to detect the infection are reported.

Pathogen	Host	Molecular biology techniques
Vibrio parahaemolyticus	*P. monodon* *L. vannamei*	PCR
Hepatobacter penaei	*P. vannamei*	PCR and qPCR
Aphanomyces astaci	Crayfish	PCR
Infectious hypodermal and haematopoietic necrosis virus	*P. stylirostris* larvae	PCR
Hepatopancreatic parvovirus	*P. monodon*	qPCR
Infectious Myonecrosis Virus	*P. vannamei*	qPCR, RT-LAMP, LFD
Macrobrachium rosenbergii nodavirus	*M. rosenbergii*	qPCR, S-ELISA
White Spot Syndrome Virus	*P. vannamei* *P. monodon* *F. chinensis*	PCR, SHERLOCK method
Yellowhead Virus	*Penaeus* spp.	GeneAmp® 5700 Sequence
Taura Syndrome Virus	*Penaeus* spp.	GeneAmp® 5700 Sequence

The Acute Hepatopancreatic Necrosis Disease (AHPND) is a bacterial disease caused by *Vibrio parahaemolyticus* and other bacteria belonging to the *Vibrio* spp. (*Vibrio owensii, Vibrio campbelli* and *Vibrio harveyi*). The infected species are specifically *P. monodon* and *L. vannamei*, which are largely cultured in Asia [*31*]. The AHPND is also known as Early Mortality Syndrome (EMS), due to the fact that the bacterial infection can cause up to the 100% of mortality in tanks containing post-larvae.

In 2013, the causative agent of AHPND was identified as *Vibrio parahaemolyticus* [*32*], an ubiquitous, opportunistic, marine pathogen carrying a virulent pVA1 plasmid with binary toxin genes similar to the Photorhabdus insect-related (Pir) toxin, PirA and PirB [*33*]. In addition to these toxin genes, the plasmid also encodes conjugative transfer genes and transposons, suggesting the possibility of plasmid mobilization into other strains or species [*34*]. The presence of specific regions of DNA such as pVA1 plasmid and the PirAB toxin genes and their exclusive sequences permit to the molecular biologist to identify of the pathogen simply by using the PCR-based amplification, so providing to the farmers an almost immediate detection tool. Another bacteria that causes the Necrotising Hepatopancreatitis (NHP) is *Hepatobacter penaei*, which is able to infect the shrimp *P. vannamei*. The detection of *H. penaei* takes place with new end-point PCR and qPCR assays, which have as a target the *flgE* involved in the flagella genesis [*35*] (Table 1).

Infectious agents that cause diseases in the crustaceans are also viruses, for instance the infectious hypodermal and haematopoietic necrosis virus (IHHNV) is one of the most diffuse virus that causes mass deaths in shrimp thar are in farms or nature. This virus affects *Penaeus stylirostris, P. monodon* and *P. vannamei*, but high mortality rates were recorded only for juveniles of *P. stylirostris* (90% of deaths). As for the bacteria this virus can be detected with the use of the PCR technique, but there is the lack of a unique primer pairs for the detection of all the different pathogen strains, hence, further studies are needed [*36*].

The Hepatopancreatic parvovirus (HPV) infects *P. monodon* and determines high mortality in the cultured larvae. Though different geographical isolates of HPV with large sequence variations, a sensitive PCR assay specific to Indian isolate was not yet reported. A sensitive SYBR Green-based and TaqMan realtime PCR were applied for the detection and quantification of the virus. The assays were found to be sensitive, specific and reproducible with a wide dynamic range for the detection and quantification of the Indian strain of HPV. These assays were compared for their sensitivity, and further applied to detect and quantify the viral load in the post-larvae of *P. monodon*. The tests were also used to screen samples of the Pacific white shrimp *L. vannamei*, freshwater prawn *M. rosenbergii* and mud crab *Scylla serrata*.

Another method is the use of RT-LAMP and LFD that in just seventy-five minutes can give the response on the presence of the virus or not [*37*]. Another viral disease is the white tail disease (WTD) detected in *M. rosenbergii*, which can cause huge losses in the farmer's economy in India. The responsible of this disease is *Macrobrachium rosenbergii nodavirus* (MrNV). This is an RNA virus, which genetic code is made up by two single strands of RN [*38,39*]. Several different methods can be used for the detection of the virus: immunedetection, nucleic acid hybridization and amplification. The WTD is usually diagnosed by the characteristic white coloration of the tail, which is not a specific symptom for this kind of disease the diagnosis and must be confirmed by molecular analysis through RT-PCR. Sometimes in absence of a specific nucleotidic sequence the virus can be detected using artificially synthesized antibodies in mice with a sandwich linked immunosorbent assay (S-ELISA). Another exploited technique applied is the use of dot-blot method, whose only constraint consisits in the use of genetic material extracted previously from other viruses. The three viruses that affect mostly the aquaculture facilities are the Taura Syndrome Virus (TSV), white spot syndrome virus (WSSV) and yellowhead virus (YHV). TSV and YHV are both RNA viruses that infect *Penaeus* spp. Both those two viruses can be detected by using a novel technique called GeneAmp® 5700 Sequence Detection System and SYBR Green chemistry. The WSSV determines the outbreak of the White Spot Syndrome (WSS) and affects different species of crustaceans, which are economically relevant for the aquaculture as *P. vannamei, P. monodon* and *Fenneropenaeus chinensi* [*40*]. Also the WSSV can be detected with the use of the PCR with specific probes. Another innovative method proposed in 2019 is the SHERLOCK (Sensitive High Efciency Reporter unlocking) method. It starts with the isothermal amplification through Recombinase Polymerase Amplification followed by T7 transcription to produce RNA from amplified copies, and lastly Cas13a detection to enable fluorescent or colorimetric reporting [*41*]. One of the most challenging studies is the development of a successful vaccine that could lead to reduced mortality in the aquaculture plans. Although there are no true traces of a specific immunity in invertebrates, some studies have demonstrated that shrimps vaccinated against WSSV had a partial protection for thirty days if the vaccine was administered with intramuscular injection and forty-five days with oral administration [*42*]. An Australian strain of infectious hypodermal and haematopoietic necrosis virus was discovered amongst cultured penaeid prawns

P. monodon on the basis of a PCR analysis. Later, an infectious type of IHHNV identified from farmed *P. monodon* was detected in Australia in April 2008 by PCR amplification from two regions of the IHHNV genome, one discriminatory for the infectious type of IHHNV and the other (from IHHNV ORF1/2) not described in the non-infectious, inserted type [*36*]. In addition, overlapping PCR primers was used to amplify all of the ORFs of the IHHNV genome (approximately 3.6 kb) from the Australian samples, except for the hairpin loop ends (GQ475529). Comparison of the maximum possible portion of this sequence with eight GenBank records of IHHNV isolates reported from Asia revealed 94–95% identity in nucleic acid sequence and 96 to 97% identity in amino acid sequence.

17.3.2 Fungal diseases

In the aquaculture not only bacterial and viral pathogens infect crustaceans, but also fungi cause damages to the cultured species: an example is provided by *Aphanomyces astaci*. This pathogen that infects the crayfish [*44*]. The generated disease is called Crayfish plague and the infection takes place firstly in the cuticle of the animal. Then it penetrates inside the cuticle expanding the site of infection also on the inside of the animal. Some early diagnostic signs are melanin plaques on the cuticle of the animal because of the melanisation process that is an attempt of the animal to marginate the infection and defeat the fungi. The melanin plaques are not diagnostic of this disease since their formation is highly aspecific. The confirmation that the aetiological agent is *A. ostaci* is given by molecular biology analyses.

A PCR-based method was used to detect *Aphanomyces astaci* in North American crayfish, causing melanised spots in their cuticle, while remaining otherwise largely unaffected. Single round and a semi-nested assay were applied by using primers designed to specifically amplify by PCR parts of the internal transcribed spacer (ITS) regions and the 5.8 rRNA gene of *A. astaci*. PCR methods are especially useful where cultivation of the pathogen is difficult and other methods of diagnosis such as clinical symptoms or pathology are not sufficient for reliable diagnosis, as is the case with *A. astaci*.

17.3.3 Sex determination in crustaceans

The order Decapoda includes approximately 17,000 species of crayfish, shrimps, crabs, and lobsters [*45*]. Some decapods are economically cultivated and some maintained in domestic aquariums, but some are invasive and are considered as pathogenic vectors. Crustacean culture represent a rapidly developing industry, due to the high economic value as a source of animal protein, employment, and foreign exchange gains. Crustaceans production reached 9.4 million tons in 2018 (6.4 and 3 million tons were provided from capture fisheries and aquaculture, respectively), with an estimated sale value of US$ 69.3 billion [*46*]. Understanding the reproductive biology and mechanism of sexual differentiation are key issues for sexual manipulation and improving the reproductive efficiency of decapods. Male and female decapods have different growth performances at different growth stages, which is important in artificial selective breeding and farming.

Two scenarios regard sex determination in decapods. The first is a steady primary sex determination system with female heterogamety (a sex gene or genes somewhere at the W-chromosome), and the second involves a hormone secreted by a gland, the androgenic gland (AG), which is at the base of differentiation between males and females (presumably, from a gene or genes at the Z-chromosome or at autosomal chromosomes). In theory, there must be a connection between these two scenarios but, until now, the molecular sex-mechanism that precedes the AG-development to elicit maleness, or that precedes and overrules a male-AG sexual pathway to elicit femaleness, remains unknown. In particular, AG, a unique crustacean endocrine organ that secretes factors such as the insulin-like androgenic gland (IAG) hormone, is a key player in crustacean sex differentiation processes. IAG expression induces masculinization, while the absence of the AG or a deficiency in IAG expression results in feminization. Therefore, by virtue of its universal role as a master regulator of crustacean sexual development, the IAG hormone may be regarded as the sexual "IAG-switch". The switch functions within an endocrine axis is governed by neuropeptides secreted from the eyestalks, and interacts downstream with specific insulin receptors at its target organs. In recent years, IAG hormones have been found and sequenced in dozens of decapod crustacean species, including crabs, prawns, crayfish and shrimps with a ring different types of reproductive strategies, such as gonochorism, hermaphroditism and parthenogenesis. The IAG-switch represents the key step in the efforts to manipulate sex developmental processes in crustaceans. Most sex manipulations were performed using AG ablation or knock-down of the IAG gene in males in order to sex reverse them into "neo-females", or using AG implantation/injecting AG extracts or cells into females to produce "neo-males" [47–49]. The sex determination system of some crustaceans can be identified by karyotype analysis. However, due to a large number of chromosomes and complex genomes, sex determination by karyotype analysis cannot be determined in most crustaceans [50]. In addition, a recent study questioned the reliability of karyotype analysis for inferring sex determination, as the centromeres could not be identified in some of the chromosomes.

Modern biological research, such as RNA interference (RNAi), permits the post-transcriptional silencing or knockdown of a gene that may play a key role in sex differentiation, allowing to analyse or to corroborate its function. Briefly, the principle is based on the *in vitro* synthesis of double-stranded RNA (dsRNA) of a knowing gene sequence (such as the insulin-like AG precursor of the AG hormone), which is injected into the organism tissues. Once in the cellular cytoplasm, dicer enzymes bind and cleavage dsRNA in small fragments (short interfering RNA, siRNA), which are recognized by the RNA induced silencing complex (RISC). RISC complexes separate dsRNA, degrade the sense strand, and recognize and align by complementarity the antisense strand into the messenger RNA transcript (mRNA). Once incorporated, a RISC protein, Argonaute, activates and cleaves the target mRNA, thus interfering with gene expression. This technique was used to demonstrate that sex hormone (CFSH) is involved in the regulation of sex differentiation of early juvenile in *Scylla paramamosain* (Table 2).

Moreover, the *SpSxl* gene was successfully identified in *S. paramamosain*, homologous to the Sex-lethal *Sxl* and characterized its expression from different

Table 2. Crustacean species, molecular techniques to detect sex determination are reported.

Species	Molecular technique
S. paramamosain	dsRNA interference
S. paramamosain	dsRNA interference
P. trituberculatus	AFLP
C. feriatus	RAD-seq
L. vannamei	BSR-seq

tissues and different ontogeny stages or from embryo to juvenile. To investigate the role of *SpSxl* in sexual development, dsRNA interference was used with *SpSxl* to determine its relationship with other sex-determining candidate genes and sex-differentiation gene IAG.

Over the last decade, several types of genetic approaches were successfully applied to identify sex-specific DNA sequences or markers in aquatic species. For example, amplified fragment length polymorphism (AFLP) was used for swimming crab *Portunus trituberculatus* [31,51] identified male-specific single nucleotide polymorphism (SNP) markers using restriction site-associated DNA sequencing (RAD-seq) in the crab *Charybdis feriatus*. This technique is genotyping-by-sequencing approach, which allows the simultaneous analysis of multiple individuals with reduced costs. In addition, RAD-seq can generate a lot of single nucleotide polymorphism (SNP) markers in tens to hundreds of individuals through cutting activity of restriction enzymes. In this study, the male-specific primers were designed according to the five male-specific SNPs, amplifying PCR products exclusively in males but not in females, confirming the PCR-based sex identification method as well working [52] performed sex-specific transcriptome sequencing of Zoea I stage in *L. vannamei* using a DNA/RNA co/extraction method, and for the first time analyzed some differentially expressed genes (DEGs) related to sex development. Two sex-linked genes were also found with fully sex-associated SNPs by bulked segregant RNA sequencing analysis (BSRseq), which is alternative method to rapidly and efficiently identify sex-linked SNPs and genes at transcriptomic level. These genes were likely located in the sex determination region of *L. vannamei*, and might participate in sex determination and differentiation.

17.4 Concluding remarks

As reported above, over the last decades global aquaculture farming is fastly growing with the aim to have a fantastic production of aquatic animal food for human consumption. Though a very competitive market, aquaculture is a global awareness regarding the use of scientific knowledge and emerging technologies to gain a better-farmed organism through sustainable production. To obtain the world's health requirements for aquaculture products, a relentless growth in production is expected for future decades. With the continual growth of global aquaculture, fish production continues to grow globally and to date only a small proportion of the aquatic animals derive from managed breeding. This trend prompted researchers to apply -omics approaches in aquaculture biology research. In fact, several evidences reported that advancing in the production efficiency and profitability of aquaculture resources is strongly dependent by the use of molecular approaches. The advanced molecular techniques have as their ultimate goals to enhance aquaculture production efficiency, sustainability, product quality and profitability in support of the commercial sector and for the benefit of consumers. Taking this into account, a thorough understanding of the genetic mechanisms and genomic structure and organization of aquaculture species can be very useful for the developing of aquaculture market. In fact, molecular biology of aquatic organisms can offers many opportunities for the breeding of a wider range of model and non-model species. The future of molecular biology in aquaculture represents a good challenge thanks to all the advanced technologies mentioned above, reducing economic losses as well as risk of diseases and/or infection among wild aquatic species. However, a strong constrain is represented from the fact that for the most farmed aquatic animals little is known about their genomes or the genes that affect important economic traits in culture. Among aquatic animals crustaceans represent the most widely cultivated and highest-yielding species all over the world. Crustacean aquaculture production are undergoing a significant global expansion, but their growth is not without problems. Many farms have experienced disease outbreaks and catastrophic production losses, often through newly emerging pathogens. A better understanding of crustacean stress and disease issues, within the wider context of aquaculture, is central to the sustainability of natural stocks and to the success of farmed commodities. Due to their extremely high industry value, genomic, genetic, proteomic and metabolomic improvements by selection breeding are key steps in aquaculture research. In particular, growth trait is considered as the most important economic trait in shrimps and the main target trait in most breeding programs. Effort to stimulate increased research on the identification of growth-related genes in crustacean species, will be devoted to underline the associations between key genes and growth.

References

1. H.M. Maqsood, and S.M. Ahmad. (2017). *Genet. Aquat. Org.* **1(1):** 27–41.
2. S. Chandhini, and V.J. Rejish Kumar. (2019). *Rev. Aquac.* **11(4):** 1379–1397.
3. H. Abdelrahman, M. ElHady, A. Alcivar-Warren, S. Allen, R. Al-Tobasei, L. Bao, and T. Zhou. (2017). *BMC Genom.* **18(1):** 1–23.
4. K. Saito, and F. Matsuda. (2010). *Annu. Rev. Plant Biol.* **61:** 463–489.
5. A. Rhoads, and K.F. Au. (2015). *Genom. Proteom. Bioinform.* **13(5):** 278–289.
6. H. Ma, S. Chen, J. Yang, X. Ji, S. Chen et al. (2010). *Environ Biol Fishes.* **88(1):** 9–14.
7. W. Jiao, X. Fu, J. Dou, H. Li, H. Su, J. Mao, Q. Yu, L. Zhang, X. Hu, X. Huang, Y. Wang, S. Wang, and Z. Bao. (2014). *DNA Res.* **21:** 85–101.
8. F. Andriantahina, X. Liu, and H. Huang. (2013). *PLoS One* **8(9):** e75206.
9. H. Shao, H. Jiang, X. Zhang, and M. Niu. (2015). *Meas. Sci. Technol.* **26(11):** 115002.
10. M. Baranski, G. Gopikrishna, A.N. Robinson, V.K. Katneni, S.M. Shekhar et al. (2014). *PLoS One* **9(1):** e85413.
11. Z. Cui, M. Hui, Y. Liu, C. Song, X. Li et al. (2015). *Heredity* **115(3):** 206–215.
12. D. Zeng, C. Yang, Q. Li, W. Zhu, X. Chen et al. (2020). *BMC Genom.* **21(1):** 1–12.
13. D. Lyu, Y. Yu, Q. Wang, Z. Luo, Q. Zhang et al. (2021). *Front. Genet.* **12:** 611570.
14. L.R. Martínez-Córdova, M. Emerenciano, A. Miranda-Baeza, and M. Martínez-Porchas. (2015). *Rev Aquac.* **7(2):** 131–148.
15. K. Kurokawa, T. Itoh, T. Kuwahara, K. Oshima, H. Toh et al. (2007). *DNA Res.* **14(4):** 169–181.
16. F. Sanger, S. Nicklen, and A.R. Coulson. (1977). *Proc. Natl. Acad. Sci. USA* **74(12):** 5463–5467.

17. D.J, Lane, B. Pace, G.J. Olsen, D.A. Stahl, M.L. Sogin et al. (1985). *Proc. Natl. Acad. Sci. USA* **82(20):** 6955–6959.
18. L. Bodrossy, and A. Sessitsch. (2004). *Curr. Opin. Microb.* **7(3):** 245–254.
19. G. Caruso. (2013). *J. Pollut. Eff. Control* **1:** 1–3.
20. T. Gollas-Galván, L.A. Avila-Villa, and M. Martínez-Porchas. (2014). *J. Rev. Aquac.* **6(4):** 256–269.
21. M. Ünlü, M.E. Morgan, and J.S. Minden. (1997). *Electrophoresis* **18(11):** 2071–2077.
22. P.M. Rodrigues, T.S. Silva, J. Dias, and F. Jessen. (2012). *J. Proteom.* **75(14):** 4325–4345.
23. M. Altaf-Ul-Amin, Y. Shinbo, K. Mihara, K. Kurokawa, and S. Kanaya. (2006). *BMC Bioinform.* **7(1):** 1–13.
24. H. Ma, S. Chen, J. Yang, X. Ji, S. Chen et al. (2010). *Environ. Biol. Fishes.* **88(1):** 9–14.
25. S. Sun, Z. Guo, H. Fu, X. Ge, J. Zhu et al. (2018). *Front Physiol.* **9:** 76.
26. X. Sun, and W. Weckwerth. (2012). *Metabolomics* **8:** 81–93.
27. M.G. Bondad-Reantaso, R.P. Subasinghe, H. Josupeit, J. Cai, and X. Zhou. (2012). *J. Invertebr. Pathol.* **110(2):** 158–165.
28. A.G. Tacon. (2020). *Rev. Fish. Sci. Aquac.* **28(1):** 43–56.
29. L. Vazquez, J. Alpuche, G. Maldonado, C. Agundis, A. Pereyra-Morales et al. (2009). *Innate Immun.* **15(3):** 179–188.
30. A. Kulkarni, S. Krishnan, D. Anand, S. Kokkattunivarthil Uthaman, and S.K. Otta. (2021). *Rev. Aquac.* **13(1):** 431–459.
31. H.C. Laia, T. Hann Ng, M. Ando, C.T. Lee, and I.T. Chen. (2015). *Fish. Shellfish Immunol.* **47(2):** 1006–1014.
32. L. Tran, L. Nunan, R.M. Redman, L.L. Mohney, C.R. Pantoja et al. (2013). *Dis. Aquat. Org.* **105(1):** 45–55.
33. T.H. Lee, N. Naitoh, and F. Yamazaki. (2004). *Fish. Scie.* **70:** 211–214.
34. R Kumar, T.H. Ng, and H.C. Wang. (2020). *Rev. Aquac.* **12(3):** 1867–1880.
35. L.F. Aranguren, and A.K. Dhar. (2018). *Dis. Aquat. Organ.* **131(1):** 49–57.
36. V. Saksmerprome, O. Puiprom, C. Noonin, and T.W. Flegel. (2010). *Aquaculture* **298(3-4):** 190–193.
37. K.P. Prasad, K.U. Shyam, H. Banu, K. Jeena, and R. Krishnan. (2017). *Aquaculture* **477:** 99–105.
38. S. NaveenKumar, M. Shekar, I. Karunasagar, and I. Karunasagar. (2013). *Virus Res.* **173(2):** 377–385.
39. J. Sri Widada, S. Durand, I. Cambournac, D. Qian, Z. Shi et al. (2003). *J. Fish. Dis.* **26(10):** 583–590.
40. L.M. Nunan, and D.V. Lightner. (2011). *J. Virol. Methods.* **171(1):** 318–321.
41. T.J. Sullivan, A.K. Dhar, R. Cruz-Flores, and A.G. Bodnar. (2019). *Scient. Rep.* **9(1):** 1–7.
44. D.J. Alderman, and J.L. Polglase. (1986). *J. Fish Dis.* **5:** 367–379.
45. S. De Grave, N.D. Pentcheff, S.T. Ahyong, T.Y. Chan, and K.A. Crandall. (2009). *Raffles Bull Zool.* **21:** 1–109.
46. FAO. (2020). The state of world fisheries and aquaculture, sustainability in action. Rome.
47. A. Sagi, and E.D. Aflalo. (2005). *Aquac. Res.* **36(3):** 231–237.
48. T. Levy, O. Rosen, B. Eilam, D. Azulay, E.D. Aflalo et al. (2016). *J. Mar. Biotechnol.* **18(5):** 554–563.
49. T. Levy, and A. Sagi. (2020). *Front. Endocrinol.* **11:** 651.
50. Z. Torrecilla, A. Martínez-Lage, A. Perina, E. González-Ortegón, and A.M. González-Tizón. (2017). *Front. Zool.* **14(1):** 1–9.
51. S. Fang, Y. Zhang, X. Shi, H. Zheng, and S. Li. (2020). *Genomics* **112(1):** 404–411.
52. Y. Wang, Y. Yu, S. Li, X. Zhang, J. Xiang et al. (2020). *J. Mar. Biotechnol.* **22(3):** 423–432.

Index

A

Abdomen iv
Abiotic 110
Acidity 205
Adaptation 135–145
Aggression 147–150
Agonistic 146–150
Algae 128
Alien species 111, 149
Allelochemicals 72
Androgenic 14
Animal welfare 261, 268–270
Anthropogenic 157, 161
Aquaculture 5, 12, 13, 15, 213–216, 218, 219, 225–235, 274–281, 283–285, 287, 288, 290, 291
Aquatic chemistry 162
Arthropods iii, iv, vi
Automatic device 226
Automation 226–230, 235

B

Bacterial disease 192, 193
Baltic 138
Barcode 166, 167, 169, 170, 172–178
Beach 134–144
Behaviour 72–74, 76, 146–155, 160, 162, 163
Biodiversity iii, v, vii, ix, 166, 168, 177–179
Biofloc 250, 251
Biogenic amine 147
Biotechnology iii, ix, 225, 228
Biotic 110
Body regions iv
Branchiopoda 204
Breeding 274, 275, 278, 279, 285, 288, 290, 291
Burrow 147, 151
Burrower 122, 128, 131

C

Calanoid 167
Calcite 161

Caligus 213–216
Carapace 160, 161
Chalimus 215
Chemical cue 72, 73, 77, 79, 80
Chemoreceptor 75, 77
Chemorecognition 74
Chemosensory proteins 73
CHH 82, 83, 86, 89, 90, 95, 96
Chromatophore 7, 10, 12
Cladocera 204
Climate change vii, 135, 138, 141, 145, 157–164
CO_2 157, 158, 161–164
Coastal environments 134, 135, 139
Color 3, 25, 28, 124, 127, 154
Communications 72–74, 76–79
Community 111, 112
Competition 121
Complementary feeding 127
Complexity 109, 110
Connectance 110, 112
Consumption 121, 122, 124, 126–131
Copepods iii, vii, 166, 167, 169, 173, 175–178, 180
Crayfish 241–255, 258, 261, 262, 264–266, 268–270
Cryptic species 166, 169, 177
Cultivation 228
Culture vii, viii
Cumulative distribution 113
Cyanobacteria 218, 219

D

Daphnia vii, 204–211
Debris 134–136
Decapods iii, vi, vii
Decaying 72
Developmental biology 19
Diapause 138
Diet 246–248, 251, 254–256, 258, 261, 262, 265
Disease vii, 185–195, 198–201, 274, 275, 285, 286, 288, 291
Disease resistance 275, 278

Diversity iii, vii, 146, 147, 150, 152, 155
DNA barcoding 166, 168, 169, 173, 176, 177, 180
Dormant egg 205

E

Ecdysis 6, 14
Ecological model 113
Ecology iii, vi, vii, ix
Ecotoxicology 85, 88, 105, 204, 205
Ecotoxigenomic 208
Ectoparasites 213, 214
Eggs 19, 20, 22, 23, 25–30, 32–35
Emamectin benzoate 216
Endocrinology 3, 4, 10, 13
Environment iii, vi, vii
Epigenetics 274
Epiphytes 122, 125–127, 130
Ethology vii
Evolution 20, 22, 33, 35, 36
Exoskeletal 161
Exposure 205, 209, 210
Extensive 247, 248, 250, 265
Eyestalk 5, 6, 8–11, 13, 15

F

Farming industry 246
Feeding source 110, 112
Feeding strategies 126
Feeds 274
Fertilization 18, 19, 21–24, 26, 28–30, 32–34
Fighting 147, 149, 154
Food item 115, 116, 118
Food webs 109, 110, 112, 115, 116, 118, 134
Freshwater 166, 167, 169, 173, 175, 177–179, 241–247, 250, 251, 254, 255, 258, 261, 262, 269
Fungal disease 191

G

Ganglia 5, 6, 9, 11–13
Genetic linkage map 278
Genital papillae 19
Genome 205–207, 209, 274–276, 278–282, 288, 289, 291
Gland 4–6, 9–15
Global change 161, 164
Glycoprotein 46
Gonochoristic 20, 22, 33
Gonopores 19, 23, 25, 27, 29–33
Great lakes 166, 167, 173, 175–177, 179

H

Haemolymph 83, 96, 100, 101
Heat shock proteins 83, 96
Hemocyanins 84, 89, 92
Hemocytes 161
Hermaphrodite 20, 23, 24, 33–35
Hierarchy 75
Hormone 3–15, 205
Host 186, 188–201, 214, 215, 222, 223
Hyperglycemic hormone 82, 90, 94

I

Immune system 82, 84, 91, 92
Infection 186–192, 195, 200
Infochemical 72, 73, 78
Integrated multi-trophic 249
Intensive 247–251, 256–258, 265
Interactions 109–112, 115
Invasion 173, 175–177, 179
Isopods iii, vii

J

Juvenile 77

K

Kelp 134, 143

L

Larva vi
Lepeophtheirus 213, 215–217, 219–222
Life history 160
Lipid transfer proteins 40–43, 52
Lipids 40–47, 50–53, 55–57, 61, 62, 66
Lipoprotein 40–48, 50–52, 57, 61, 63–65
Lipovitellin 43–45
Litter 122–125, 127–131
Live food 229, 231, 232
Lobster 241, 246, 247, 268–270

M

Malacostracan 46, 47, 52, 53, 55–57, 60, 62
Mean-field approximation 112
Mediterranean sea 121, 129
Melanin engaging proteins 41
Mesograzer 121–123
Mesoherbivore 121
Metabolomics 274–276, 282–284, 291
Metagenomics 274–276, 279, 280
Microsomal triglyceride transfer protein 41, 50, 54

Microsporidia 192, 195–197
Migrations 124, 127
Mitochondrial cytochrome oxidase 166
Model 111, 112, 118
Model organism 146, 155, 225–227, 232, 237, 239
Molecular iii, iv, viii
Monitoring 228–231, 233–235, 237, 239
Monophyletic iv, vi
Morphology 146, 147, 151
Mortality 226
Moult 6, 8, 11, 13
MTP 41–44, 47, 50–52, 54
Multi-omic 206
Multixenobiotic resistance 83
Mycotic infection 191

N

Natural drug 213, 217, 218
Nauplius 214, 215
Near-shore 176
Neuro-hormone 4, 5, 9–14
Neurosecretion 5, 12

O

Omics vii
Opportunistic 124
Ostracods 166
Ovary 22, 26, 27, 29, 33
Oxidative stress 84, 86, 88, 89, 91, 92, 95–98

P

Pain 268–270
Pancrustacea iv
Parasite 147, 148, 185, 192, 195, 197–201, 213, 214, 216, 218, 223
Parasitic copepod 197, 199, 200
Parthenogenesis 20, 25, 36
Pathogen 185–188, 191, 192, 201
Penaeid 185–188
Peroxidation 84, 96, 100
Pesticide 217, 218
pH 157–164
Pheromone 72, 73, 75–79, 148
Phospholipid 44–46, 51, 52, 62, 66
Phylogenetic 41, 53–56
Physiology iii, vi, viii, ix, 161, 163, 164, 225, 226
Phytotherapy 188
Planktonic 166, 167, 169
Plant-animal 121
Plasma membrane 83
Pleon iv
Pollutant 83, 84, 86, 92, 97

Pollution 274
Polyserine 47, 66
Posture 148, 153, 154
Prawn 187–190, 241, 253, 268
Predation 137
Prey escape 137
Production 244–251, 254, 258, 261, 265
Production cost 227, 238, 239
Productivity 228
Proteomics 274–276, 280–282, 291

R

Reactive oxygen species 84
Recirculation 226
Relative abundance 112, 114, 118
Reproduction vi–viii, 18–21, 23, 25, 26, 28, 31, 34, 138, 139, 141, 241, 248, 251, 252, 258
Rhythms iii, vii

S

Salinity 205
Salmo salar 214, 216
Salmonid 213–216
Sand-hopper 136, 137, 142
Seagrass 121–126
Seagrass detritus 128
Sediment 134–136, 139–142
Semi-automatic culture 227
Sensors iii, iv, vii
Sex 3–5, 7, 8, 14
Sex determination 275, 278, 285, 288–290
Sex differentiation 18
Sex ratio 138, 139, 141
Sexual biology vi
Sexual selection 146, 150, 151
Shellfish 245
Signaling cascade 207
Sinus gland 82, 101
Somites iv
Spatial skills 147
Species diversity 110, 112, 114, 115, 118
Spermatozoan 32
Steroid 4–8, 12, 13
Stress iii, vii
Stressor 205–211

T

Tactile cue 75
Taxonomy iii
Temperature 158, 159, 161
Testis 33
Toxicity 218–220, 222, 223
Transcriptome 42, 53, 55, 56, 63, 65

Transmission 187, 188, 191, 192, 195
Trophic level 74, 78

U

Ultraviolet 162
Uridine 76

V

Virus 186–190
Vitellogenin 40, 41, 43, 44, 46–48, 50–55, 57, 58,
    61, 65–67

W

Water flea 204–206, 209, 210
Water waste 228
Welfare viii

X

Xenobiotic 85
X-organ 5, 11, 13, 14

Y

Yolk 40–42, 46, 47, 52, 53, 57, 59, 60, 63, 66
yolk protein vii
Y-organ 4–6, 8, 11–14

Z

Zooplankton 166, 167, 169, 173

For Product Safety Concerns and Information please contact our EU
representative GPSR@taylorandfrancis.com
Taylor & Francis Verlag GmbH, Kaufingerstraße 24, 80331 München, Germany